VIROLOGY

VIROLOGY

PRINCIPLES AND APPLICATIONS

John B. Carter and **Venetia A. Saunders**

School of Biomolecular Sciences,
Liverpool John Moores University, UK

BICENTENNIAL
1807
WILEY
2007
BICENTENNIAL

John Wiley & Sons, Ltd

Other Wiley Editorial Offices

John Wiley & Sons Inc., 111 River Street, Hoboken, NJ 07030, USA

Jossey-Bass, 989 Market Street, San Francisco, CA 94103-1741, USA

Wiley-VCH Verlag GmbH, Boschstr. 12, D-69469 Weinheim, Germany

John Wiley & Sons Australia Ltd, 42 McDougall Street, Milton, Queensland 4064, Australia

John Wiley & Sons (Asia) Pte Ltd, 2 Clementi Loop #02-01, Jin Xing Distripark, Singapore 129809

John Wiley & Sons Canada Ltd, 6045 Freemont Blvd, Mississauga, Ontario, L5R 4J3, Canada

Wiley also publishes its books in a variety of electronic formats. Some content that appears in print may not be available in electronic books.

Library of Congress Cataloging-in-Publication Data:

Carter, John, 1944–
    Virology : principles and applications / John Carter and Venetia
Saunders.
          p. ; cm.
    Includes bibliographical references and index.
    ISBN 978-0-470-02386-0 (cloth)
  1. Virology. 2. Viruses. 3. Virus diseases. I. Saunders, Venetia
A., 1949– II. Title.
    [DNLM: 1. Viruses. 2. Virus Diseases. QW 160 C323v2007]
    QR360.C36 2007
    616.9′101 – dc22
                                            2007017896

British Library Cataloguing in Publication Data

A catalogue record for this book is available from the British Library

ISBN: 978-0-470-02386-0 (HB)
        978-0-470-02387-7 (PB)

Typeset in 10/12pt Times by Laserwords Private Limited, Chennai, India
Printed and bound by Printer Trento Srl., Trento, Italy
This book is printed on acid-free paper.

To Myra, Robert, Jon and Mark

Contents

Preface

Virology is a fascinating and rapidly developing subject, and is worthy of study purely because viruses are interesting! Furthermore, virology is a branch of science that is of immense relevance to mankind for a host of reasons, not least of which are the threats to human health caused by viruses, such as HIV, hepatitis B virus, papillomaviruses, measles and influenza viruses, to mention just a few. There is a continuing need for trained virologists, and it is hoped that this book will play a small role in helping to fulfil this need. To a large extent the material in the book is based on virology taught at Liverpool John Moores University.

This is not a textbook of fundamental virology, medical virology, veterinary virology, plant virology or of bacteriophages, but a bit of each of these! The general pattern of the book is that principles of virology are covered earlier and applications are covered later. There is no strict demarcation between the two, however, so the reader may be made aware of important applications while principles are being introduced.

The first 10 chapters cover basic aspects of virology. A chapter on methods used in virology comes early in the book, but could be skimmed to gain an overview of its contents and thereafter used for reference. There is one chapter on each of the seven Baltimore classes, concentrating mainly on animal viruses. There is a chapter devoted entirely to HIV and an extended chapter on phages, reflecting the renewed interest in their biology and applications. After a chapter on origins and evolution of viruses, there follow five chapters covering various aspects of applied virology, including vaccines and antiviral drugs. The final chapter is on prions, which are not viruses but are often considered along with the viruses.

Each chapter starts with 'At a glance', a brief summary with the dual aim of giving a flavour of what is coming up and providing a revision aid. Each chapter ends with a list of learning outcomes and a guide to further reading in books and journals. The references are mainly from the 21st century, but there is a selection of important papers from the last century.

The book has a web site (www.wiley.com/go/carter), where you can find

- many references additional to those in the book;

- links to the journal references (to the full text where this is freely available, otherwise to the abstract);

- links to virology web sites;

- self-assessment questions and answers for each chapter, to reinforce and extend concepts developed in the book.

A key feature of our book is a standard colour code to differentiate various types of nucleic acid and protein molecule in the diagrams. The colour code is explained on page xxiii. It is appreciated that colour coding may be of limited value to individuals who have difficulty in differentiating colours, so we have also labelled many of the molecules.

A number of virus replication cycles are described and the reader should be aware that these are models based on evidence to date; the models may have to be modified in the light of future evidence. We present the virus replication cycles as fitting within a general framework of seven steps:

(1) Attachment of a virion to a cell

(2) Entry into the cell

(3) Transcription of virus genes into mRNAs

(4) Translation of virus mRNAs into virus proteins

(5) Genome replication

(6) Assembly of the virus proteins and genomes into virions

(7) Exit of the virions from the cell.

We hope that this helps in appreciating how virus replication fits into a general pattern and in comparing the replication cycles of different types of virus. For some groups of viruses the framework has to be modified and we make clear when this is the case.

If you come across an unfamiliar term, please consult the 'Virologists' vocabulary' at the back of the book. This glossary includes not only virology-specific terms, but also a selection of terms from cell biology, molecular biology, immunology and medicine.

A list of the abbreviations that are used in this book appears on the following pages.

We wish to thank the many people who have made the production of this book possible. We thank all those who supplied images and those who gave permission for the use of their images; we are especially grateful to David Bhella, Tom Goddard, Kathryn Newton and Jean-Yves Sgro. Thanks also to Robert Carter for assistance with images. We acknowledge the contributions of the many students who have acted as guinea-pigs for our teaching materials and who have provided us with feedback. Grateful thanks also to those who

reviewed material for the book and provided valuable feedback. We are sorry that we were unable to include all the topics suggested, but if we had done so the book would have run to several volumes! Many thanks to Rachael Ballard, Robert Hambrook and all at John Wiley & Sons, Ltd who helped the book come to fruition. Finally, thanks to our families for their support and for their patience during those many hours we spent ensconced in the study.

We hope you find the book useful and we would be interested to hear what you think of it. We have tried to ensure that there are no errors, but it is probable that some have slipped through; if you come across any errors please inform us.

John B. Carter
J.B.Carter@ljmu.ac.uk

Venetia A. Saunders
V.A.Saunders@ljmu.ac.uk

School of Biomolecular Sciences,
Liverpool John Moores University,
Byrom Street,
Liverpool L3 3AF, UK

Abbreviations used in this book

(+) DNA	plus strand (positive strand) DNA
(−) DNA	minus strand (negative strand) DNA
(+) RNA	plus strand (positive strand) RNA
(−) RNA	minus strand (negative strand) RNA
A	adenine
ADP	adenosine diphosphate
AIDS	acquired immune deficiency syndrome
AP-1	activator protein 1
ATP	adenosine triphosphate
b	base(s)
BL	Burkitt's lymphoma
bp	base pair(s)
BSE	bovine spongiform encephalitis
C	cytosine
C terminus	carboxy terminus
cccDNA	covalently closed circular DNA
CD	cluster of differentiation
cDNA	copy DNA
CJD	Creutzfeldt-Jakob disease
cos	cohesive end
CP	coat protein
CPE	cytopathic effect
DIP	defective interfering particle
DNA	deoxyribose nucleic acid
ds	double-stranded
DTR	direct terminal repeat
E	early
EBV	Epstein-Barr virus
EF	elongation factor

ELISA	enzyme-linked immunosorbent assay
ERV	endogenous retrovirus
E. coli	*Escherichia coli*
Ff	F-specific filamentous
G	guanine
GFP	green fluorescent protein
HAV	hepatitis A virus
HBV	hepatitis B virus
HBsAg	hepatitis B surface antigen
HCV	hepatitis C virus
HIV	human immunodeficiency virus
HPV	human papillomavirus
HSV	herpes simplex virus
HTLV-1	human T-lymphotropic virus 1
IC_{50}	50% inhibitory concentration
ICTV	International Committee on Taxonomy of Viruses
ICTVdB	International Committee on Taxonomy of Viruses database
IE	immediate early
IG	intergenic
IRES	internal ribosome entry site
ITR	inverted terminal repeat
kb	kilobase(s)
kbp	kilobase pair(s)
kD	kiloDalton(s)
KSHV	Kaposi's sarcoma-associated herpesvirus
LDI	long distance interaction
LE	left end
LIN	lysis inhibition
LPS	lipopolysaccharide
LTR	long terminal repeat
Mbp	megabase pair(s)
MHC	major histocompatibility complex
MJ	Min Jou
m.o.i.	multiplicity of infection
MP	movement protein
mRNA	messenger RNA
N terminus	amino terminus
NF-κB	nuclear factor kappa B
NK cell	natural killer cell

nm	nanometre(s) (10^{-9} metre)
NPC	nasopharyngeal carcinoma
NSP	non-structural protein
O	operator
ORF	open reading frame
ori	origin (replication)
P	promoter
PBS	primer binding site
PCR	polymerase chain reaction
pfu	plaque-forming unit
phage	bacteriophage
PS	packaging signal
RBS	ribosome binding site
RE	right end
RF	replicative form
RI	replicative intermediate
RNA	ribose nucleic acid
RNAi	RNA interference
RNAse H	ribonuclease H
rRNA	ribosomal RNA
RT-PCR	reverse transcriptase–polymerase chain reaction
S	Svedberg unit
SARS	severe acute respiratory syndrome
S-D	Shine-Dalgarno
SI	selectivity index
SIV	simian immunodeficiency virus
Sp1	stimulatory protein 1
ss	single-stranded
ssb	single-stranded binding
T	thymine
$TCID_{50}$	virus dose that infects 50% of tissue cultures
TK	thymidine kinase
tRNA	transfer RNA
TSE	transmissible spongiform encephalitis
U	uracil
UV	ultra-violet
vCJD	variant Creutzfeldt-Jakob disease
VP	virus protein
VPg	virus protein, genome linked
VSV	vesicular stomatitis virus

Greek letters used in this book

α	alpha
β	beta
γ	gamma
ε	epsilon
θ	theta
κ	kappa
λ	lambda
σ	sigma
φ	phi
ψ	psi

Colour coding for molecules

With the aim of maximizing clarity of the diagrams the following standard colour code is used to depict molecules.

Virus Molecules

DNA

dsDNA

Where there are distinct (+) and (−) strands these are coded: **(+) DNA**
(−) DNA

RNA

primary transcripts and mRNA

(+) RNA (other than primary transcripts and mRNA)

(−) RNA

A cap on the 5' end of an RNA molecule is indicated by the symbol ⎛.

Protein

Some virus proteins are in this colour
Some virus proteins are in this colour

N.B. The colours of capsid images derived from X-ray crystallography and cryo-electron microscopy do not conform to the colour coding in the drawings.

Cell Molecules

DNA
tRNA
protein

1

Viruses and their importance

At a glance

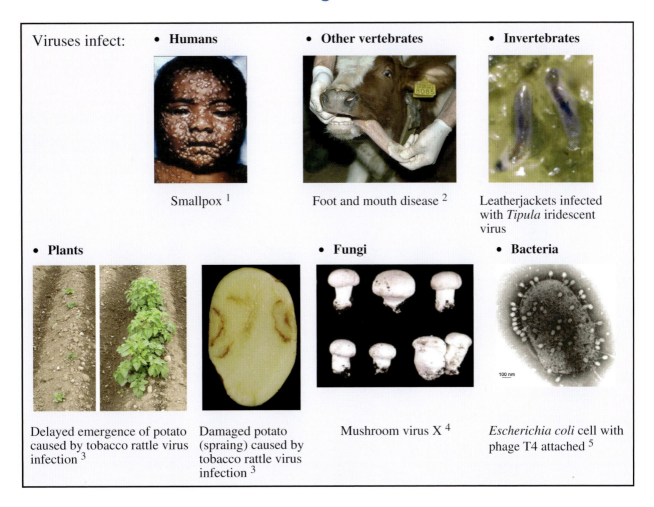

Viruses infect:

- **Humans**

Smallpox [1]

- **Other vertebrates**

Foot and mouth disease [2]

- **Invertebrates**

Leatherjackets infected with *Tipula* iridescent virus

- **Plants**

Delayed emergence of potato caused by tobacco rattle virus infection [3]

Damaged potato (spraing) caused by tobacco rattle virus infection [3]

- **Fungi**

Mushroom virus X [4]

- **Bacteria**

100 nm

Escherichia coli cell with phage T4 attached [5]

Virology: Principles and Applications   John B. Carter and Venetia A. Saunders
© 2007 John Wiley & Sons, Ltd   ISBNs: 978-0-470-02386-0 (HB); 978-0-470-02387-7 (PB)

At a glance (continued)

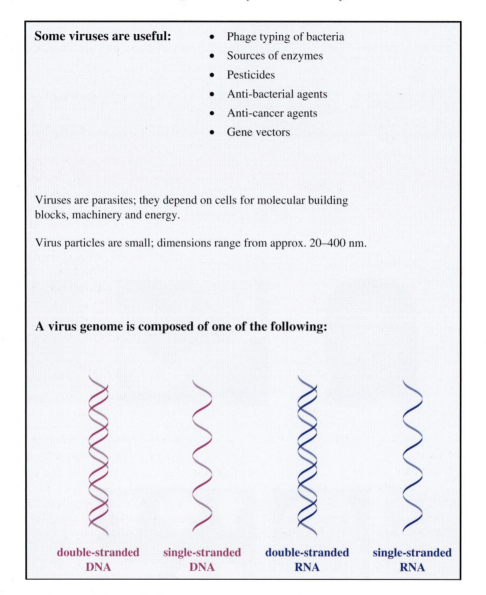

Some viruses are useful:

- Phage typing of bacteria
- Sources of enzymes
- Pesticides
- Anti-bacterial agents
- Anti-cancer agents
- Gene vectors

Viruses are parasites; they depend on cells for molecular building blocks, machinery and energy.

Virus particles are small; dimensions range from approx. 20–400 nm.

A virus genome is composed of one of the following:

| double-stranded DNA | single-stranded DNA | double-stranded RNA | single-stranded RNA |

Photographs reproduced with permission of

[1] World Health Organisation.

[2] Animal Sciences Group, Wageningen UR.

[3] MacFarlane and Robinson (2004) Chapter 11 *Microbe-Vector Interactions in Vector-Borne Diseases*, 63rd Symposium of the Society for General Microbiology, Cambridge University Press. Reprinted with permission.

[4] Warwick HRI.

[5] Cornell Integrated Microscopy Center.

1.1 Viruses are ubiquitous on Earth

Viruses infect all cellular life forms: eukaryotes (vertebrate animals, invertebrate animals, plants, fungi) and prokaryotes (bacteria and archaea). The viruses that infect prokaryotes are often referred to as bacteriophages, or phages for short.

The presence of viruses is obvious in host organisms showing signs of disease. Many healthy organisms, however, are hosts of non-pathogenic virus infections, some of which are active, while some are quiescent. Furthermore, the genomes of many organisms contain remnants of ancient virus genomes that integrated into their host genomes long ago. As well being present within their hosts, viruses are also found in soil, air and water. Many aqueous environments contain very high concentrations of viruses that infect the organisms that live in those environments.

There is a strong correlation between how intensively a species is studied and the number of viruses found in that species. Our own species is the subject of most attention as we have a vested interest in learning about agents and processes that affect our health. It is not surprising that there are more viruses known that infect mankind than any other species, and new human viruses continue to be found. The intestinal bacterium *Escherichia coli* has also been the subject of much study and many viruses have been found in this species. If other species received the same amount of attention it is likely that many would be found to be hosts to similar numbers of viruses.

It is undoubtedly the case that the viruses that have been discovered represent only a tiny fraction of the viruses on the Earth. Most of the known plants, animals, fungi, bacteria and archaea have yet to be investigated for the presence of viruses, and new potential hosts for viruses continue to be discovered. Furthermore, the analysis of DNA from natural environments points to the existence of many bacterial species that have not yet been isolated in the laboratory; it is likely that these 'non-cultivable bacteria' are also hosts to viruses.

1.2 Reasons for studying viruses

1.2.1 Some viruses cause disease

Viruses are important agents of many human diseases, ranging from the trivial (e.g. common colds) to the lethal (e.g. rabies), and viruses also play roles in the development of several types of cancer. As well as causing individuals to suffer, virus diseases can also affect the well-being of societies. Smallpox had a great impact in the past and AIDS is having a great impact today.

There is therefore a requirement to understand the nature of viruses, how they replicate and how they cause disease. This knowledge permits the development of effective means for prevention, diagnosis and treatment of virus diseases through the production of vaccines, diagnostic reagents and techniques, and anti-viral drugs. These medical applications therefore constitute major aspects of the science of virology.

Veterinary virology and plant virology are also important because of the economic impact of the many viruses that cause disease in domestic animals and crop plants: foot and mouth disease virus and rice yellow mottle virus are just two examples. Another area where viruses can cause economic damage is in the dairy industry, where phages can infect the lactic acid bacteria that are responsible for the fermentations that produce cheese, yogurt and other milk products.

1.2.2 Some viruses are useful

Some viruses are studied because they have useful current or potential applications.

- *Phage typing of bacteria.* Some groups of bacteria, such as some *Salmonella* species, are classified into strains on the basis of the spectrum of phages to which they are susceptible. Identification of the phage types of bacterial isolates can provide useful epidemiological information during outbreaks of disease caused by these bacteria.

- *Sources of enzymes.* A number of enzymes used in molecular biology are virus enzymes. Examples include reverse transcriptases from retroviruses and RNA polymerases from phages.

- *Pesticides.* Some insect pests are controlled with baculoviruses and myxoma virus has been used to control rabbits.

- *Anti-bacterial agents.* In the mid-20th century phages were used to treat some bacterial infections of humans. Interest waned with the discovery of antibiotics, but has been renewed with the emergence of antibiotic-resistant strains of bacteria.

- *Anti-cancer agents.* Genetically modified strains of viruses, such as herpes simplex virus and vaccinia virus, are being investigated for treatment of cancers. These strains have been modified so that they are able to infect and destroy specific tumour cells, but are unable to infect normal cells.

- *Gene vectors for protein production.* Viruses such as certain baculoviruses and adenoviruses are used as vectors to take genes into animal cells growing in culture. This technology can be used to insert into cells genes encoding useful proteins, such as vaccine components, and the cells can then be used for mass production of the proteins.

- *Gene vectors for treatment of genetic diseases.* Children with severe combined immunodeficiency (baby in the bubble syndrome) have been successfully treated using retroviruses as vectors to introduce into their stem cells a non-mutated copy of the mutated gene responsible for the disease (Section 16.5).

1.2.3 Virus studies have contributed to knowledge

Much of the basic knowledge of molecular biology, cell biology and cancer has been derived from studies with viruses. Here are a few examples.

- A famous experiment carried by Alfred Hershey and Martha Chase, and published in 1952, used phage T2 and *E. coli* to provide strong evidence that genes are composed of DNA.

- The first enhancers to be characterized were in genes of simian virus 40 (SV40).

- The first transcription factor to be characterized was the transplantation (T) antigen of SV40.

- The first nuclear localization signal of a protein was identified in the T antigen of SV40.

- Introns were discovered during studies of adenovirus transcription.

- The role of the cap structure at the 5′ end of eukaryotic messenger RNA was discovered during studies with vaccinia virus and a reovirus.

- The first internal ribosomal entry site to be discovered was found in the RNA of poliovirus.

- The first RNA pseudoknot to be discovered was that in the genome of turnip yellow mosaic virus.

1.3 The nature of viruses

1.3.1 Viruses are small particles

Evidence for the existence of very small infectious agents was first provided in the late 19th century by two scientists working independently: Martinus Beijerinck in Holland and Dimitri Ivanovski in Russia. They made extracts from diseased plants, which we now know were infected with tobacco mosaic virus, and passed the extracts through fine filters. The filtrates contained an agent that was able to infect new plants, but no bacteria could be cultured from the filtrates. The agent remained infective through several transfers to new plants, eliminating the possibility of a toxin. Beijerinck called the agent a 'virus' and the term has been in use ever since.

At around the same time, Freidrich Löeffler and Paul Frosch transmitted foot and mouth disease from animal to animal in inoculum that had been highly diluted. A few years later Walter Reed and James Carroll demonstrated that the causative agent of yellow fever is a filterable agent.

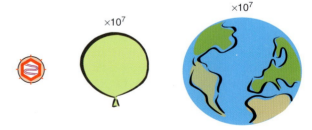

Figure 1.1 *Comparative sizes of a herpesvirus particle, a balloon and the Earth.* A large balloon is about ten million times larger than a herpesvirus particle, while the Earth is larger than the balloon by the same factor.

Figure 1.2 *Transmission electron microscope.* This is a microscope in which the image is formed by electrons transmitted through the specimen. Photograph courtesy of Kathryn Newton.

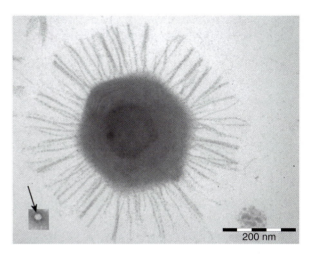

200 nm

Figure 1.3 *Virions of mimivirus, one of the largest viruses, and a parvovirus (arrowed), one of the smallest viruses.* Electron micrograph of mimivirus courtesy of Prof. D. Raoult, Unité des Rickettsies, Marseille, France. Electron micrograph of parvovirus from Walters *et al.* (2004) *Journal of Virology*, **78**, 3361. Reproduced by permission of the American Society for Microbiology.

Figure 1.1 gives some indication of the size of these agents, which are known as virus particles or virions. The virion of a herpesvirus, which is a fairly large virus, is about ten million times smaller than a large balloon, while the balloon is smaller than the Earth by the same factor. The virions of most viruses are too small to be seen with a light microscope and can be seen only with an electron microscope (Figure 1.2).

The units in which virions are normally measured are nanometres (1 nm $= 10^{-9}$ m). Although virions are very small, their dimensions cover a large range. Amongst the smallest are parvoviruses, with diameters about 20 nm, while the *mi*crobe-*mi*micking virus (mimivirus), isolated from an amoeba, is amongst the largest (Figure 1.3).

Virology is therefore concerned with very small particles, though often with very large numbers of those particles! A concentrated suspension of virions might contain 10^{12} virions/ml. A single virus-infected cell might produce 10^5 virions. A person infected with HIV might produce 10^{11} virions in a day.

Virions are not cells. They do not contain organelles, except for the virions of the arenaviruses, which contain cell ribosomes that were packaged when the virions were assembled.

1.3.2 Viruses have genes

The virion contains the genome of the virus. Whereas the genomes of cells are composed of double-stranded DNA, there are four possibilities for a virus genome:

- double-stranded DNA

- single-stranded DNA

- double-stranded RNA

- single-stranded RNA.

The genome is enclosed in a protein coat known as a capsid. The genome plus the capsid, plus other components in many cases, constitute the virion. The functions of the virion are to protect the genome and to deliver it into a cell in which it can replicate.

Generally, virus genomes are much smaller than cell genomes and the question arises as to how viruses encode all their requirements in a small genome. Viruses achieve this in a number of ways.

- *Viruses use host cell proteins.* The genomes of large viruses duplicate some of the functions of the host cell, but the small viruses rely very heavily on functions of the host cell. There is, however, one function that an RNA virus must encode, no matter how small its genome. That function is an RNA polymerase, because cells do not encode enzymes that can replicate virus RNA. A significant proportion of the genome of an RNA virus is taken up with the gene for an RNA polymerase.

- *Viruses code efficiently.* There may be overlapping genes and genes encoded within genes. The small genome of hepatitis B virus is a good example (see Section 18.6).

- *Many virus proteins are multifunctional.* A virus protein may have several enzyme activities.

1.3.3 Viruses are parasites

Viruses differ from cells in the way in which they multiply. A new cell is always formed directly from a pre-existing cell, but a new virion is never formed directly from a pre-existing virion. New virions are formed by a process of replication, which takes place inside a host cell and involves the synthesis of components followed by their assembly into virions.

Viruses are therefore parasites of cells, and are dependent on their hosts for most of their requirements, including

- building-blocks such as amino acids and nucleosides;

- protein-synthesizing machinery (ribosomes);

- energy, in the form of adenosine triphosphate.

A virus modifies the intracellular environment of its host in order to enhance the efficiency of the replication process. Modifications might include production of new membranous structures, reduced expression of cell genes or enhancement of a cell process. Some large phages encode proteins that boost photosynthesis in the cells of their photosynthetic bacterial hosts, thereby probably boosting the yields of virus from the cells.

A point has now been reached where the nature of viruses can be summarized in a concise definition (see the box).

Virus definition

A virus is a very small, non-cellular parasite of cells. Its genome, which is composed of either DNA or RNA, is enclosed in a protein coat.

1.3.4 Are viruses living or nonliving?

'Viruses belong to biology because they possess genes, replicate, evolve, and are adapted to particular hosts, biotic habitats, and ecological niches. However, . . .they are nonliving infectious entities that can be said, at best, to lead a kind of borrowed life.'
   Marc van Regenmortel and Brian Mahy (2004)

'It's life, Jim, but not as we know it!'
   Dr. McCoy speaking to Captain Kirk of the Starship Enterprise, *Star Trek*

There is an ongoing debate as to whether viruses are living or nonliving; the view taken depends on how life is defined. Viruses have genes and when they infect cells these genes are replicated, so in this sense viruses are living. They are, however, very different to cellular life forms, so Dr. McCoy's stock phrase (see the box) on finding new life forms in the galaxy could be applied to viruses. When viruses are outside their host cells they exist as virus particles (virions), which are inert, and could be described as nonliving, but viable bacterial spores are inert and are not considered to be nonliving. You might form your own view as to whether viruses are living or nonliving as you progress through this book.

When Beijerinck selected the word 'virus' he chose the Latin word for poison. This term has now been in use for over a century and virology has developed into a huge subject. More recently, the term virus has acquired further meanings. Computers

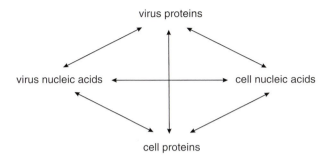

Figure 1.4 *Interactions between virus molecules and cell molecules.*

are threatened by *infection* with viruses that can be found *in the wild* once they have been *released* by their authors. These viruses are *specific* for certain file types. *Infected* files may be put on several web sites and a virus *epidemic* may ensue. Another use of the term virus is exemplified in John Humphrys' book 'Lost For Words', in which he talks about the *deadly* virus of management-speak *infecting* language. All the italicized terms in this paragraph are also used in the context of the viruses that are the subject of this book.

1.4 The remainder of the book

Having outlined the nature of viruses and why they are important, the remainder of the book will examine many aspects of fundamental and applied virology. The early chapters cover principles, such as the structure of virions, virus replication and the classification of viruses. There are then nine chapters devoted to reviews of particular groups of viruses, where both principles and applications of virology are covered. Towards the end of the book we consider specific applications of virology, including viral vaccines and anti-viral drugs. The final chapter is devoted to prions, which are not viruses!

It is important to point out that much of virology is concerned with characteristics of the proteins and nucleic acids of viruses, and with interactions between these molecules and the proteins and nucleic acids of cells (Figure 1.4). Most of these interactions rely on specific binding between the molecules. We shall also be discussing cellular structures, and processes such as transcription, translation and DNA replication. A good background in molecular biology and cell biology is therefore essential; some useful sources of information for plugging any gaps can be found under *Sources of further information*.

Learning outcomes

By the end of this chapter you should be able to

- discuss reasons for studying viruses;
- explain how viruses differ from other organisms;
- define the term 'virus'.

Sources of further information

Cell biology and molecular biology books

Alberts B. *et al.* (2004) *Essential Cell Biology*, 2nd edition, Garland

Brown W. M. and Brown P. M. (2002) *Transcription*, Taylor and Francis

Cooper G. M. and Hausman R. E. (2004) *The Cell: a Molecular Approach*, 3rd edition, ASM Press

Drlica K. (2004) *Understanding DNA and Gene Cloning*, 4th edition, Wiley

Lodish H. *et al.* (2004) *Molecular Cell Biology*, 5th edition, Freeman

Pollard T. D. and Earnshaw W. C. (2004) *Cell Biology*, Saunders

Reece R. J. R. (2004) *Analysis of Genes and Genomes*, Wiley

Watson J. D. *et al.* (2004) *Molecular Biology of the Gene*, 5th edition, Addison-Wesley

Weaver R. F. (2005) *Molecular Biology*, 3rd edition, McGraw-Hill

Historical paper

Hershey A. D. and Chase M. (1952) Independent functions of viral protein and nucleic acid in growth of bacteriophage *Journal of General Physiology*, **36**, 39–56

Recent papers

Breitbart M. and Rohwer F. (2005) Here a virus, there a virus, everywhere the same virus? *Trends In Microbiology*, **13**, 278–284

Chinen J. and Puck J. M. (2004) Successes and risks of gene therapy in primary immunodeficiencies *Journal of Allergy and Clinical Immunology*, **113**, 595–603

Suttle C. A. (2005) Viruses in the sea *Nature*, **437**, 356–361

van Regenmortel M. H. V. and Mahy B. W. J. (2004) Emerging issues in virus taxonomy *Emerging Infectious Diseases*, **10**(1) http://www.cdc.gov/ncidod/eid/vol10no1/03-0279.htm

Young L. S. *et al.* (2006) Viral gene therapy strategies: from basic science to clinical application *Journal of Pathology*, **208**, 299–318

2

Methods used in virology

At a glance

Virus Isolation and Culture

Animal virus plaques
in a cell culture [1]

Phage plaques
in a lawn of
bacterial cells [2]

Density Gradient Centrifugation

Separation of virus particles
in a density gradient [3]

Fluorescence Microscopy

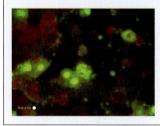
Virus-infected cells
detected using a
virus-specific antibody
labelled with a
fluorescent dye [4]

Confocal Microscopy

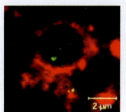
An endosome (labelled red)
containing virus protein
(labelled green)
in an infected cell [5]

Virology: Principles and Applications   John B. Carter and Venetia A. Saunders
© 2007 John Wiley & Sons, Ltd   ISBNs: 978-0-470-02386-0 (HB); 978-0-470-02387-7 (PB)

At a glance (continued)

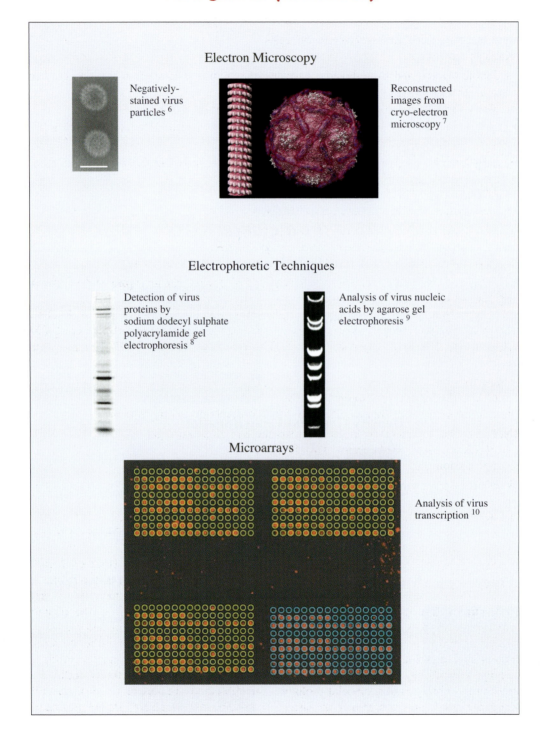

Electron Microscopy

Negatively-stained virus particles [6]

Reconstructed images from cryo-electron microscopy [7]

Electrophoretic Techniques

Detection of virus proteins by sodium dodecyl sulphate polyacrylamide gel electrophoresis [8]

Analysis of virus nucleic acids by agarose gel electrophoresis [9]

Microarrays

Analysis of virus transcription [10]

[1]Plaques formed by porcine reproductive and respiratory syndrome virus in a cell culture
From Lee and Yoo (2005) *Journal of General Virology*, **86**, 3091.
Reproduced by permission of the Society for General Microbiology and the author.

[2]Plaques formed by phage MS2 in *Escherichia coli* cells
Courtesy of Kathryn Newton.

[3]Separation of double-layered and triple-layered particles of rotavirus in a caesium chloride gradient
From López *et al.* (2005) *Journal of Virology*, **79**, 184.
Reproduced by permission of the American Society for Microbiology and the author.

[4]Light micrograph of cells infected with influenza A virus treated with antibody labelled with a fluorescent dye
Reproduced by permission of Argene SA, France.

[5]An endosome (labelled with a red fluorescent dye) containing rabies virus protein (labelled with green fluorescent protein) in the cytoplasm of an infected cell
From Finke and Conzelmann (2005) *Virus Research*, **111**, 120.
Reproduced by permission of Elsevier Limited and the author.

[6]Negatively stained particles of an orbivirus. The bar represents 50 nm.
From Attoui *et al.* (2005) *Journal of General Virology*, **86**, 3409.
Reproduced by permission of the Society for General Microbiology and the author.

[7]Reconstructed images from cryo-electron microscopy: measles virus nucleocapsid (left), echovirus type 12 bound to a fragment of its cell receptor (right)
Courtesy of Dr. David Bhella (MRC Virology Unit, Glasgow). Reinterpretations of data in Bhella *et al.* (2004) *Journal of Molecular Biology*, **340**, 319 (by permission of Elsevier Limited) and Bhella *et al.* (2004) *Journal of Biological Chemistry*, **279**, 8325 (by permission of The American Society for Biochemistry and Molecular Biology).

[8]Rotavirus proteins from infected cells
From López *et al.* (2005) *Journal of Virology*, **79**, 184.
Reproduced by permission of the American Society for Microbiology and the author.

[9]Orbivirus RNA segments separated by electrophoresis through an agarose gel
From Attoui *et al.* (2005) *Journal of General Virology*, **86**, 3409.
Reproduced by permission of the Society for General Microbiology and the author.

[10]Analysis of varicella-zoster virus transcription using microarrays
From Kennedy *et al.* (2005) *Journal of General Virology*, **86**, 2673.
Reproduced by permission of the Society for General Microbiology and the author.

2.1 Introduction to methods used in virology

Methods used in virology are introduced early in this book in order to provide an appreciation of the nature of the techniques that have been used to achieve our current level of knowledge and understanding of viruses. Virology is a huge subject and uses a wide range of methods. Many of the techniques of molecular biology and cell biology are used, and constraints on space permit us to mention only some of them. Much of the focus of this chapter is on methods that are unique to virology. Many of these methods are used, not only in virus research, but also in the diagnosis of virus diseases of humans, animals and plants.

Initially this chapter could be skimmed to gain an overview of its contents and thereafter used for reference. Details of the methods outlined here, and of other methods important in virology, fill many volumes, some of which are listed at the end of the chapter.

2.2   Cultivation of viruses

Virologists need to be able to produce the objects of their study, so a wide range of procedures has been developed for cultivating viruses. Virus cultivation is also referred to as propagation or growth, all terms borrowed from horticulture! A few techniques have been developed for the cultivation of viruses in cell-free systems, but in the vast majority of cases it is necessary to supply the virus with appropriate cells in which it can replicate.

Phages are supplied with bacterial cultures, plant viruses may be supplied with specially cultivated plants or with cultures of protoplasts (plant cells from which the cell wall has been removed), while animal viruses may be supplied with whole organisms, such as mice, eggs containing chick embryos (Figure 2.1) or insect larvae. For the most part, however, animal viruses are grown in cultured animal cells.

2.2.1 Animal cell culture

Animal cell culture techniques are well developed and most of the cells used are from continuous cell lines derived from humans and other animal species. Continuous cell lines consist of cells that have been immortalized, either in the laboratory or in the body (Figure 2.2); they can be subcultured indefinitely. The HeLa cell line is a widely used continuous cell line that was initiated in the middle of the 20th century from cells taken from a cervical carcinoma.

Sometimes it is difficult to find a cell line in which a virus can replicate. For many years no suitable cell culture system could be found for hepatitis C virus, but eventually a human hepatoma cell line was found to support replication of an isolate of the virus.

Cells are cultured in media that provide nutrients. Most media are supplemented with animal serum, which contains substances that promote the growth of many cell lines. Other important roles for the medium are the maintenance of optimum osmotic pressure and pH for the cells. Viruses can be cultivated in cells growing on the surface of a variety of plastic vessels (Figure 2.3) with the cells bathed in the growth medium. Most cells grow on a plastic or glass surface as a single layer of cells, known as a monolayer. Alternatively the cells can be suspended in the medium, which is stirred to keep them in suspension.

Contamination with bacteria and fungi can cause major problems in cell culture work; in order to minimize these problems work is normally done in a sterile cabinet (Figure 2.4) and most media contain antibiotics. Many cell types require a relatively high concentration of carbon dioxide, which can be supplied in a special incubator.

Sites into which viruses can be inoculated   Inoculation of an egg

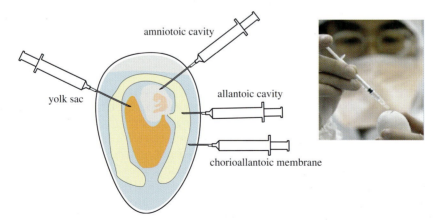

amniotoic cavity

yolk sac

allantoic cavity

chorioallantoic membrane

Figure 2.1   *Cultivation of viruses in eggs containing chick embryos.* Photograph courtesy of World Health Organisation.

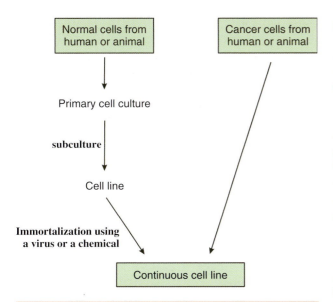

Figure 2.2 *Derivation of continuous cell lines of human and animal cells.* Most types of cell taken from the body do not grow well in culture. If cells from a primary culture can be subcultured they are growing as a cell line. They can be subcultured only a finite number of times unless they are immortalized, in which case they can be subcultured indefinitely as a continuous cell line. Cancer cells are already immortalized, and continuous cell lines may be established from these without further treatment.

2.3 Isolation of viruses

Many viruses can be isolated as a result of their ability to form discrete visible zones (plaques) in layers of host cells. If a confluent layer of cells is inoculated with virus at a concentration so that only a small proportion of the cells is infected, then plaques may form where areas of cells are killed or altered by the virus infection. Each plaque is formed when infection spreads radially from an infected cell to surrounding cells.

Plaques can be formed by many animal viruses in monolayers if the cells are overlaid with agarose gel to maintain the progeny virus in a discrete zone (Figure 2.5). Plaques can also be formed by phages in lawns of bacterial growth (Figure 2.6).

It is generally assumed that a plaque is the result of the infection of a cell by a single virion. If this is the case then all virus produced from virus in the plaque should be a clone, in other words it should be genetically identical. This clone can be referred to as an isolate, and if it is distinct from all other isolates it can be referred to as a strain. This is analogous to the derivation of a bacterial strain from a colony on an agar plate.

There is a possibility that a plaque might be derived from two or more virions so, to increase the probability that a genetically pure strain of virus has been obtained, material from a plaque can be inoculated onto further monolayers and virus can be derived from an individual plaque. The virus is said to have been plaque purified.

When a virus is first isolated it may replicate poorly in cells in the laboratory, but after it has gone through a number of replication cycles it may replicate more efficiently. Each time the virus is 'sub-cultured' (to borrow a term from bacteriology) it is said to have been passaged. After a number of passages the virus may be genetically different to the original wild strain, in which case it is now a laboratory strain.

2.4 Centrifugation

After a virus has been propagated it is usually necessary to remove host cell debris and other contaminants before the virus particles can be used for laboratory studies, for incorporation into a vaccine, or for some other purpose. Many virus purification procedures involve centrifugation; partial purification can be achieved by differential centrifugation and a higher degree of purity can be achieved by some form of density gradient centrifugation.

2.4.1 Differential centrifugation

Differential centrifugation involves alternating cycles of low-speed centrifugation, after which most of the virus is still in the supernatant, and high-speed centrifugation, after which the virus is in the pellet (Figure 2.7).

2.4.2 Density gradient centrifugation

Density gradient centrifugation involves centrifuging particles (such as virions) or molecules (such as nucleic

Figure 2.3   *Cell culture flasks, dishes and plates.* Photographs of TPP cell culture products courtesy of MIDSCI.

Figure 2.4   *Cell culture work.* Precautions to avoid contamination include working in a sterile cabinet and wearing gloves and mask. Photograph courtesy of Novartis Vaccines.

Plaques formed by West Nile virus in a cell monolayer

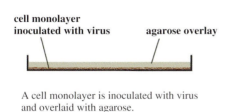

cell monolayer
inoculated with virus   agarose overlay

A cell monolayer is inoculated with virus and overlaid with agarose.

Both flasks were inoculated with virus. Plaque formation in the flask on the right has been inhibited by West Nile virus-specific antibody. Courtesy of Dr. Elieen Ostlund, National Veterinary Services Laboratories, US Department of Agriculture.

Figure 2.5 *Method for production of plaques by animal viruses.*

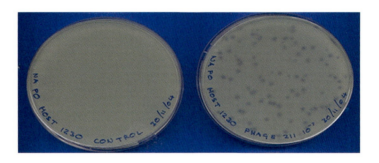

Figure 2.6 *Plaques formed by a phage in a bacterial lawn.* The control plate on the left was inoculated with only the bacterial host. The plate on the right was inoculated with phage and bacterial host. Photograph courtesy of Philip O'Grady.

acids) in a solution of increasing concentration, and therefore density. The solutes used have high solubility: sucrose is commonly used. There are two major categories of density gradient centrifugation: rate zonal and equilibrium (isopycnic) centrifugation (Figure 2.8).

In rate zonal centrifugation a particle moves through the gradient at a rate determined by its sedimentation coefficient, a value that depends principally on its size. Homogeneous particles, such as identical virions, should move as a sharp band that can be harvested after the band has moved part way through the gradient.

In equilibrium centrifugation a concentration of solute is selected to ensure that the density at the bottom of the gradient is greater than that of the particles/molecules to be purified. A particle/molecule suspended in the gradient moves to a point where the gradient density is the same as its own density. This technique enables the determination of the buoyant densities of nucleic acids and of virions. Buoyant densities of virions determined in gradients of caesium chloride are used as criteria in the characterization of viruses.

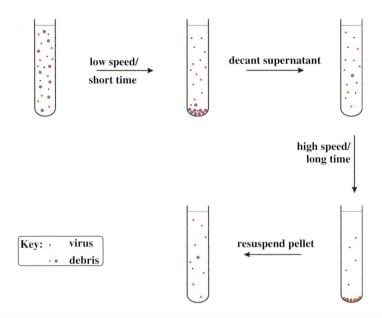

Figure 2.7 *Partial purification of virions by differential centrifugation.* A crude preparation of virus containing host debris is subjected to low-speed/short-time centrifugation (e.g. 10 000 *g*/20 minutes) followed by high-speed/long-time centrifugation (e.g. 100 000 *g*/2 hours). This cycle can be repeated to obtain a higher degree of purity. The final pellet containing partly purified virus is resuspended in a small volume of fluid.

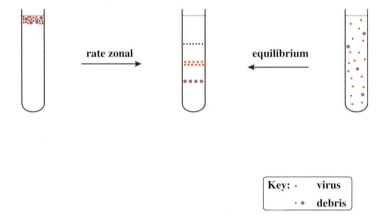

Figure 2.8 *Purification of virions by density gradient centrifugation.* A partly purified preparation of virus is further purified in a density gradient. Rate zonal centrifugation involves layering the preparation on top of a pre-formed gradient. Equilibrium centrifugation can often be done starting with a suspension of the impure virus in a solution of the gradient material; the gradient is formed during centrifugation.

2.5 Structural investigations of cells and virions

2.5.1 Light microscopy

The sizes of most virions are beyond the limits of resolution of light microscopes, but light microscopy has useful applications in detecting virus-infected cells, for example by observing cytopathic effects (Section 2.7.2) or by detecting a fluorescent dye linked to antibody molecules that have bound to a virus antigen (fluorescence microscopy).

Confocal microscopy is proving to be especially valuable in virology. The principle of this technique is the use of a pinhole to exclude light from out-of-focus regions of the specimen. Most confocal microscopes scan the specimen with a laser, producing exceptionally clear images of thick specimens and of fluorescing specimens. Furthermore, 'optical slices' of a specimen can be collected and used to create a three-dimensional representation. The techniques can be used with live cells and can be applied to investigations of protein trafficking, with the virus or cell protein under investigation carrying a suitable label, e.g. green fluorescent protein (a jellyfish protein).

2.5.2 Electron microscopy

Many investigations of the structure of virions or of virus-infected cells involve electron microscopy. Large magnifications are achievable with a transmission electron microscope but the specimen, whether it is a suspension of virions or an ultrathin section of a virus-infected cell, must be treated so that details can be visualized.

Negative staining techniques generate contrast by using heavy-metal-containing compounds, such as potassium phosphotungstate and ammonium molybdate. In electron micrographs of virions the stains appear as dark areas around the virions, allowing the overall virion shape and size to be determined. Further structural detail may be apparent if the stains penetrate any crevices on the virion surface or any hollows within the virion. Negative staining techniques have generated many high quality electron micrographs, but the techniques have limitations, including structural distortions resulting from drying.

Cryo-electron microscopy techniques are more recent. In these a wet specimen is rapidly cooled to a temperature below $-160\,^\circ$C, freezing the water as a glasslike material, as in the method outlined in Figure 2.9. The images are recorded while the specimen is frozen. They require computer processing in order to extract maximum detail, and data from multiple images are processed to reconstruct three-dimensional images of virus particles. This may involve averaging many identical copies or combining images into three-dimensional density maps (tomograms).

2.5.3 X-ray crystallography

X-ray crystallography is another technique that is revealing detailed information about the three-dimensional structures of virions (and DNA, proteins and DNA−protein complexes). This technique requires the production of a crystal of the virions or molecules under study. The crystal is placed in a beam of X-rays, which are diffracted by repeating arrangements of molecules/atoms in the crystal. Analysis of the diffraction pattern allows the relative positions of these molecules/atoms to be determined.

Other techniques that are providing useful information about the structure of viruses are nuclear magnetic resonance and atomic force microscopy.

2.6 Electrophoretic techniques

Mixtures of proteins or nucleic acids can be separated by electrophoresis in a gel composed of agarose or polyacrylamide. In most electrophoretic techniques each protein or nucleic acid forms a band in the gel. Electrophoresis can be performed under conditions where the rate of movement through the gel depends on molecular weight. The molecular weights of the protein or nucleic acid molecules can be estimated by comparing the positions of the bands with positions of bands formed by molecules of known molecular weight electrophoresed in the same gel. The technique for estimating molecular weights of proteins is polyacrylamide gel electrophoresis in the presence of the detergent sodium dodecyl sulphate (SDS-PAGE; Figure 2.10).

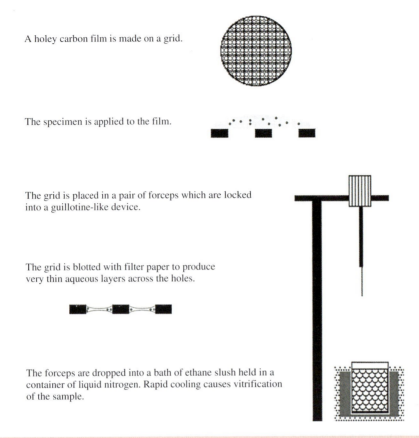

A holey carbon film is made on a grid.

The specimen is applied to the film.

The grid is placed in a pair of forceps which are locked into a guillotine-like device.

The grid is blotted with filter paper to produce very thin aqueous layers across the holes.

The forceps are dropped into a bath of ethane slush held in a container of liquid nitrogen. Rapid cooling causes vitrification of the sample.

Figure 2.9 *Preparation of a specimen for cryo-electron microscopy.* Modified, with the permission of the authors and the American Society for Microbiology, from Baker, Olson and Fuller (1999) *Microbiology and Molecular Biology Reviews,* **63**, 862.

The patterns of nucleic acids and proteins after electrophoretic separation may be immobilized by transfer (blotting) onto a membrane. If the molecules are DNA the technique is known as Southern blotting, named after Edwin Southern; if the molecules are RNA the technique is known as northern blotting, and if the molecules are protein the technique is known as western blotting!

2.7 Detection of viruses and virus components

A wide range of techniques has been developed for the detection of viruses and virus components and many of them are used in laboratories involved with diagnosis of virus diseases. The techniques can be arranged in four categories: detection of (1) virions, (2) virus infectivity, (3) virus antigens and (4) virus nucleic acids.

2.7.1 Detection of virions

Specimens can be negatively stained (Section 2.5) and examined in an electron microscope for the presence of virions. Limitations to this approach are the high costs of the equipment and limited sensitivity; the minimum detectable concentration of virions is about 10^6/ml. An example of an application of this technique is the examination of faeces from a patient with gastroenteritis for the presence of rotavirus particles.

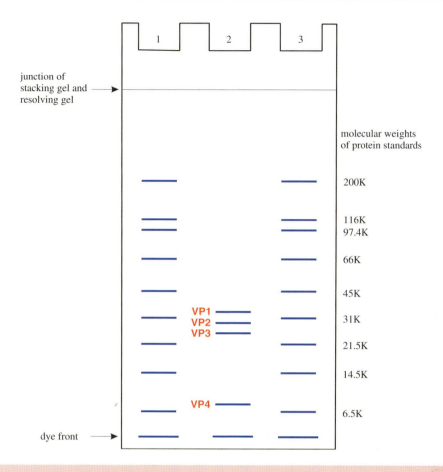

Figure 2.10 *Separation of proteins and estimation of their molecular weights using SDS-PAGE.* Lanes 1 and 3 contain standard proteins of known molecular weight. Lane 2 contains the four capsid proteins of a picornavirus.

2.7.2 Detection of infectivity using cell cultures

Not all virions have the ability to replicate in host cells. Those virions that do have this ability are said to be 'infective', and the term 'infectivity' is used to denote the capacity of a virus to replicate. Virions may be non-infective because they lack part of the genome or because they have been damaged.

To determine whether a sample or a specimen contains infective virus it can be inoculated into a culture of cells, or a host organism, known to support the replication of the virus suspected of being present. After incubation of an inoculated cell culture at an appropriate temperature it can be examined by light microscopy for characteristic changes in the appearance of the cells resulting from virus-induced damage. A change of this type is known as a cytopathic effect (CPE); examples of CPEs induced by poliovirus and herpes simplex virus can be seen in Figure 2.11. The poliovirus-infected cells have shrunk and become rounded, while a multi-nucleated giant cell known as a syncytium (plural *syncytia*) has been formed by the fusion of membranes of herpes simplex virus-infected cells.

The quantity of infective virus in a specimen or a preparation can be determined (Section 2.8).

2.7.3 Detection of virus antigens

Virus antigens can be detected using virus-specific antisera or monoclonal antibodies. In most techniques

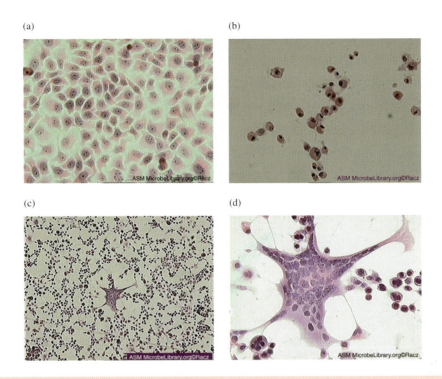

Figure 2.11 *Cytopathic effects caused by replication of poliovirus and herpes simplex virus in cultures of Vero (monkey kidney) cells.* (a) Uninfected Vero cells. (b) Infected with poliovirus. (c), (d) Infected with herpes simplex virus. (a), (b) and (d) were viewed at × 400 magnification; (c) was viewed at × 100 magnification. The cells were stained with haematoxylin and eosin. Reproduced by permission of Dr. Maria-Lucia Rácz (University of Sao Paulo) and the American Society for Microbiology MicrobeLibrary.

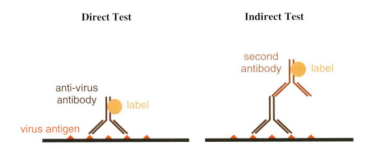

Figure 2.12 *Principles of tests to detect virus antigens.* The specimen is treated with anti-virus antibody. In an indirect test a labelled second antibody detects any anti-viral antibody that has bound to antigen.

positive results are indicated by detecting the presence of a label, which may be attached either to the anti-virus antibody (direct tests) or to a second antibody (indirect tests) (Figure 2.12). The anti-virus antibody is produced by injecting virus antigen into one animal species and the second antibody is produced by injecting immunoglobulin from the first animal species into a second animal species.

Antibodies can have many types of label attached and the labels can be detected using a range of methods. Some types of label and some methods for detecting them are listed in Table 2.1.

2.7.4 Detection of virus nucleic acids

2.7.4.a Hybridization

Virus genomes or virus messenger RNAs (mRNAs) may be detected using sequence-specific DNA probes carrying appropriate labels (Figure 2.13). Some of the labels that are used for antibody detection can be used to label the probes (Table 2.1).

Hybridization may take place on the surface of a membrane after Southern blotting (DNA) or northern blotting (RNA) (Section 2.6). Thin sections of tissue may be probed for the presence of specific nucleic acids, in which case the technique is known as *in situ* hybridization.

Table 2.1 Molecules used to label antibodies (and nucleic acids) and techniques used to detect them

Label	Detection technique
Enzyme	Enzyme-linked immunosorbent assay
Fluorescent	Fluorescence microscopy Fluorimetry
Gold	Electron microscopy
Radio-active	Autoradiography

2.7.4.b Polymerase chain reaction (PCR)

When a sample is likely to contain a low number of copies of a virus nucleic acid the probability of detection can be increased by amplifying virus DNA using a PCR, while RNA can be copied to

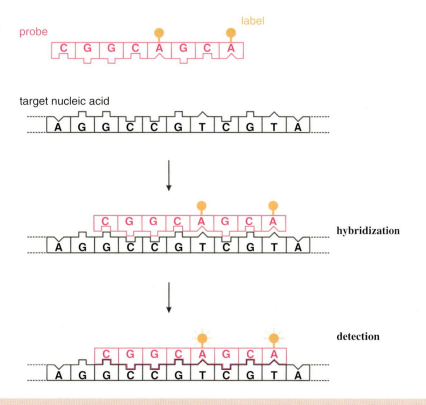

Figure 2.13 *Detection of a specific nucleic acid (DNA or RNA) using a labelled DNA probe.*

DNA and amplified using a RT (reverse transcriptase)-PCR. The procedures require oligonucleotide primers specific to viral sequences. An amplified product can be detected by electrophoresis in an agarose gel, followed by transfer to a nitrocellulose membrane, which is incubated with a labelled probe.

There are also PCR techniques available for determining the number of copies of a specific nucleic acid in a sample. Real-time PCR is commonly used for this purpose. In this technique the increase in DNA concentration during the PCR is monitored using fluorescent labels; the larger the initial copy number of DNA, the sooner a significant increase in fluorescence is observed. The PCR cycle at which the fluorescent signal passes a defined threshold is determined and this gives an estimate of the starting copy number. An example is given in Figure 2.14.

2.8   Infectivity assays

An infectivity assay measures the titre (the concentration) of infective virus in a specimen or a preparation.

Samples are inoculated into suitable hosts, in which a response can be observed if infective virus is present. Suitable hosts might be animals, plants or cultures of bacterial, plant or animal cells. Infectivity assays fall into two classes: quantitative and quantal.

2.8.1 Quantitative assays

Quantitative assays are those in which each host response can be any one of a series of values, such as a number of plaques (Section 2.3). A plaque assay can be carried out with any virus that can form plaques, giving an estimate of the concentration of infective virus in plaque-forming units (pfu).

The assay is done using cells grown in Petri dishes, or other appropriate containers, under conditions that will allow the formation of plaques. Cells are inoculated with standard volumes of virus dilutions. After incubation, dishes that received high dilutions of virus may have no plaques or very few plaques, while dishes that received low dilutions of virus may have very large numbers of plaques or all the cells may have lysed.

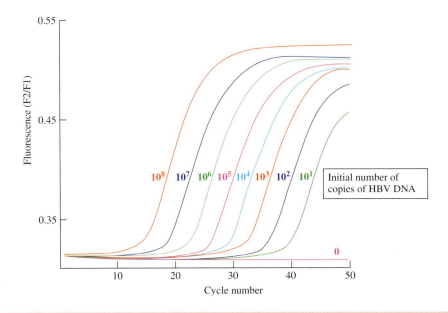

Figure 2.14 *Quantification of hepatitis B virus (HBV) DNA by real-time PCR.* When there are 10^8 copies of HBV DNA in the sample, fluorescence starts to increase at an early cycle number. The increase starts at progressively later cycles with decreasing numbers of HBV DNA copies. Data from Ho *et al.* (2003) *Journal of Medical Microbiology,* **52**, 397; by permission of the Society for General Microbiology and the authors.

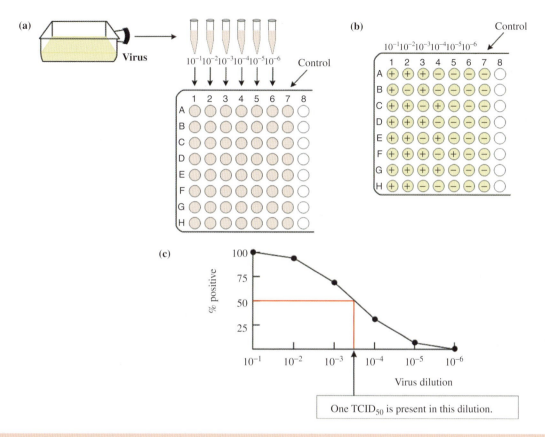

Figure 2.15 *Example of a TCID$_{50}$ assay.* (a) Tenfold dilutions of the virus were inoculated into cell cultures grown in the wells of a plate. Each well received 1 ml virus suspension, except for the control wells, which each received 1 ml diluent. (b) After incubation the cell culture in each well was scored '+' or '−' for CPE. (c) The results of the assay.

Dishes are selected that have plaque numbers within a certain range, e.g. 30–300. The plaques are counted and the concentration of virus in the sample (pfu/ml) is calculated.

2.8.2 Quantal assays

In a quantal assay each inoculated subject either responds or it does not; for example, an inoculated cell culture either develops a CPE or it does not; an inoculated animal either dies or it remains healthy. The aim of the assay is to find the virus dose that produces a response in 50 per cent of inoculated subjects. In cell culture assays this dose is known as the TCID$_{50}$ (the dose that infects 50 per cent of inoculated tissue cultures). In animal assays the dose is known as the ID$_{50}$, and where the virus infection kills the animal the

dose is known as the LD$_{50}$ (the dose that is lethal for 50 per cent of inoculated animals).

The outcome of a TCID$_{50}$ determination can be used to estimate a virus titre in pfu, or vice versa, using the formula

$$1 \text{ TCID}_{50} = 0.7 \text{ pfu}$$

This formula applies only if the method used to visualize plaques does not alter the yield of virus.

An example of a TCID$_{50}$ determination is depicted in Figure 2.15.

It is clear that the dilution of virus that contains one TCID$_{50}$ lies between 10^{-3} and 10^{-4}. Several methods have been developed for estimating this end point; one of these, the Reed-Muench method, is shown in the box.

The Reed-Muench method for estimating $TCID_{50}$

Virus dilution	Infections per number inoculated	Observed values		Cumulative values[1]		Infection ratio[2]	% infection[3]
		Positive	Negative	Positive	Negative		
10^{-1}	8/8	8	0	24	0	24/24	100
10^{-2}	7/8	7	1	16	1	16/17	94
10^{-3}	5/8	5	3	9	4	9/13	69
10^{-4}	3/8	3	5	4	9	4/13	31
10^{-5}	1/8	1	7	1	16	1/17	6
10^{-6}	0/8	0	8	0	24	0/24	0

[1]The cumulative values are derived by adding up the observed values in the direction of the arrows.
[2]The infection ratio is the number of positives for the cumulative value out of the total for the cumulative value.
[3]The % infection is the infection ratio converted to a percentage.

It had already been determined that the dilution of virus that contains one $TCID_{50}$ lies between 10^{-3} and 10^{-4}, so the end point can be expressed as $10^{-(3+x)}$, where x is the value to be estimated.

$$x = log_{10} \text{ dilution factor} \left(\frac{\% \text{ infection at next dilution above } 50\% - 50}{\% \text{ infection at next dilution above } 50\% - \% \text{ infection at next dilution below } 50\%} \right)$$

$$= 1 \left(\frac{69 - 50}{69 - 31} \right)$$

$$= 0.5$$

End point $= 10^{-(3+0.5)} = 10^{-3.5}$.
i.e. 1 ml of a $10^{-3.5}$ dilution contains one $TCID_{50}$ of virus
i.e. 1 ml of a 1/3200 dilution contains one $TCID_{50}$ of virus.
(3.2 is the antilogarithm of 0.5)
The concentration of virus in the undiluted suspension is 3.2×10^3 $TCID_{50}$/ml.

A similar approach can be used to assay the infectivity of any virus that causes the formation of discrete lesions. A number of animal viruses, including smallpox virus and other poxviruses, can cause the formation of pocks after inoculation onto the chorioallantoic membrane of a chick embryo (Figure 2.16). Similarly, many plant viruses cause the formation of lesions after they have been rubbed on the surface of leaves (Figure 2.17), though the titration is not directly equivalent to a plaque assay as the number of sites on a leaf through which a virus can enter is limited.

2.8.3 One-step growth curve

A common type of experiment that involves assaying virus infectivity is a single-burst experiment, which provides data for a one-step growth curve (Figure 2.18).

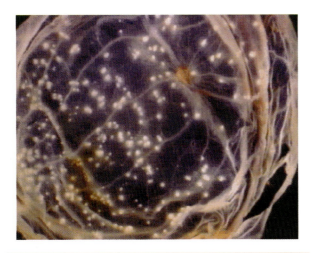

Figure 2.16 *Virus-induced pocks on the chorioallan-toic membrane of a chick embryo.* From Buxton and Fraser (1977) *Animal Microbiology*, Volume 2, Blackwell Scientific Publications. Reproduced by permission of the publisher.

The first one-step growth curves produced were for phages, but they have since been produced for other viruses.

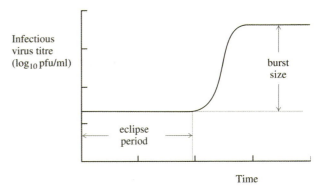

Figure 2.18 *One-step growth curve.* The eclipse period is the period during which no intracellular infectious virus can be recovered; infectious nucleic acid might be recoverable in some cases. The burst size is the average yield of infectious virus per cell.

For a period (the eclipse period), the titre of infective virus remains constant. During the eclipse period the virus is replicating, but no virus has been released from infected cells, so each infected cell gives rise to one plaque. As infected cells begin to die and release virus

Figure 2.17 *Leaf lesions resulting from infection with potato virus Y.* The leaf on the right is a control inoculated with sterile buffer. Copyright INRA, France/Didier Spire; reproduced with permission.

the titre begins to rise, and when all the infected cells have lysed the titre levels off.

The procedure for deriving a one-step growth curve is now outlined.

- A suspension of virions is mixed with a suspension of host cells. It is necessary to ensure that, as far as possible, all the cells are infected simultaneously, so the number of virions must greatly exceed the number of cells. The ratio of virions to cells is known as the multiplicity of infection (m.o.i.). In a one-step growth experiment an m.o.i. between 5 and 10 is commonly used.

- Adsorption of virions to cells is allowed to proceed for a suitable period (e.g. 2 minutes), and then adsorption is stopped by, for example, greatly diluting the mixture and/or by adding antivirus antiserum.

- A sample is taken from the suspension at intervals until lysis is complete. For phages the experiment is usually complete in less than one hour. For animal viruses, however, the timescale is measured in hours or days, reflecting the much slower growth rates of eukaryotic cells and the longer times taken for replication of animal viruses.

- A plaque assay is carried out on each sample.

- The logarithm of the virus titre is plotted against time to obtain the one-step growth curve (Figure 2.18).

This type of experiment provides valuable information about the replication cycle of a virus in a particular host cell system. The average yield of infectious virus per cell (the burst size) can be calculated from the formula

$$\text{burst size} = \frac{\text{final virus titre}}{\text{initial virus titre}}$$

It should be noted that the burst size is an average of what may be a significant variable as there is usually much variation in burst size between cells. One factor that may affect the burst size is the physiological state of the cells. In one study with a phage, a burst size of 170 was obtained when growing bacteria were used, while a value of 20 was obtained with resting bacteria.

2.9 Virus genetics

2.9.1 Genome sequencing

Determination of the sequence of bases in a DNA molecule is usually done using the dideoxy chain termination method developed by Fred Sanger and his colleagues at Cambridge. The sequence of the genome of phage φX174 DNA was determined by this group six years after the first determination of a DNA sequence, which itself was from a virus: 12 base pairs (bp) of phage lambda DNA (Figure 2.19).

With the help of computer programmes such as Artemis and BLAST much useful information can be derived from virus genome sequences.

- Open reading frames (ORFs) can be found. The ORFs can be translated into the amino acid sequences of the virus proteins, and these may allow characteristics and functions of the proteins to be deduced. Hydrophobic sequences may indicate membrane association; the sequence Gly–Asp–Asp may indicate RNA-dependent RNA polymerase activity.

- Sequences that regulate gene expression, e.g. promoters and enhancers, can be identified.

- During an outbreak of virus disease, such as foot and mouth disease or influenza, sequences of virus isolates can be compared. This can provide useful epidemiological information such as the source of the virus responsible and whether more than one strain is responsible for the outbreak.

- Phylogenetic trees can be constructed (Section 10.2.1).

2.9.2 Genome manipulation

The wide range of techniques available for the manipulation of nucleic acids can be applied to virus genomes. These techniques include the isolation of specific fragments of genomes using restriction endonucleases, the cloning of fragments in bacterial plasmids and the introduction of site-specific mutations into virus genomes. The natural processes of recombination and

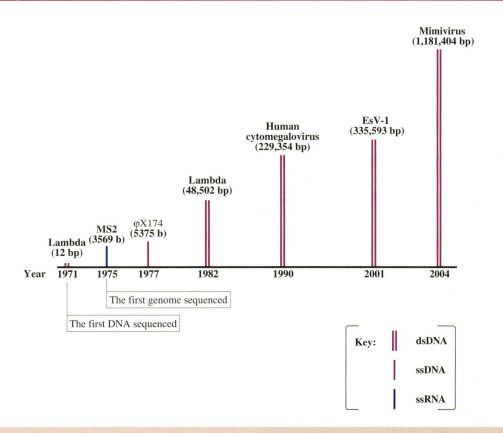

Figure 2.19 *Time line depicting some landmarks in viral genome sequencing.* Lambda, MS2 and φX174 are phages. The host of EsV-1 is the marine filamentous brown alga *Ectocarpus siliculosus*. The host of mimivirus is an amoeba. The first DNA ever sequenced was 12 bp of the genome of phage lambda. Eleven years later the complete genome sequence of this virus was published. Not to scale.

reassortment (see Chapter 20) that produce new combinations of virus genes can be harnessed to produce new viral genotypes in the laboratory.

2.9.3 Investigation of gene function and expression

The function of a gene may be deduced if its expression is blocked, and many studies of gene function involve the creation of a virus with a mutated gene. Many of the mutants are unable to replicate under conditions in which the wild-type virus replicates and are known as conditional lethal mutants. Most studies have used temperature-sensitive mutants, which are unable to replicate at temperatures at which the wild-type virus replicates (non-permissive temperatures), but are able to replicate at other temperatures (permissive temperatures).

Techniques for introducing mutations into DNA are well established and have been applied to the study of DNA viruses for some time. More recently, techniques for manipulating the genomes of RNA viruses have been developed. These techniques are referred to as reverse genetics, and they involve the reverse transcription of the RNA genome to DNA, mutation of the DNA, then transcription back to RNA. Reverse genetics are discussed further in Chapters 14 and 15.

Another way in which the expression of a gene can be blocked is by RNA silencing, or RNA interference (RNAi). This approach exploits a cell defence mechanism (described in Section 9.2.3) that can destroy virus mRNAs after induction by specific dsRNA molecules. Short sequences of dsRNA can be used to inhibit the expression of virus genes, and hence to investigate the functions of those genes. Yet another approach to the investigation of gene function involves the introduction of the gene into cells in which it is transiently expressed and changes in the cells are monitored.

As well as the production of virus mutants there are other ways in which viruses can be genetically modified in order to aid their study. A sequence encoding a tag, such as green fluorescent protein, can be appended to the gene for a virus protein so that the distribution of the protein in infected cells can be monitored (Figure 2.20).

The expression of virus genes in infected cells can be monitored with the aid of DNA microarrays. A DNA microarray consists of a substrate, such as a glass slide, with hundreds, thousands or even millions of small spots of DNA attached. Each spot contains a specific DNA probe and specific DNA or RNA molecules in a sample are detected by hybridization to specific spots.

Microarray technology allows investigators to monitor the expression of hundreds or thousands of genes, and is thus ideal for studies of viruses with large genomes such as the herpesviruses and the large DNA phages. Virus mRNAs are detected by copying RNA from infected cells to DNA using a reverse transcriptase, amplifying the DNA with a PCR, labelling it with a fluorescent dye and adding it to a microarray. The probes that bind DNA from the sample are detected by scanning with a laser at a wavelength that excites the fluorescent dye.

DNA microarrays can also be used to investigate the expression of host cell genes that are relevant to viruses. Cell surface molecules used by viruses as receptors have been identified using microarrays.

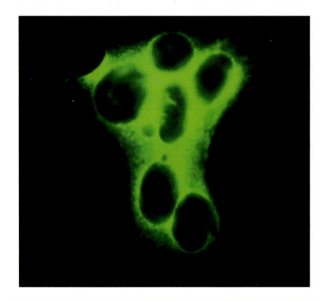

Figure 2.20 *A syncytium (giant cell) infected with a genetically modified HIV.* The gene for enhanced green fluorescent protein has been inserted into the virus genome. The distribution of the fluorescence indicates the distribution of the labelled virus protein. From Müller *et al.* (2004) *Journal of Virology*, **78**, 10803; reproduced by permission of the American Society for Microbiology and the authors.

Learning outcomes

By the end of this chapter you should be able to

- outline methods for

    o cultivation of viruses;

    o purification of viruses;

    o detection of viruses and their components;

    o assay of virus infectivity;

    o investigation of virus gene function;

- assess the value of virus genome sequencing.

Sources of further information

Books

Cann A. J., editor (1999) *Virus Culture: a Practical Approach*, Oxford University Press

Coleman W. B. and Tsongalis G. J. (2006) *Molecular Diagnostics for the Clinical Laboratorian*, 2nd edition, Humana

Collier L. and Oxford J. (2006) *Human Virology*, 3rd edition, Chapter 36, Oxford University Press

Edwards K., Logan J. and Saunders N., editors (2004) *Real-Time PCR: an Essential Guide*, Horizon

Freshney R. I. (2005) *Culture of Animal Cells: a Manual of Basic Technique*, 5th edition, Wiley-Liss

Payment P. (2002) Cultivation and assay of animal viruses. Chapter 8 in Hurst C. J. *et al.*, editors, *Manual of Environmental Microbiology*, 2nd edition, ASM Press

Ye S. and Day I. N. M., editors (2003) *Microarrays and Microplates: Applications in Biomedical Sciences*, BIOS

Historical paper

Ellis E. L. and Delbruck M. (1939) The growth of bacteriophage *Journal of General Physiology*, **22**, 365–384

Recent papers

Curry A, Appleton H. and Dowsett B. (2006) Application of transmission electron microscopy to the clinical study of viral and bacterial infections: present and future *Micron*, **37**, 91–106

DeFilippis V. *et al.* (2003) Functional genomics in virology and antiviral drug discovery *Trends in Biotechnology*, **21**, 452–457

Domiati-Saada R. and Scheuermann R. H. (2006) Nucleic acid testing for viral burden and viral genotyping *Clinica Chimica Acta*, **363**, 197–205

Kennedy P. G. E. *et al.* (2005) Transcriptomal analysis of varicella-zoster virus infection using long oligonucleotide-based microarrays *Journal of General Virology*, **86**, 2673–2684

Kuznetsov Yu. G. *et al.* (2001) Imaging of viruses by atomic force microscopy *Journal of General Virology*, **82**, 2025–2034

Natarajan P. *et al.* (2005) Exploring icosahedral virus structures with VIPER *Nature Reviews Microbiology*, **3**, 809–817

Rossmann M. G. *et al.* (2005) Combining X-ray crystallography and electron microscopy *Structure*, **13**, 355–362

Vernet G. (2004) Molecular diagnostics in virology *Journal of Clinical Virology*, **31**, 239–247

Watzinger F., Ebner K. and Lion T. (2006) Detection and monitoring of virus infections by real-time PCR *Molecular Aspects of Medicine*, **27**, 254–298

3

Virus structure

At a glance

Main types of virion structure		Genomes			
		dsDNA	ssDNA	dsRNA	ssRNA
Icosahedral, naked		✓	✓	✓	✓
Icosahedral, enveloped		✓		✓	✓
Helical, naked		✓	✓		✓
Helical, enveloped					✓

Virology: Principles and Applications   John B. Carter and Venetia A. Saunders
© 2007 John Wiley & Sons, Ltd   ISBNs: 978-0-470-02386-0 (HB); 978-0-470-02387-7 (PB)

3.1   Introduction to virus structure

Outside their host cells, viruses survive as virus particles, also known as virions. The virion is a gene delivery system; it contains the virus genome, and its functions are to protect the genome and to aid its entry into a host cell, where it can be replicated and packaged into new virions. The genome is packaged in a protein structure known as a capsid.

Many viruses also have a lipid component, generally present at the surface of the virion forming an envelope, which also contains proteins with roles in aiding entry into host cells. A few viruses form protective protein occlusion bodies around their virions. Before looking at these virus structures we shall consider characteristics of the nucleic acid and protein molecules that are the main components of virions.

3.2   Virus genomes

A virion contains the genome of a virus in the form of one or more molecules of nucleic acid. For any one virus the genome is composed of either RNA or DNA. If a new virus is isolated, one way to determine whether it is an RNA virus or a DNA virus is to test its susceptibility to a ribonuclease and a deoxyribonuclease. The virus nucleic acid will be susceptible to degradation by only one of these enzymes.

Each nucleic acid molecule is either single-stranded (ss) or double-stranded (ds), giving four categories of virus genome: dsDNA, ssDNA, dsRNA and ssRNA. The dsDNA viruses encode their genes in the same kind of molecule as animals, plants, bacteria and other cellular organisms, while the other three types of genome are unique to viruses. It interesting to note that most fungal viruses have dsRNA genomes, most plant viruses have ssRNA genomes and most prokaryotic viruses have dsDNA genomes. The reasons for these distributions presumably concern diverse origins of the viruses in these very different host types.

A further categorization of a virus nucleic acid can be made on the basis of whether the molecule is linear, with free 5′ and 3′ ends, or circular, as a result of the strand(s) being covalently closed. Examples of each category are given in Figure 3.1. In this figure, and indeed throughout the book, molecules of DNA and RNA are colour coded. Dark blue and light blue depict (+) RNA and (−) RNA respectively; these terms are explained in Section 6.2.

It should be noted that some linear molecules may be in a circular conformation as a result of base pairing between complementary sequences at their ends (see Figure 3.7 below). This applies, for example, to the DNA in hepadnavirus virions and to the RNA in influenza virions.

3.2.1 Genome size

Virus genomes span a large range of sizes. Porcine circovirus (ssDNA) and hepatitis delta virus (ssRNA) each have a genome of about 1.7 kilobases (kb), while at the other end of the scale there are viruses with dsDNA genomes comprised of over 1000 kilobase pairs (kbp). The maximum size of a virus genome is subject to constraints, which vary with the genome category. As the constraints are less severe for dsDNA all of the large virus genomes are composed of dsDNA. The largest RNA genomes known are those of some coronaviruses, which are 33 kb of ssRNA.

The largest virus genomes, such as that of the mimivirus, are larger than the smallest genomes of cellular organisms, such as some mycoplasmas (Figure 3.2).

3.2.2 Secondary and tertiary structure

As well as encoding the virus proteins (and in some cases untranslated RNAs) to be synthesized in the infected cell, the virus genome carries additional information, such as signals for the control of gene expression. Some of this information is contained within the nucleotide sequences, while for the single-stranded genomes some of it is contained within structures formed by intramolecular base pairing.

In ssDNA complementary sequences may base pair through G–C and A–T hydrogen bonding; in ssRNA weaker G–U bonds may form in addition to G–C and A–U base pairing. Intramolecular base pairing results in regions of secondary structure with stem-loops and bulges (Figure 3.3(a)). In some ssRNAs intramolecular base pairing results in structures known

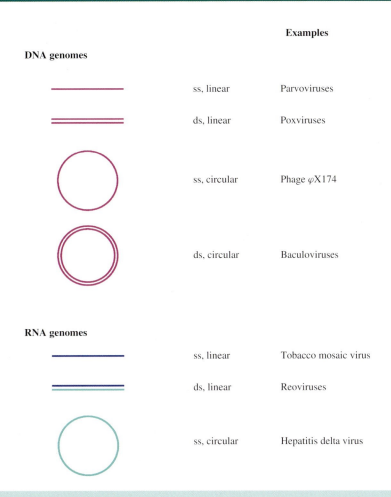

Examples

DNA genomes

ss, linear   Parvoviruses

ds, linear   Poxviruses

ss, circular   Phage φX174

ds, circular   Baculoviruses

RNA genomes

ss, linear   Tobacco mosaic virus

ds, linear   Reoviruses

ss, circular   Hepatitis delta virus

Figure 3.1 *Linear and circular viral genomes.*
ss: single-stranded
ds: double-stranded
There are no viruses known with circular dsRNA genomes

as pseudoknots, the simplest form of which is depicted in Figure 3.3(b).

Regions of secondary structure in single-stranded nucleic acids are folded into tertiary structures with specific shapes, many of which are important in molecular interactions during virus replication. For an example see Figure 14.6, which depicts the 5′ end of poliovirus RNA, where there is an internal ribosome entry site to which cell proteins bind to initiate translation. Some pseudoknots have enzyme activity, while others play a role in ribosomal frameshifting (Section 6.4.2).

3.2.3 Modifications at the ends of virus genomes

It is interesting to note that the genomes of some DNA viruses and many RNA viruses are modified at one or both ends (Figures 3.4 and 3.5). Some genomes have a covalently linked protein at the 5′ end. In at least some viruses this is a vestige of a primer that was used for initiation of genome synthesis (Section 7.3.1).

Some genome RNAs have one or both of the modifications that occur in eukaryotic messenger RNAs (mRNAs): a methylated nucleotide cap at the 5′ end

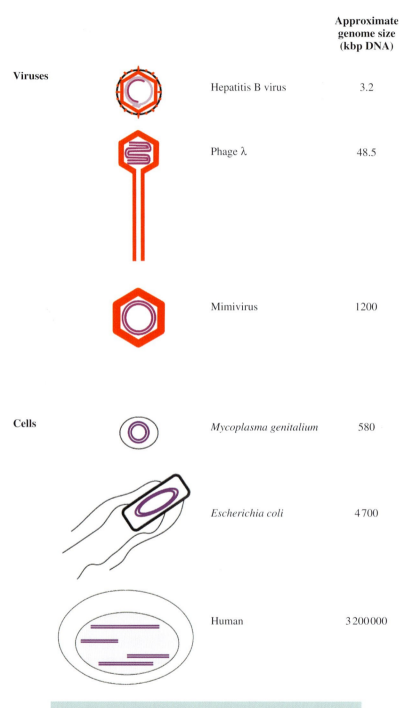

Figure 3.2 *Genome sizes of some DNA viruses and cells.*

(a) (b)

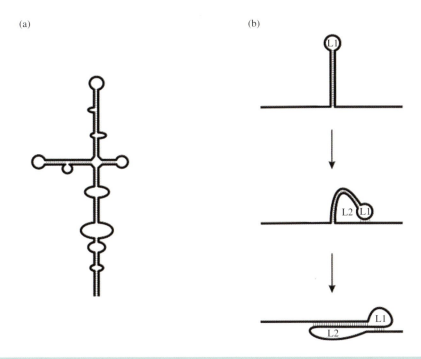

Figure 3.3 *Secondary structures resulting from intramolecular base-pairing in single-stranded nucleic acids.* (a) Stem-loops and bulges in ssRNA and ssDNA. (b) Formation of a pseudoknot in ssRNA. A pseudoknot is formed when a sequence in a loop (L1) base-pairs with a complementary sequence outside the loop. This forms a second loop (L2).

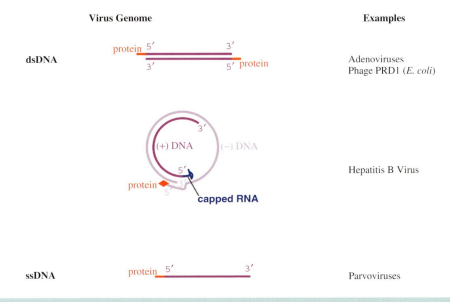

Figure 3.4 *DNA virus genomes with one or both ends modified.* The 5′ end of some DNAs is covalently linked to a protein. One of the hepatitis B virus DNA strands (the (+) strand) is linked to a short sequence of RNA with a methylated nucleotide cap.

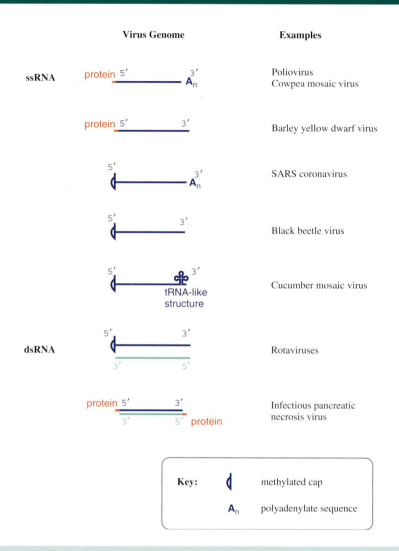

Figure 3.5 *RNA virus genomes with one or both ends modified.* The 5′ end may be linked to a protein or a methylated nucleotide cap. The 3′ end may be polyadenylated or it may be folded like a transfer RNA.

(Section 6.3.4) and a sequence of adenosine residues (a polyadenylate tail; poly(A) tail) at the 3′ end (Section 6.3.5).

The genomes of many RNA viruses function as mRNAs after they have infected host cells. A cap and a poly(A) tail on a genome RNA may indicate that the molecule is ready to function as mRNA, but neither structure is essential for translation. All the ssRNAs in Figure 3.5 function as mRNAs, but not all have a cap and a poly(A) tail.

The genomes of some ssRNA plant viruses are base paired and folded near their 3′ ends to form structures similar to transfer RNA. These structures contain sequences that promote the initiation of RNA synthesis.

3.2.4 Proteins non-covalently associated with virus genomes

Many nucleic acids packaged in virions have proteins bound to them non-covalently. These proteins have

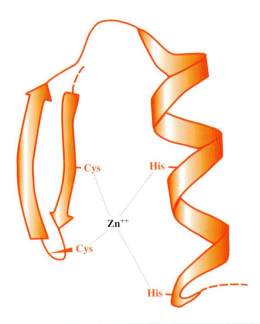

Figure 3.6 *A zinc finger in a protein molecule.* A zinc finger has recurring cysteine and/or histidine residues at regular intervals. In this example there are two cysteines and two histidines.

regions that are rich in the basic amino acids lysine, arginine and histidine, which are negatively charged and able to bind strongly to the positively charged nucleic acids.

Papillomaviruses and polyomaviruses, which are DNA viruses, have cell histones bound to the virus genome. Most proteins associated with virus genomes, however, are virus coded, such as the HIV-1 nucleocapsid protein that coats the virus RNA; 29 per cent of its amino acid residues are basic. As well as their basic nature, nucleic-acid-binding proteins may have other characteristics, such as zinc fingers (Figure 3.6); the HIV-1 nucleocapsid protein has two zinc fingers.

In some viruses, such as tobacco mosaic virus (Section 3.4.1), the protein coating the genome constitutes the capsid of the virion.

3.2.5 Segmented genomes

Most virus genomes consist of a single molecule of nucleic acid, but the genes of some viruses are encoded in two or more nucleic acid molecules. These segmented genomes are much more common amongst RNA viruses than DNA viruses. Examples of ssRNA viruses with segmented genomes are the influenza viruses (see Figure 3.20 below), which package the segments in one virion, and brome mosaic virus, which packages the segments in separate virions. Most dsRNA viruses, such as members of the family *Reoviridae* (Chapter 13), have segmented genomes.

The possession of a segmented genome provides a virus with the possibility of new gene combinations, and hence a potential for more rapid evolution (Section 20.3.3.c). For those viruses with the segments packaged in separate virions, however, there may be a price to pay for this advantage. A new cell becomes infected only if all genome segments enter the cell, which means that at least one of each of the virion categories must infect.

3.2.6 Repeat sequences

The genomes of many viruses contain sequences that are repeated. These sequences include promoters, enhancers, origins of replication and other elements that are involved in the control of events in virus replication. Many linear virus genomes have repeat sequences at the ends (termini), in which case the sequences are known as terminal repeats (Figure 3.7). If the repeats are in the same orientation they are known as direct terminal repeats (DTRs), whereas if they are in the opposite orientation they are known as inverted terminal repeats (ITRs). Strictly speaking, the sequences referred to as 'ITRs' in single-stranded nucleic acids are not repeats until the second strand is synthesized during replication. In the single-stranded molecules the 'ITRs' are, in fact, repeats of the complementary sequences (see ssDNA and ssRNA (−) in Figure 3.7).

3.3 Virus proteins

The virion of tobacco mosaic virus contains only one protein species and the virions of parvoviruses contain two to four protein species. These are viruses with small genomes. As the size of the genome increases, so the number of protein species tends to increase; 39 protein species have been reported in the virion of

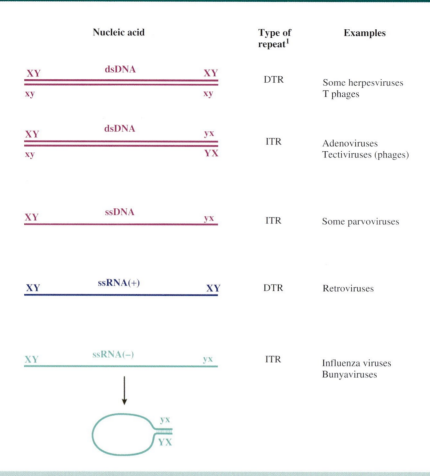

Figure 3.7 *Terminal repeats in virus genomes.*

> [1] DTR: direct terminal repeat
> ITR: inverted terminal repeat
>
> X and x represent complementary sequences.
> Y and y represent complementary sequences.
>
> ssRNA (+) has the same sequence as the virus mRNA.
> ssRNA (−) has the sequence complementary to the virus mRNA.
>
> The RNAs of single-stranded RNA viruses with ITRs can circularize; a 'panhandle' is formed by base pairing between the complementary sequences at the termini.

herpes simplex virus 1, and over 100 in the virion of the algal virus *Paramecium bursaria Chlorella virus 1*.

Proteins that are components of virions are known as structural proteins. They have to carry out a wide range of functions, including

- protection of the virus genome
- attachment of the virion to a host cell (for many viruses)
- fusion of the virion envelope to a cell membrane (for enveloped viruses).

Virus proteins may have additional roles, some of which may be carried out by structural proteins, and some by non-structural proteins (proteins synthesized by the virus in an infected cell but they are not virion components). These additional roles include

- enzymes, e.g. protease, reverse transcriptase

- transcription factors

- primers for nucleic acid replication

- interference with the immune response of the host.

Nomenclature of virus proteins

There is no standard system of nomenclature for virus proteins, with different systems having evolved for different groups of viruses. For quite a number of viruses the following system has been adopted, the proteins being numbered in decreasing order of size:

- structural proteins   VP1, VP2, VP3, . . .   (VP = virus protein)

- non-structural proteins: NSP1, NSP2, NSP3, ...

Many virus proteins are known by an abbreviation of one or two letters, which may indicate

- a structural characteristic   G (glycoprotein)
  P (phosphoprotein)

- or a function   F (fusion)
  P (polymerase)
  RT (reverse transcriptase).

In a virion the virus genome is enclosed in a protein coat, known as a capsid. For some viruses the genome and the capsid constitute the virion, while for other viruses there are additional components. There may be an envelope at the surface of the virion, in which case there may be protein between the envelope and the capsid, or there may be an internal lipid membrane. A few viruses produce protein occlusion bodies in which virions become embedded. We shall consider each of these components in turn.

3.4 Capsids

Virus genomes removed from their capsids are more susceptible to inactivation, so a major function of the capsid is undoubtedly the protection of the genome. A second major function of many capsids is to recognize and attach to a host cell in which the virus can be replicated. Although the capsid must be stable enough to survive in the extracellular environment, it must also have the ability to alter its conformation so that, at the appropriate time, it can release its genome into the host cell.

For many viruses the capsid and the genome that it encloses constitute the virion. For other viruses a lipid envelope (Section 3.5.1), and sometimes another layer of protein, surrounds this structure, which is referred to as a nucleocapsid.

Capsids are constructed from many molecules of one or a few species of protein. The individual protein molecules are asymmetrical, but they are organized to form symmetrical structures. Some examples of symmetrical structures are shown in Figure 3.8. A symmetrical object, including a capsid, has the same appearance when it is rotated through one or more angles, or when it is seen as a mirror image. For the

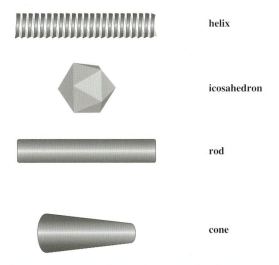

helix

icosahedron

rod

cone

Figure 3.8 *Symmetrical structures.* All these types of symmetry are seen amongst viruses. The most common are helical and icosahedral symmetries.

vast majority of viruses the capsid symmetry is either helical or icosahedral.

3.4.1 Capsids with helical symmetry

The capsids of many ssRNA viruses have helical symmetry; the RNA is coiled in the form of a helix and many copies of the same protein species are arranged around the coil (Figure 3.9(a), (b)). This forms an elongated structure, which may be a rigid rod if strong bonds are present between the protein molecules in successive turns of the helix, or a flexible rod (Figure 3.9(c)) if these bonds are weak. The length of the capsid is determined by the length of the nucleic acid.

For many ssRNA viruses, such as measles and influenza viruses, the helical nucleic acid coated with protein forms a nucleocapsid, which is inside an envelope (see Figure 3.20 below). The nucleocapsid may be coiled or folded to form a compact structure.

The virions of some plant viruses that have helical symmetry (e.g. tobacco mosaic virus) are hollow tubes; this allows the entry of negative stain, making the centre of the virion appear dark in electron micrographs. The rod-shaped tobacco rattle virus has a segmented genome with two RNAs of different sizes packaged in separate virions, resulting in two lengths of virion.

The virions of a few DNA viruses, such as the filamentous phages (Section 19.4.2), also have helical symmetry.

3.4.2 Capsids with icosahedral symmetry

Before proceeding further, a definition of the term 'icosahedron' is required.

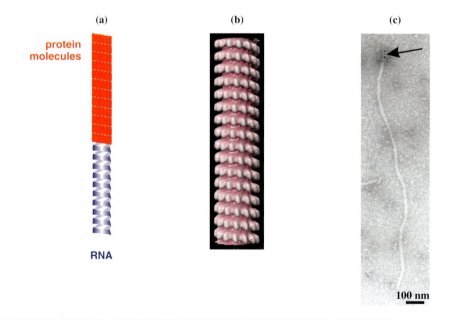

Figure 3.9 *Helical symmetry.* (a) Structure of a capsid with helical symmetry. The ssRNA coils are coated with repeated copies of a protein. (b) Part of measles virus nucleocapsid. The complete nucleocapsid is folded and enclosed within an envelope. Reconstructed image from cryo-electron microscopy, courtesy of Dr. David Bhella (MRC Virology Unit, Glasgow). Reinterpretation of data in Bhella *et al.* (2004) *Journal of Molecular Biology,* **340**, 319 (by permission of Elsevier Limited). (c) Beet yellows virus particle. The virion is a long flexible rod, at one end of which there is a 'tail' (arrow) composed of a minor capsid protein, detected here by specific antibodies labelled with gold. Image courtesy of Professor Valerian Dolja, originally published in Alzhanova *et al.* (2001) *The EMBO Journal,* **20**, 6997. Reproduced by permission of Nature Publishing Group.

An icosahedron is an object with

- 20 faces, each an equilateral triangle;
- 12 vertices, each formed where the vertices of five triangles meet;
- 30 edges, at each of which the sides of two triangles meet.

An icosahedron has five-, three- and two-fold axes of rotational symmetry (Figure 3.10).

Capsids with icosahedral symmetry consist of a shell built from protein molecules that appear to have been arranged on scaffolding in the form of an icosahedron. They have less contact with the virus genome than the capsid proteins of viruses with helical symmetry.

To construct an icosahedron from identical protein molecules the minimum number of molecules required is three per triangular face, giving a total of 60 for the icosahedron (Figure 3.11(a)). The capsid of satellite tobacco mosaic virus is constructed in this way (Figure 3.11(b)).

In capsids composed of more than 60 protein molecules it is impossible for all the molecules to be arranged completely symmetrically with equivalent bonds to all their neighbours. In 1962 Donald Caspar and Aaron Klug proposed a theory of quasi-equivalence, where the molecules do not interact equivalently with one another, but nearly equivalently.

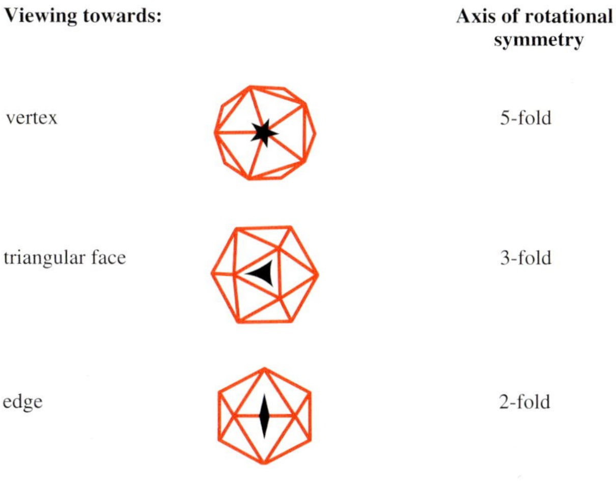

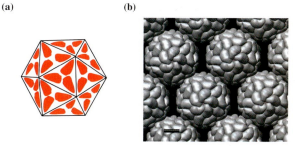

(a)   **(b)**

Figure 3.11 *Capsid constructed from sixty protein molecules.* (a) Arrangement of protein molecules, with three per triangular face. (b) Virions of satellite tobacco mosaic virus. The bar represents 5 nm. Image created with the molecular graphics program UCSF Chimera from the Resource for Biocomputing, Visualization, and Informatics at the University of California, San Francisco. Courtesy of Tom Goddard.

Hence, the capsid of a virus built from 180 identical protein molecules, such as tomato bushy stunt virus, contains three types of bonding between the molecules.

The capsids of many icosahedral viruses are composed of more than one protein species. That of cowpea mosaic virus is composed of two proteins (Figure 3.12): one is present as 'pentamers' at the vertices of the icosahedron ($12 \times 5 = 60$ copies) and the other is present as 'hexamers' on the faces. Each 'hexamer' is composed of three copies of a protein with two domains. The arrangement is similar to that of the panels on the surface of the football in Figure 3.12.

It was pointed out in Section 3.2.1 that there is a huge range in the sizes of virus genomes, with all the large genomes being dsDNA. There is also a huge range in the sizes of icosahedral capsids. The satellite tobacco mosaic virus capsid is about 17 nm in diameter, whereas the diameter of the *Paramecium bursaria Chlorella* virus 1 capsid is about ten times greater than this (Figure 3.13) and the mimivirus capsid is about 300 nm in diameter (Figure 1.3).

3.4.2.a Capsid shapes

It is clear from the images in Figure 3.13 that capsid surfaces vary in their topography; there may be canyons, hollows, ridges and/or spikes present. It

Viewing towards:		Axis of rotational symmetry
vertex		5-fold
triangular face		3-fold
edge		2-fold

Figure 3.10 *The three axes of symmetry of an icosahedron.*

Cowpea mosaic virus capsid

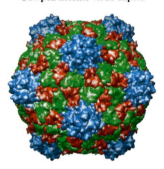

Figure 3.12 *Capsid constructed from two protein species.* The cowpea mosaic virus capsid is constructed from one protein species (blue) that forms 12 'pentamers', and from a second protein species with two domains (green and red) that forms 20 'hexamers'. The football is similarly constructed from 12 'pentamers' and 20 'hexamers'. The cowpea mosaic virus image is from the VIPER database (Shepherd *et al.*, 2006). The image was reconstructed using the data of Lin *et al.* (1999) *Virology*, **265**, 20. Reproduced by permission of Elsevier Limited.

is also clear that some capsids actually have the shape of an icosahedron, such as that of *Paramecium bursaria Chlorella* virus 1, which is 165 nm across when measured along the two- and three-fold axes and 190 nm across when measured along the five-fold axes. Capsids that have an icosahedral shape have an angular outline in electron micrographs (Figure 3.14).

An icosahedral shape is not an inevitable outcome of icosahedral symmetry; the football in Figure 3.12 is constructed in the form of icosahedral symmetry, but the structure is spherical. Many small viruses that have capsids with icosahedral symmetry appear to be spherical, or almost spherical, and their virions are often described as isometric, such as those of densoviruses and foot and mouth disease virus (Figure 3.13).

Some capsids with icosahedral symmetry are elongated. The capsids of geminiviruses (plant viruses) are formed from two incomplete icosahedra. Another plant virus, alfalfa mosaic virus, has four sizes of virion; all are 19 nm diameter, but three are elongated as a result of insertions of a protein lattice

between a half icosahedral structure at each end of the capsid.

3.4.2.b Capsomeres

The capsids of some viruses, such as papillomaviruses (Figure 3.15), are clearly constructed from discrete structures. These structures are called capsomeres and each is built from several identical protein molecules.

The capsids of papillomaviruses are constructed from 72 capsomeres, which are all identical, but the capsids of some viruses are constructed from two types of capsomere: pentons, which are found at the vertices of the icosahedron, and hexons, which make up the remainder of the capsid. In these viruses there are always 12 pentons (one at each vertex), but the number of hexons varies; for example, the capsids of herpesviruses and adenoviruses contain 150 and 240 hexons, respectively.

3.4.2.c Structures at capsid vertices

Some icosahedral viruses have a structure such as a knob, projection or fibre at each of the 12 vertices of the capsid. For example, the virions of some phages (e.g. G4; Figure 3.13) have projections, while the adenovirus virion has a fibre, with a knob attached, at each of the 12 pentons (Figure 3.16). These structures at the capsid vertices are composed of distinct proteins that are involved in attachment of the virion to its host cell and in delivery of the virus genome into the cell.

3.4.2.d Tailed bacteriophages

The majority of the known phages are constructed in the form of a tail attached to a head, which contains the virus genome. All of these phages have dsDNA genomes. The head has icosahedral symmetry and may be isometric as in phage lambda (λ), or elongated as in phage T4. The tail, which is attached to one of the vertices of the head via a connector, may be long as in phage λ, or short as in phage T7. Attached to the tail there

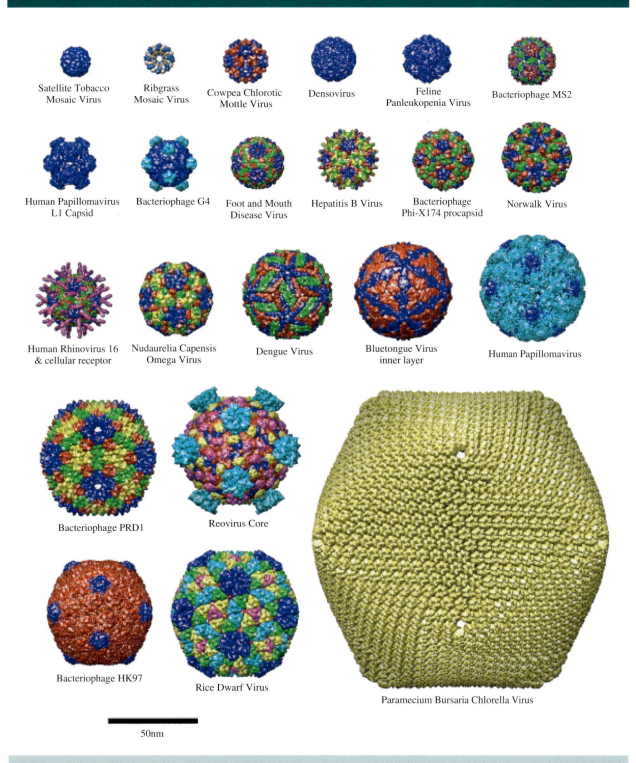

Figure 3.13 *Capsids with icosahedral symmetry.* Some of the wide ranges of capsid architectures and sizes are illustrated. The images were created with the molecular graphics program UCSF Chimera using data from cryo-electron microscopy and X-ray diffraction. From Goddard *et al.* (2005) *Structure*, **13**, 473. Reproduced by permission of Elsevier Limited.

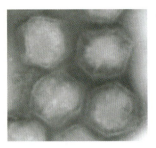

Figure 3.14 *Transmission electron micrograph of negatively stained virions of* Tipula *iridescent virus.*

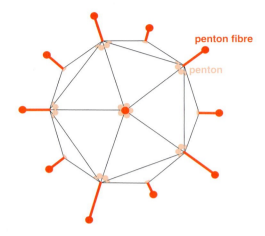

Figure 3.16 *Adenovirus virion.* At each of the 12 vertices of the virion there is a penton, and attached to each penton there is a protein fibre with a knob at the end. The rest of the capsid is constructed from hexons. Please see Figure 12.1 for an electron micrograph of an adenovirus.

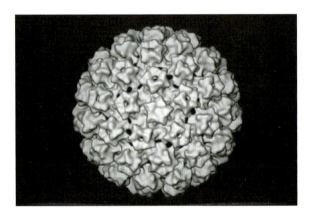

Figure 3.15 *Papillomavirus capsid reconstruction.* From Trus *et al.* (1997) *Nature Structural Biology,* **4**, 413, with the permission of the authors and Nature Publishing Group.

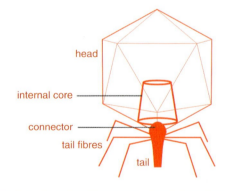

Figure 3.17 *Structure of phage T7.* Each of the components is composed of one or more distinct proteins.

may be specialized structures such as fibres and/or a baseplate.

Some of the tailed phages have been objects of intensive study and a lot of the detail of their structures has been uncovered. One such phage is T7 (Figure 3.17).

Inside the head of phage T7 is a cylindrical structure (the internal core) around which the DNA is wound. The connector has a wider region inserted into one of the vertices of the head and a narrower region to which the tail is attached. The tail is very short and tapers from the connector to the tip; attached to the tail are six tail fibres. Further details about the structure of tailed phages are given in Section 19.5.

3.4.3 Conical and rod-shaped capsids

HIV-1 and baculoviruses have capsids that are conical and rod shaped, respectively (Figure 3.18). Inside each capsid is a copy of the virus genome coated in a highly basic protein. Both of these viruses have enveloped virions (Section 3.5.1).

HIV-1 Capsids

Baculovirus Virions

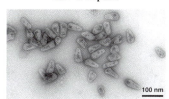
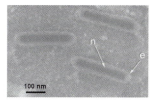

Figure 3.18 *Conical and rod-shaped capsids.*
   e: envelope
   n: nucleocapsid
HIV-1 capsids from Kotov *et al.* (1999) *Journal of Virology*, **73**, 8824. Reproduced by permission of the American Society for Microbiology. Baculovirus virions are those of *Aglais urticae* nucleopolyhedrovirus. From Harrap (1972) *Virology*, **50**, 124. Reproduced by permission of Elsevier Limited

3.5   Virion membranes

Many viruses have a lipid membrane component. In most of these viruses the membrane is at the virion surface and is associated with one or more species of virus protein. This lipid–protein structure is known as an envelope and it encloses the nucleocapsid (nucleic acid plus capsid). The virions of most enveloped viruses, such as herpesviruses, are spherical or roughly spherical, but other shapes exist (Figure 3.19). Some viruses have a membrane located not at the virion surface, but within the capsid.

3.5.1 Enveloped virions

Many animal viruses are enveloped, including all those with helical symmetry, e.g. influenza viruses, and a significant number of those with icosahedral symmetry, e.g. herpesviruses. Enveloped virions are much less common amongst the viruses that infect plants (e.g. potato yellow dwarf virus) and they are extremely rare amongst the viruses that infect prokaryotes (e.g. *Pseudomonas* phage φ6).

Associated with the membranes of an enveloped virus are one or more species of protein. Most of these proteins are integral membrane proteins and most are O- and/or N-glycosylated (Section 6.4.3.a). Many of the glycoproteins in virion envelopes are present as multimers. The envelope of influenza A virus (Figure 3.20), for example, contains two glycoprotein species: a haemagglutinin present as trimers and a

neuraminidase present as tetramers. There is also a third protein species (M2) that is not glycosylated.

Some envelope proteins span the membrane only once, but some, such as those of hepatitis B virus (Section 18.3.4), span the membrane several times. The polypeptide chain is highly hydrophobic at each membrane anchor.

Some surface glycoproteins of enveloped viruses perform the function of fusing the virion membrane to a cell membrane during the infection process (Section 5.2.4.b). These fusion proteins have an additional hydrophobic region that plays a major role in the fusion process.

Many enveloped viruses, including influenza viruses (Figure 3.20) and retroviruses, have a layer of protein between the envelope and the nucleocapsid. This protein is often called a matrix protein. In some viruses, however, such as yellow fever virus, there is no such layer and the nucleocapsid interacts directly with the internal tails of the integral membrane protein molecules.

3.5.2 Virions with internal membranes

There are several viruses that have a lipid membrane within the virion rather than at the surface (Figure 3.19). These membranes have proteins associated with them and are present in the virions of, for example, *Paramecium bursaria Chlorella* virus 1 (Figure 3.13), the iridoviruses (Figure 3.14) and the tectiviruses (phages).

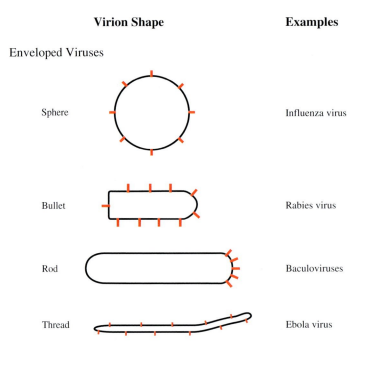

Virion Shape   **Examples**

Enveloped Viruses

Sphere   Influenza virus

Bullet   Rabies virus

Rod   Baculoviruses

Thread   Ebola virus

Viruses with an Internal Lipid Membrane

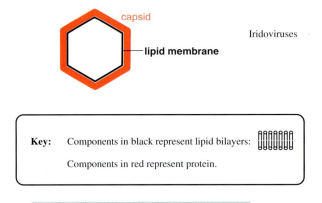

capsid

lipid membrane   Iridoviruses

Key: Components in black represent lipid bilayers:

Components in red represent protein.

Figure 3.19 *Lipid-containing viruses.*

3.5.3 Membrane lipids

Most virion membranes are derived from host cell membranes that undergo modification before incorporation into virions. For example, the HIV-1 envelope is derived from the plasma membrane of the host cell, but the virus envelope contains more cholesterol and sphingomyelin, and less phosphatidylcholine and phosphatidylinositol.

Some viruses are able to replicate in more than one kind of host cell; for example, alphaviruses replicate in both mammalian cells and insect cells. When progeny

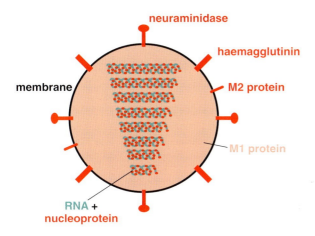

neuraminidase

haemagglutinin

membrane

M2 protein

M1 protein

RNA +
nucleoprotein

Figure 3.20 *Influenza A virus.* The envelope contains two species of glycoprotein (haemagglutinin and neuraminidase) and a non-glycosylated protein (M2). Underlying the envelope is a layer of M1 (membrane or matrix) protein, which encloses the helical nucleocapsid (eight segments of ssRNA coated with nucleoprotein) and several other proteins.

virions are released from cells the lipid composition of the envelope may reflect that of the cell. Virions of Semliki Forest virus produced in hamster kidney cells contain about five times more cholesterol than virions produced in mosquito cells.

3.6 Occlusion bodies

Some viruses provide added protection to the virions whilst outside their hosts by occluding them in protein crystals. These occlusion bodies, as they are known, are produced by many of the viruses that infect invertebrates, including most baculoviruses. There are two major types of occlusion body in which baculoviruses embed their rod-shaped virions (Figure 3.18); the granuloviruses form small granular occlusion bodies, generally with a single virion in each (Figure 3.21(a)), while the nucleopolyhedroviruses form large occlusion bodies with many virions in each (Figure 3.21(b), (c)).

3.7 Other virion components

3.7.1 Virus RNA in DNA viruses

Virus RNA is present in the virions of a number of DNA viruses. The hepadnaviruses and the caulimoviruses have short RNA sequences covalently attached to their DNAs. These RNAs have functioned as primers for DNA synthesis and remain attached to the genome in the mature virion.

There is evidence that virus mRNAs are packaged in the virions of herpesviruses, but the roles of these molecules are not clear.

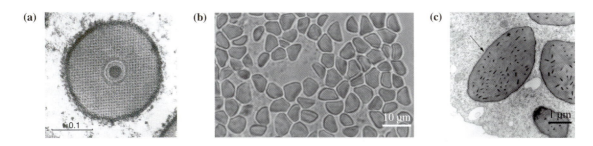

(a)   (b)   (c)

Figure 3.21 *Baculovirus occlusion bodies.* (a) Transverse section through an occlusion body of a granulovirus of the Indian meal moth (*Plodia interpunctella*). The nucleocapsid surrounded by its membrane can be seen at the centre, surrounded by the crystalline protein that forms the occlusion body. The bar represents 0.1 μm. From: Arnott and Smith (1967) *Journal of Ultrastructure Research,* **21,** 251. Reproduced by permission of Elsevier Limited. (b), (c) Nucleopolyhedrovirus of the leatherjacket (*Tipula paludosa*). Courtesy of Dr. Liz Boslem. (b) Light micrograph of occlusion bodies. (c) Electron micrograph of a thin section of an occlusion body. The virions are randomly embedded in the crystalline protein.

3.7.2 Cell molecules

The incorporation of cell lipids into virions has already been discussed. Other cell molecules that become incorporated into virions include the following.

- *Transfer RNA molecules.* These are present in the virions of retroviruses.

- *Proteins.* When a virion is assembled some cell protein may be incorporated. There are reports of HIV-1 incorporating several cell proteins, including cyclophilin A in association with the capsid and human leukocyte antigens in the envelope.

- *Polyamines.* Spermidine and other polyamines have been reported in a variety of viruses.

Cations, such as Na^+, K^+, Zn^{2+} and Mg^{2+} have also been reported as components of virions. One likely role for polyamines and cations is the neutralization of negative charges on the virus genome.

Learning outcomes

By the end of this chapter you should be able to

- describe the components of virions;

- illustrate the variety of virus genomes;

- outline the functions of virus structural and non-structural proteins;

- define the terms 'helical symmetry' and 'icosahedral symmetry';

- describe the virions of a selection of naked and enveloped viruses.

Sources of further information

Books

Chiu W. *et al.*, editors (1997) *Structural Biology of Viruses*, Oxford University Press

Fischer W. B., editor (2005) *Viral Membrane Proteins: Structure, Function and Drug Design*, Kluwer–Plenum

Harrison S. C. (2001) Principles of virus structure. Chapter 3 in *Fields Virology*, editors-in-chief Knipe D. M. and Howley P. M., 4th edition, Lippincott, Williams and Wilkins

Historical paper

Caspar D. L. D. and Klug A. (1962) Physical principles in the construction of regular viruses *Cold Spring Harbor Symposia Quantitative Biology*, **27**, 1–24

Recent papers

Amos L. A. and Finch J. T. (2004) Aaron Klug and the revolution in biomolecular structure determination *Trends In Cell Biology*, **14**, 148–152

Baker T. S., Olson N. H. and Fuller S. D. (1999) Adding the third dimension to virus life cycles: three-dimensional reconstruction of icosahedral viruses from cryo-electron micrographs *Microbiology and Molecular Biology Reviews*, **63**, 862–922

Chiu W. and Rixon F. J. (2002) High resolution structural studies of complex icosahedral viruses: a brief overview *Virus Research*, **82**, 9–17

Klug A. (1999) The tobacco mosaic virus particle: structure and assembly *Philosophical Transactions of the Royal Society of London B*, **354**, 531–535

Lander G. C. *et al.* (2006) The structure of an infectious P22 virion shows the signal for headful DNA packaging *Science*, **312**, 1791–1795

Lee K. K. and Johnson J. E. (2003) Complementary approaches to structure determination of icosahedral viruses *Current Opinion in Structural Biology*, **13**, 558–569

Shepherd C. M. *et al.* (2006) VIPERdb: a relational database for structural virology *Nucleic Acids Research*, **34** (Database Issue), D386–D389

4

Virus transmission

At a glance

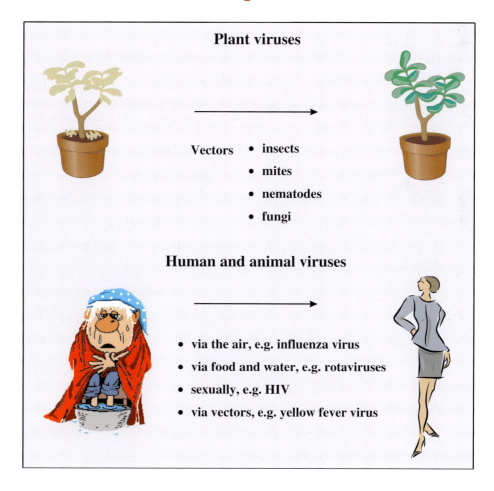

Plant viruses

Vectors
- insects
- mites
- nematodes
- fungi

Human and animal viruses

- via the air, e.g. influenza virus
- via food and water, e.g. rotaviruses
- sexually, e.g. HIV
- via vectors, e.g. yellow fever virus

Virology: Principles and Applications   John B. Carter and Venetia A. Saunders
© 2007 John Wiley & Sons, Ltd   ISBNs: 978-0-470-02386-0 (HB); 978-0-470-02387-7 (PB)

4.1 Introduction to virus transmission

A minimum proportion of the virions produced in infected hosts must be transmitted to new hosts in which more virions can be manufactured. If this does not happen the virus will die out. The only other possibility for the survival of virus genes is for them to be maintained in cells as nucleic acids, which are replicated and passed on to daughter cells when the cells divide.

Viruses of bacteria and other microbial hosts are released from infected cells into the environment of the host, where further susceptible cells are likely to be present. These viruses are dependent upon chance encounters with susceptible cells, to which they may bind if receptors on the surface of those cells come into contact with virus attachment sites.

Viruses of multicellular animals and plants must also find new cells to infect. An infection may spread to adjacent cells, or to cells in a distant part of the host after transport in the blood of an animal or in the phloem of a plant, but ultimately a virus must find new hosts to infect if it is to survive.

Some virus infections modify the behaviour of their hosts in order to increase the probability of transmission. Mammals infected with rabies virus often become aggressive; this change in behaviour increases the likelihood of the host biting another individual and transmitting virus in the saliva. Some plant-feeding insect larvae infected with baculoviruses become more mobile in the late stages of infection, thus aiding virus dispersal. If the virus-infected insect migrates towards the top of a plant then lower leaves, where other larvae will feed, will become contaminated when the host dies and putrefies.

On their journeys between hosts viruses may have to survive adverse conditions in an environment such as air, water or soil; the survival of virus infectivity outside the host is discussed in Chapter 23. There are some viruses, however, that can be transmitted to new hosts without 'seeing the light of day', in other words without exposure to the outside environment. These are viruses that can be transmitted directly from host to host, for example during kissing or sexual intercourse, and viruses that are transmitted via vectors. Also included in this category are viruses that can be transmitted directly from a parent to members of the next generation. Transmission in these cases is said to be vertical; otherwise it is described as horizontal.

Viruses may be moved over long distances in a variety of ways. Rivers and winds can move viruses to new areas. The foot and mouth disease outbreak on the Isle of Wight, UK, in 1981 was initiated by virus that had spread via the air from Brittany, more than 250 km away. Viruses of birds, fishes, humans and other hosts are transported within their hosts to other parts of the planet as a result of migration, travel and animal export, for example

- bird migration (avian influenza viruses)

- human travel (SARS virus)

- animal export (monkeypox virus).

Furthermore, viruses can be transported on inanimate materials, for example, foot and mouth disease virus on straw and farm vehicles.

Once virions have entered a multicellular organism they may have further to travel before suitable host cells are encountered. During this final stage of their journey the virions will encounter hazards in the form of host defence mechanisms, which must be survived if the virions are to remain infective when they reach their destination.

In theory, a single virion can initiate an infection, but in practice it is often found that a host must be inoculated with a minimum number of virions in order for that host to become infected. The reasons for this are probably many and varied, and may include some virions being defective and some being inactivated by the host's immune systems. This minimum amount of virus required for infection of a host is known as the minimum infective dose.

In this chapter we shall look at the ways that viruses of plants and animals are transmitted from host to host, but first we must look at some general aspects of virus transmission by vectors.

4.1.1 Transmission via vectors: general principles

Many viruses of plants and animals are transmitted between hosts by organisms that feed on them; these organisms act as vectors. There are several concepts concerning vector transmission that are common to viruses of both plants and animals so, before discussing these host categories separately, it is useful to consider these concepts.

Most vectors of viruses are arthropods (insects, mites, ticks); the arthropod-transmitted viruses of vertebrates are sometimes referred to as arboviruses (*ar*thropod-*bo*rne viruses).

The principle of vector transmission is that the vector acquires a virus when it feeds on an infected host and subsequently transmits the virus to one or more new hosts. Some viruses are transmitted after virions have become attached to the mouthparts of their vectors during feeding. Transmission in this way may occur within seconds or minutes of the vector acquiring virus. Many vector-transmitted viruses, however, cross the gut wall of the vector and enter its circulatory system. The virus ultimately reaches the salivary glands and is secreted into the saliva, which may transport virus into new hosts when the vector feeds. This mode of transmission is said to be circulative, and transmission does not occur until hours or days after the vector has acquired the virus.

Some circulative viruses replicate in one or more tissues and organs of their vectors (Figure 4.1); thus there are viruses that can replicate in both invertebrates and plants, and viruses that can replicate in both invertebrates and vertebrates. It is fascinating to contemplate the existence of viruses with such broad host ranges. Many (but not all) of the invertebrates appear to suffer little or no harm when they are infected, so it has been hypothesized that many of the modern 'plant viruses' and 'vertebrate viruses' are descended from viruses of invertebrates that later extended their host ranges to plants or vertebrates.

If the reproductive organs of the vector are infected there may be possibilities for vector-to-vector transmission. Some viruses are sexually transmitted (male to female and vice versa) and some are transmitted to the next generation within the egg; the latter is known as transovarial transmission.

There is a high degree of specificity between most vectors and the viruses that they transmit. Generally specificity is greater between plant viruses and their vectors than between vertebrate viruses and their vectors.

4.2 Transmission of plant viruses

Plant cells are surrounded by thick cell walls that present significant barriers to virus entry; most plant viruses are carried across these barriers by vectors. A wide variety of organisms use plants as sources of nutrition and some of these organisms, especially invertebrates, act as virus vectors (Figure 4.2). Many

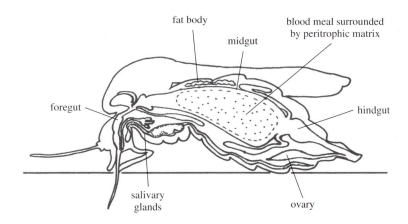

Figure 4.1 *Cross-section of a mosquito.* Some of the organs and tissues that may become infected by a virus acquired in a blood meal are indicated. From Higgs (2004) Chapter 7, 63rd Symposium of the Society for General Microbiology, Cambridge University Press, by permission of the author and the Society for General Microbiology.

Examples of Plant Virus Vectors

Examples of Viruses Transmitted

Insects   Aphids

Potato virus Y
Cauliflower mosaic virus

Beet yellows virus
Bean yellow mosaic virus

Leafhoppers

Rice dwarf virus

Whiteflies

Tomato yellow leaf curl virus

Beetles

Maize chlorotic mottle virus

Mites

Ryegrass mosaic virus

Nematodes

Grapevine fanleaf virus

Figure 4.2

of the vectors (e.g. aphids, nematodes) feed by piercing cell walls and ingesting the contents, while beetles feed by biting.

The most common vectors of plant viruses are aphids, which, as we have already pointed out, feed by ingesting the contents of cells. The reader may question how an invertebrate that feeds by removing the contents of plant cells can introduce virus into cells that subsequently support replication of the virus. The answer appears to be that these vectors probe a number of cells before selecting one on which to feed, so virus may be transmitted into cells that are probed, but not significantly damaged by the vector.

The nematodes that transmit viruses are soil-dwelling animals that pierce root cells and then ingest their contents. Virus transmission ceases after a nematode moults, indicating that virus does not move from its gut into its body. An interesting feature of virus-vector specificity concerns transmission of viruses with different shapes by different types of nematode. Tobraviruses have rod-shaped virions and are transmitted by nematodes in the family *Trichodoridae*, while nepoviruses have isometric virions and are transmitted by nematodes in the family *Longidoridae*.

The basis of some cases of plant virus-vector specificity lies in specific amino acid sequences in capsid proteins. In other cases important roles are played by virus-coded non-structural proteins (helper factors) that are synthesized in the infected plant cell. Virions may bind specifically to structures in the mouthparts of their vectors via specific sequences on the surfaces of capsids and/or helper factors.

Some plant-parasitic fungi can also act as virus vectors, for example *Spongospora subterranea*, which infects potato causing powdery scab disease, is a vector of potato mop-top virus. If a plant is infected with both a fungus and a virus, then virions may be taken into developing fungal spores. The virus may survive in a spore for months or years until it germinates on a new host, which then becomes infected, not only with the fungus but also with the virus. Some fungus-transmitted plant viruses have been classified in a genus, the name of which reflects their mode of transmission and the virion shape: the genus *Furovirus* (*fu*ngus-transmitted *ro*d-shaped).

About 20 per cent of plant viruses can be transmitted vertically; in other words, seed can be infected, leading to infection of the next generation. Most seed-transmitted viruses are carried in the embryo, which may have acquired its infection from either an infected ovule or an infected pollen grain. Examples of viruses that can be transmitted via seed include the nepoviruses and the tobraviruses, two groups of nematode-transmitted viruses mentioned above. Many plant viruses can also be transmitted by artificial means; for example, grafting material from a virus-infected plant can introduce virus into a new host.

Figure 4.2   *Invertebrate vectors of plant viruses.* The examples shown are

- aphids   – upper: *Myzus persicae*, courtesy of Magnus Gammelgaard;
           – lower: *Aphis fabae*, courtesy of Robert Carter;

- leafhopper (*Nephotettix cincticeps*), courtesy of Keiji Morishima;

- whitefly (*Bemisia tabaci*), courtesy of Dr. J.K. Brown, University of Arizona;

- beetle (*Diabrotica virgifera virgifera*), courtesy of Austrian Agency of Health and Food Safety;

- mite (*Abacarus hystrix*), courtesy of Dr. Anna Skoracka, Adam Mickiewicz University, Poland;

- nematode (*Xiphinema index* feeding on a root), from MacFarlane and Robinson (2004) Chapter 11, 63rd Symposium of the Society for General Microbiology, Cambridge University Press, reproduced by permission of the Society for General Microbiology

4.3 Transmission of vertebrate viruses

4.3.1 Non-vector transmission of vertebrate viruses

Many viruses of vertebrates (including humans) infect their hosts via the mucous membranes of the upper respiratory tract. An infected individual may shed virus-containing droplets into the air during sneezing, speaking and coughing, and new hosts may become infected by inhaling this material. Virus present in genital secretions can be transmitted during sexual contact, while viruses infecting the intestinal tract are shed in faeces and may enter new hosts ingesting faeces-contaminated food or water.

Some viruses may be released from lesions, for example foot and mouth disease virus from lesions on the feet and in the mouth, papillomaviruses from warts, and herpes simplex virus from lesions on the lips. These viruses could be transmitted directly by contact between an infected host and a non-infected host, or they could be transmitted indirectly through contamination of the environment. Transmission of rabies virus, for example from dog to human, requires

a bite from an infected mammal to introduce virus-containing saliva into the body.

In most of the cases discussed so far the host becomes infected when epithelial cells and/or lymphoid cells at a body surface become infected. Some infections (e.g. common cold, rotavirus) remain limited to these tissues, while others (e.g. measles, polio, HIV) cross the epithelial surface and spread to other organs and tissues.

Modes of vertebrate virus transmission not involving vectors are summarized in Table 4.1, along with some examples.

4.3.2 Vector transmission of vertebrate viruses

Vectors that transmit vertebrate viruses are mainly blood-feeding arthropods that acquire their viruses when they take blood meals from infected animals (Figure 4.3). Obviously for this to occur virus must be present in the blood of infected animals, a situation known as viraemia.

Some arthropod parasites that acquire virus infections from the vertebrates on which they feed remain infected for life. This is the case with ticks that

Table 4.1 Non-vector transmission of vertebrate viruses

Transmission route	Examples of viruses transmitted
Horizontal transmission	
Respiratory tract	Influenza viruses (mammals)
	Common cold viruses
	Measles virus
Intestinal tract	Influenza viruses (birds)
	Rotaviruses
Abrasions and wounds	Papillomaviruses
	Rabies virus
Genital tract	HIV
	Papillomaviruses
Vertical transmission	
Mother to foetus via the placenta	Rubella virus
Mother to baby via milk	HIV

Examples of Vertebrate Virus Vectors		Examples of Viruses Transmitted	Virus Hosts
Living Vectors			
Mosquitoes		Yellow fever virus West Nile virus	Humans
Midges		Bluetongue virus	Sheep
Ticks		Louping ill virus	Sheep
Inanimate Vectors			
Syringes and Needles	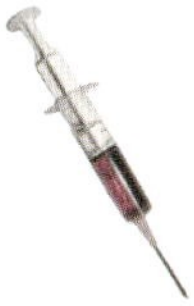	Hepatitis B virus HIV	Humans

Figure 4.3 *Vectors of vertebrate viruses.* Photograph credits: mosquito (*Aedes aegypti*) courtesy of James Gathany and the US Centers for Disease Control and Prevention; midge (*Culicoides* sp.) from Defra website. Crown copyright material, reproduced with the permission of the Controller of HMSO and Queen's Printer for Scotland; tick (*Ixodes ricinus*) courtesy Dr. František Dusbábek, Institute of Parasitology, Academy of Sciences of the Czech Republic

become infected with tick-borne encephalitis virus. Transovarial transmission has been demonstrated for a number of viruses, including yellow fever virus in mosquitoes.

Viruses in blood or blood products can be transmitted during medical procedures by inanimate 'vectors' such as blood transfusion equipment and syringes and needles. Viruses can also be transmitted when needles for drug injection are shared, when needlestick injuries occur and when haemophiliacs are injected with blood factors containing viruses.

4.3.3 Virus survival in the vertebrate host

Once a virus has reached a potential host it must evade several defence systems if it is to initiate an infection.

A virus in the lower respiratory tract must avoid removal by the mucociliary escalator, while a virus in the gastro-intestinal tract must avoid inactivation by extremes of pH. The mucosal surfaces of the respiratory, intestinal and urinogenital tracts present significant barriers to infection. The surfaces are bathed in fluids that contain antiviral substances and the fluids are viscous, limiting virus access to the cells.

The antiviral substances include complement proteins, which can be triggered by the presence of certain viruses to undergo modifications that help to protect the host from infection. Complexes of modified complement proteins can damage enveloped virions by insertion into their membranes. Some modified complement proteins can coat virions, which then attach to phagocytes via receptors on the cell surfaces.

Viruses may be phagocytosed by a range of cell types including macrophages, which are large cells. Viruses that are introduced into the blood by a vector may be phagocytosed by neutrophils, which constitute the majority of the white blood cells.

These aspects of host immunity are not virus specific; they are components of the innate immune system. The virus may also have to contend with antibodies and other components of the adaptive immune system (Section 9.2.2) if the host has previously been infected with, or vaccinated against, the virus.

4.4 Transmission of invertebrate viruses

We have discussed the replication of 'plant' and 'vertebrate' viruses in invertebrate vectors, and it could be argued that these viruses are as much 'invertebrate' viruses as they are viruses of these other host types. In addition to these viruses, invertebrates are hosts to many other viruses that replicate only in invertebrates; most of the known invertebrate viruses have insect hosts.

Many of these viruses have evolved occlusion bodies, which are large protein structures in which virions become embedded in the infected cell (Section 3.6). Viruses that produce occlusion bodies include cypoviruses (Section 13.1) and baculoviruses. Occlusion bodies may be expelled in the faeces of a virus-infected insect or they may remain in the host until it is ingested by a predator or it dies as a result of the virus infection. They are robust structures, able to survive putrefaction of the host and providing protection for virions in the outside environment. New insects become infected when occlusion bodies become ingested along with food, which is often plant material. Enzymes and high pH in the gut break down occlusion bodies, releasing the virions. The transmission of a baculovirus via occlusion bodies is depicted in Figure 4.4.

Vertical transmission and vector transmission of a number of invertebrate viruses have been reported. Vertical transmission may occur either within eggs (transovarially) or on the surface of eggs. Vectors involved in virus transmission between insect hosts include parasitic wasps.

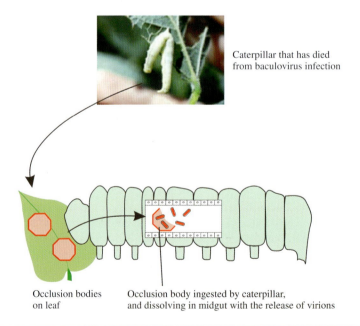

Caterpillar that has died from baculovirus infection

Occlusion bodies on leaf

Occlusion body ingested by caterpillar, and dissolving in midgut with the release of virions

Figure 4.4 *Occlusion bodies as vehicles for baculovirus transmission.* Putrefaction of insects that have been killed by virus infection results in the deposit of occlusion bodies on leaves. Occlusion bodies ingested by insects are dissolved in the midgut, releasing virions. Photograph courtesy of Dr. Flavio Moscardi.

4.5 Permissive cells

In order to replicate itself, a virus must gain access to a permissive cell, meaning a cell that will permit the replication of the virus. For viruses that bind to the host cell surface as the first step in infection the cell must have appropriate receptors that the virus can bind to. Furthermore, all the requirements of the virus must either be present in the cell or they must be inducible; these requirements include cell proteins, such as transcription factors and enzymes.

Some viruses are restricted to a narrow range of permissive cell types, for example hepatitis B viruses are restricted almost exclusively to hepatocytes. Many other viruses are much less specific and, as we have seen in this chapter, some can replicate in both animal and plant hosts.

Experimentally, the range of permissive cell types may be much wider than the cell types found infected in nature. Flock house virus is an insect virus that was isolated from the New Zealand grass grub *Costelytra zealandica*. As well as replicating in other insect species after artificial inoculation, the virus can also replicate in a range of plant species.

Some viruses of eukaryotes need the host cell to be in a particular phase of the cell cycle (Figure 4.5). For example, retroviruses require access to the cell nucleus and most can enter the nucleus only when the nuclear envelope has broken down in the M (mitosis) phase. Parvoviruses are DNA viruses that use the DNA replicating enzymes of the host to replicate their own DNA, so they need a cell in which those enzymes are present, i.e. a cell in the S phase.

A further requirement of a permissive cell is that it must lack defences against the virus, or the virus must have the ability to overcome the defences of the cell. Cells have evolved many mechanisms to defend themselves against infection with viruses and cellular microbes. Prokaryotic cells produce restriction endonucleases to degrade foreign DNA, including that of phages. (The cells modify their own DNA so that it is unaffected by the restriction enzymes.) The cells of vertebrate animals respond to the presence of dsRNA, which is produced during infection with many viruses,

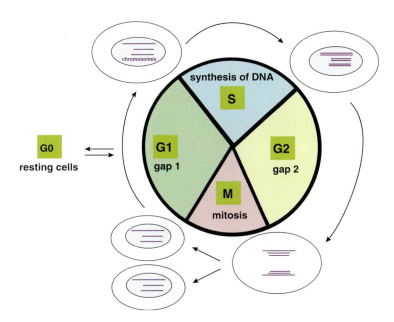

Figure 4.5 *The eukaryotic cell cycle.* The cell synthesizes DNA and histones during the S (synthesis) phase and divides during the M (mitotic) phase. These phases are separated by two gap phases (G1 and G2). Non-dividing cells suspend the cycle in the G1 phase and are said to be in the G0 state.

by synthesizing cytokines, including interferons, interleukins and tumour necrosis factor. These proteins can trigger a range of anti-viral defences.

In order to successfully infect a cell a virus must deliver its genome into this cell, the genome must survive intact and all attempts by the host to terminate virus replication must be thwarted. In the next four chapters we shall consider the various stages that must be completed before a virus has successfully replicated itself.

Learning outcomes

By the end of this chapter you should be able to

- describe the modes of transmission of plant viruses and animal viruses;

- evaluate the roles of vectors in virus transmission;

- discuss the immune mechanisms encountered by an animal virus when it enters the body of a new host.

Sources of further information

Books

Gillespie S. H., Smith G. L. and Osbourn A. (2004) *Microbe–Vector Interactions in Vector-Borne Diseases*, 63rd Symposium of the Society for General Microbiology, Cambridge University Press

Taylor C. E. and Brown D. J. F. (1997) *Nematode Vectors of Plant Viruses*, CAB

Journals

Bergelson J. M. (2003) Virus interactions with mucosal surfaces: alternative receptors, alternative pathways *Current Opinion in Microbiology*, **6**, 386–391

Pereira L. *et al.* (2005) Insights into viral transmission at the uterine–placental interface *Trends in Microbiology*, **13**, 164–174

5

Attachment and entry of viruses into cells

At a glance

Overview Of Virus Replication

1. **A**ttachment

2. **E**ntry

3. **T**ranscription

4. **T**ranslation

5. **G**enome replication

6. **A**ssembly

7. **E**xit

Animal Viruses

All animal viruses must cross the **plasma membrane**.

Some are transported in the cytoplasm via **microtubules**.

Some must cross the **nuclear envelope,** usually via a **nuclear pore**.

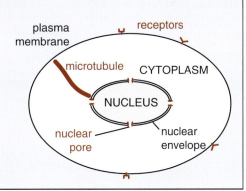

Virology: Principles and Applications   John B. Carter and Venetia A. Saunders
© 2007 John Wiley & Sons, Ltd   ISBNs: 978-0-470-02386-0 (HB); 978-0-470-02387-7 (PB)

At a glance (continued)

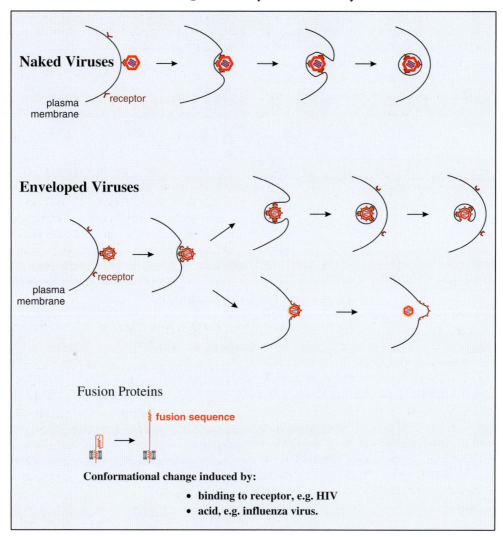

5.1   Overview of virus replication

The aim of a virus is to replicate itself, and in order to achieve this aim it needs to enter a host cell, make copies of itself and get the new copies out of the cell. In general the process of virus replication can be broken down into seven steps:

1. Attachment of a virion to a cell

2. Entry into the cell

3. Transcription of virus genes into messenger RNA molecules (mRNAs)

4. Translation of virus mRNAs into virus proteins

5. Genome replication

6. Assembly of proteins and genomes into virions

7. Exit of the virions from the cell.

The first letter from each step gives the acronym AETTGAE, which may provide a memory aid.

When trying to understand the modes of replication of different types of virus these steps provide a useful template. It is, however, a general template, because not all of the seven steps are relevant to all viruses, the steps do not always occur in the same order and some viruses have an additional step! In the later stages of replication several steps occur concurrently. For many viruses, transcription, translation, genome replication, virion assembly and exit can all be in progress at the same time.

> In the dialogue between the cell and the intruder, the cell provides critical cues that allow the virus to undergo molecular transformations that lead to successful internalization, intracellular transport, and uncoating.
>
> Smith and Helenius (2004)

In this chapter we discuss the first two steps of this generalized replication cycle: attachment and entry into the host cell. We concentrate mainly on the mechanisms used by animal viruses to gain entry into host cells, but bacteriophages are also considered. The following three chapters deal with the remaining five steps of the generalized replication cycle.

The first step, attachment of a virion to a cell, applies to viruses that infect animal and bacterial hosts. Before these viruses can cross the bounding membrane or wall of the host cell they must first bind to specific molecules on the cell surface. Most plant viruses, on the other hand, are delivered directly into a cell by a vector (Section 4.2).

5.2 Animal viruses

5.2.1 Cell receptors and co-receptors

A virion attaches via one or more of its surface proteins to specific molecules on the surface of a host cell. These cellular molecules are known as receptors and the recognition of a receptor by a virion is highly specific, like a key fitting in its lock. It has been found that some viruses need to bind to a second type of cell surface molecule (a co-receptor) in order to infect a cell. In at least some cases binding to a receptor causes a conformational change in the virus protein that enables it to bind to the co-receptor.

Receptors and co-receptors are cell surface molecules, usually glycoproteins, with a wide range of functions that include

- acting as receptors for chemokines and growth factors;

- mediating cell-to-cell contact and adhesion.

Some examples of cell receptors are given in Table 5.1.

A number of receptors are glycoproteins that are folded into domains similar to those found in immunoglobulin molecules. Three examples are shown in Figure 5.1; in each molecule the binding site for the virus is located in the outermost domain.

Many of the cell surface molecules used by viruses as receptors are in regions of the plasma membrane that are coated on the inner surface with one of the proteins clathrin or caveolin.

5.2.1.a Evidence that a cell surface molecule is a virus receptor

A search for the receptor for a particular virus might begin with the production of a panel of monoclonal antibodies against cell surface proteins. If one of the antibodies blocks virus binding and infectivity then this is strong evidence that the corresponding antigen is the receptor.

Further evidence that a molecule is a virus receptor might come from experiments that show one or more of the following.

- Soluble derivatives of the molecule block virus binding/infectivity.

- The normal ligand for the molecule blocks virus binding/infectivity.

- Introduction of the gene encoding the molecule into virus-resistant cells, and expression of that gene, makes those cells susceptible to infection.

5.2.2 Virus attachment sites

Each virion has multiple sites that can bind to receptors, and each site is made up of regions of one or more

Table 5.1 Examples of cell receptors, virus proteins involved in attachment, and (for enveloped viruses) fusion proteins

Virus	Cell receptor	Virus protein(s) involved in	
		attachment to receptor	fusion
Naked viruses			
Approx. 90% of human rhinoviruses	Intercellular adhesion molecule-1 (ICAM-1)	VP1 + VP3	
Approx. 10% of human rhinoviruses	Low-density lipoprotein receptors	VP1	
Poliovirus	CD155	VP1	
Enveloped viruses			
Murine leukaemia viruses	Mouse cationic amino acid transporter	SU (surface glycoprotein)	TM (transmembrane glycoprotein)
HIV-1	CD4	gp120	gp41
Influenza viruses A & B	Sialic-acid-containing glycoproteins	Haemagglutinin	Haemagglutinin
Measles virus	Signalling lymphocyte activation molecule (CD150)	Haemagglutinin	Fusion

protein molecules. Some examples of proteins that bear virus attachment sites are given in Table 5.1.

The virus attachment sites of naked viruses are on the capsid surface, sometimes within depressions (e.g. poliovirus) and sometimes on ridges (e.g. foot and mouth disease virus). Poliovirus and foot and mouth disease virus are picornaviruses, and it has been demonstrated for picornaviruses in general that binding to receptors induces major structural changes in the virion (Section 14.4.1).

The virus attachment sites of some naked viruses are on specialized structures, such as the fibres and knobs of adenoviruses (Section 3.4.2.c) and the spikes of rotaviruses (Chapter 13), while the virus attachment sites of enveloped viruses are on the surface glycoproteins.

Some virion surface proteins that bear the virus attachment sites are able to bind strongly to red blood cells of various species and cause them to clump, a phenomenon known as haemagglutination. The proteins responsible for haemagglutination are called haemagglutinins. Examples of viruses that can haemagglutinate are influenza viruses (Figure 3.22) and measles virus.

5.2.3 Attachment of virions to receptors

The forces that bind a virus attachment site to a receptor include hydrogen bonds, ionic attractions and van der Waals forces. Sugar moieties, when present on the receptor and/or on the virus attachment site, are commonly involved in the forces that bind the two.

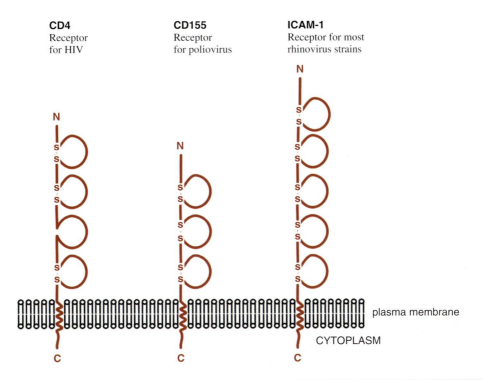

| CD4 | CD155 | ICAM-1 |
| Receptor for HIV | Receptor for poliovirus | Receptor for most rhinovirus strains |

plasma membrane

CYTOPLASM

Figure 5.1 *Cell receptors with immunoglobulin-like domains.* Each loop indicates an immunoglobulin-like domain. Most of the domains are stabilized by one or two disulphide (–S–S–) bonds. In each receptor the virus binding site is located in the outermost domain.

No covalent bonds are formed between virions and receptors.

Initially, a virion is weakly bound to a cell at only one or a few receptors. At this stage the attachment is reversible and the virion may detach, but if it remains attached there are opportunities for more virus attachment sites to bind to more receptors, and if sufficient bind then the attachment to the cell becomes irreversible.

5.2.4 Entry of animal viruses into cells

After binding to receptors animal viruses must cross the plasma membrane to gain entry to the host cell. They may do this either at the cell surface or they may cross the membrane of an endosome, which is a vesicle formed by part of the plasma membrane pinching off into the cytoplasm. This process (endocytosis) is used by cells for a variety of functions, including nutrient uptake and defence against pathogens. There are a number of endocytic mechanisms, including clathrin-mediated endocytosis and caveolin-mediated endocytosis; most animal viruses hi-jack one or more of these mechanisms in order to gain access to their host cells.

If a virion binds to a region of the plasma membrane coated with clathrin or caveolin these protein molecules force the membrane to bend around the virion. Many viruses, such as adenoviruses, are endocytosed at clathrin-coated regions of the plasma membrane. The virions end up in clathrin-coated endosomes, from which the clathrin is soon lost. Some viruses, such as simian virus 40, are endocytosed at caveolin-coated regions of the plasma membrane and the virions end up in caveolin-coated endosomes. Other viruses are taken up by mechanisms that are independent of clathrin and caveolin.

An endosome may fuse with other vesicles such as lysosomes, which have a pH of 4.8–5.0, thus lowering the pH within the vesicle. The pH may be further lowered by a process that pumps hydrogen ions across the membrane; the process involves the hydrolysis of ATP to ADP. This acidification of the environment of the virion is important for those enveloped viruses that need to carry out acid-triggered fusion of the envelope with the vesicle membrane (Section 5.2.4.b).

5.2.4.a Entry of naked viruses

It is possible that some naked viruses deliver their genomes into their host cells through a pore formed in the plasma membrane, but for most naked viruses irreversible attachment of the virion to the cell surface leads to endocytosis. The plasma membrane 'flows' around the virion, more receptors bind, and eventually the virion is completely enclosed in membrane, which pinches off as an endosome (Figure 5.2). The endosome contents, however, are part of the external environment and the virus is not yet in the cytoplasm. The mechanisms by which virions, or their genomes, are released from endosomes are not fully understood.

5.2.4.b Entry of enveloped viruses

Reversible attachment of an enveloped virion may lead to irreversible attachment, as for naked viruses. There are then two processes whereby infection of the cell may occur: either fusion of the virion envelope with the plasma membrane, or endocytosis followed by fusion of the virion envelope with the endosome membrane (Figure 5.3).

Both processes involve the fusion of the virion envelope with a cell membrane, either the plasma membrane or a vesicle membrane. Lipid bilayers do not fuse spontaneously and each enveloped virus has a specialized glycoprotein responsible for membrane fusion (Figure 5.4). Some examples of these fusion proteins are given in Table 5.1.

Many fusion proteins are synthesized as part of a larger protein, which is cleaved after synthesis. The cleavage products remain linked, either through non-covalent bonds (e.g. retroviruses) or through a disulphide bond (e.g. influenza viruses), and are found in the virion envelope as dimers or trimers, depending on the protein. Each monomer has at least two hydrophobic sequences: a transmembrane sequence and

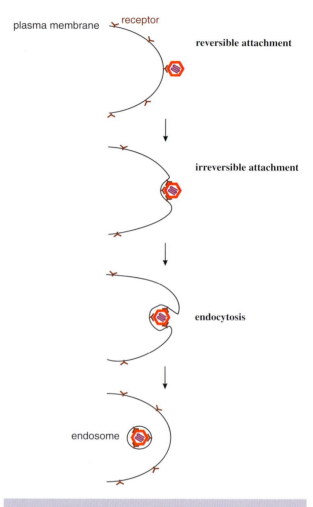

Figure 5.2 *Attachment and entry of a naked virion.* As more receptors bind to the virion its attachment becomes irreversible. The virion is taken into an endosome, the membrane of which is formed by pinching off from the plasma membrane.

a fusion sequence that is responsible for membrane fusion by inserting into a cell membrane.

The fusion sequence normally lies hidden and must become exposed in order for fusion to take place. Conformational changes to the fusion protein that expose the fusion sequence may come about either as a result of the virus binding to a receptor or they may be induced by the low pH within an endosome.

If the fusion sequence becomes exposed while the virion is at the cell surface, then infection could

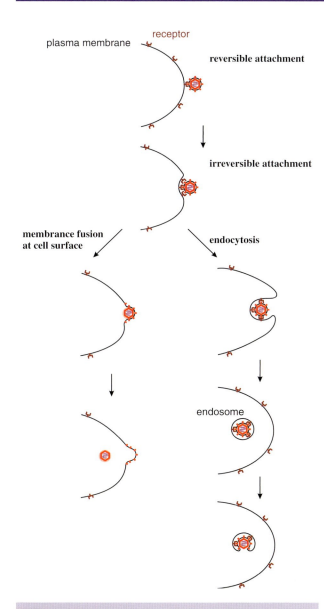

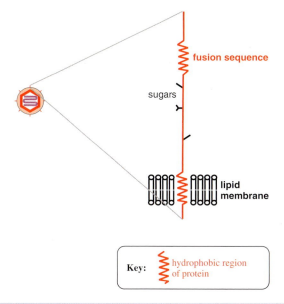

Figure 5.4 *Characteristics of fusion proteins of enveloped viruses.* Fusion proteins are envelope glycoproteins that have an external hydrophobic sequence that can be inserted into the target membrane.

- pH-independent fusion, e.g. herpesviruses, HIV

- acid-triggered fusion, e.g. influenza viruses, rhabdoviruses.

Once the fusion sequence is exposed it can insert into the target membrane (plasma membrane or endosome membrane). A further conformational change in the fusion protein pulls the two membranes together and then mediates their fusion. This process involves a release of energy from the fusion protein, which irreversibly changes shape.

5.2.5 Intracellular transport

Once in the cell the virus, or at least its genome, may have to be delivered to a particular location, such as the nucleus. For some viruses the destination is reached using one of the transport systems of the cell, such as the microtubules.

Most RNA viruses of eukaryotes replicate in the cytoplasm; the majority encode all the enzymes for replication of their genomes and they have no requirement for the enzymes of the nucleus. The

Figure 5.3 *Attachment and entry of an enveloped virion.* Some enveloped viruses can enter a host cell both by membrane fusion at the cell surface and by endocytosis. Others can enter only by endocytosis.

occur either by fusion with the plasma membrane or by endocytosis. If, however, exposure to low pH is required to expose the fusion sequence, then endocytosis is the only option. There are therefore two categories of membrane fusion:

influenza viruses, however, are exceptions as they require the cell splicing machinery, so their genomes must be delivered into the nucleus.

Retroviruses too are RNA viruses that replicate their genomes in the nucleus. They copy their genomes to DNA in the cytoplasm, then most retroviruses must wait in the cytoplasm until mitosis begins. During mitosis the nuclear envelope is temporarily broken down and the virus DNA (with associated proteins) is able to enter the nuclear compartment. These viruses therefore can replicate only in cells that are dividing. The DNA (with associated proteins) of a group of retroviruses, however, can be transported into an intact nucleus. This group (the lentiviruses, which includes HIV) can therefore replicate in non-dividing cells.

Some DNA viruses, such as iridoviruses and poxviruses, replicate in the cytoplasm of eukaryotic cells, but most DNA viruses replicate in the nucleus. For these viruses (and the influenza viruses and the lentiviruses) the virus genome must be transported to the nuclear envelope and then across it. The structural proteins of some of these viruses have sequences that allow them to attach to microtubules.

Microtubules are components of the cytoskeleton, providing support for various components of the cell and acting as tracks for the transport of materials, such as certain organelles, to particular sites in the cell. Microtubules are hollow cylinders, 25 nm in diameter, and are composed of the proteins α- and β-tubulin. The ends of each microtubule are designated as plus and minus. In most animal cells the plus ends are located near the plasma membrane, while the minus ends are attached to a structure called the centrosome (also referred to as the microtubule organizing centre) near the nucleus (Figure 5.5).

Proteins, known as motor proteins, move themselves and any cargo along the microtubules. A number of viruses (including herpesviruses, adenoviruses, parvoviruses and retroviruses) exploit this transport system to take their nucleocapsid, or a structure derived from it, from the periphery of the cytoplasm to a location close to the nucleus. The virus structures are transported at speeds of 1–4 μm per second.

The nuclear envelope is composed of two membranes (Figure 5.6), each of which is a lipid bilayer. Within the nuclear envelope are nuclear pores, which number between 3000 and 5000 in a typical growing cell. The nuclear pores act as gatekeepers, controlling the transport of materials in and out of the nucleus. Traffic into the nucleus includes ribosomal proteins and histones, while traffic out of the nucleus includes ribosomes, mRNAs and transfer RNAs.

Each nuclear pore is constructed from a complex of more than 50 protein species that control the movement of materials in and out of the nucleus. The protein molecules are assembled into discrete structures, including a 'basket' protruding into the nucleus and eight filaments that protrude into the cytoplasm (Figure 5.6). At the centre of the pore is a channel through which molecules can diffuse, though most molecules and particles pass through the pore complexed with specific carrier proteins (importins and exportins). The channel can open to allow the passage of particles up to 25 nm across, and there have been reports of larger particles being moved through nuclear pores. Small nucleocapsids/virions, such as the parvovirus virion, can pass through, but larger viruses must either shed some of their load to form slimmer structures or uncoat at a nuclear pore.

5.2.6 Genome uncoating

Uncoating can be defined as the complete or partial removal of the capsid to release the virus genome. Depending on the virus, the process can take place

- at the cell surface, the capsid remaining on the exterior surface of the cell;

- within the cytoplasm;

- at a nuclear pore;

- within the nucleus.

It should be noted that successful entry of a virion into a cell is not always followed by virus replication. The host's intracellular defences, such as lysosomal enzymes, may inactivate infectivity before or after uncoating. In some cases the virus genome may initiate a latent infection (Section 9.3.1) rather than a complete replication cycle.

In at least some cases, however, provided that the uncoated virus genome survives intact, and is at the correct location in the cell, then its transcription (or in some cases translation) can begin. These aspects are dealt with in the next chapter.

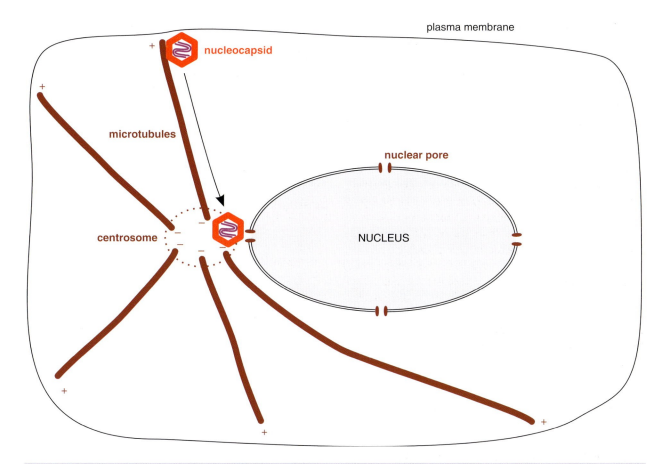

Figure 5.5 *Transport of a nucleocapsid along a microtubule.* The nucleocapsid is transported from the plus end of a microtubule near the plasma membrane to the centrosome, which is located close to the nucleus.

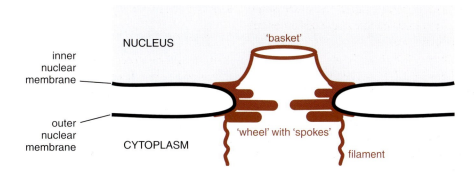

Figure 5.6 *Structure of the nuclear envelope and a nuclear pore.*

5.3   Bacteriophages

Like animal viruses, phages bind specifically to cell surface molecules that function as receptors and co-receptors. For many phages these molecules are on the surface of the host cell wall, but for some phages they are on the surface of other structures (pili, flagella or capsules). Many of the virus attachment sites are on particular virion structures, such as the tail fibres of phage T4. As with animal viruses, the attachment of a phage to its host is initially reversible, and becomes irreversible after attachment to further receptors and/or co-receptors.

Infection of a cell with a phage involves entry into the cell of the virus genome and perhaps a few associated proteins, the capsid and any associated appendages remaining at the cell surface. This is in contrast to most animal and plant viruses, where the entire virion, or at least the nucleocapsid, enters the host cell. Unless the host is a mycoplasma (mycoplasmas do not have cell walls), delivery of a phage genome into a cell requires penetration of a cell wall and perhaps also a slime layer or capsule; the virions of many phages carry enzymes, such as lysozymes, to aid this process.

The attachment and entry of a number of phages are discussed in Chapter 19.

Learning outcomes

By the end of this chapter you should be able to

- outline a generalized scheme of virus replication involving seven steps;

- describe how animal viruses attach to and enter their host cells;

- differentiate between the entry mechanisms of naked and enveloped animal viruses;

- describe the roles of cell components in the delivery of some viral genomes to the nucleus;

- outline the infection mechanisms of phages.

Sources of further information

Books

Marsh M., editor (2005) *Membrane Trafficking in Viral Replication*, Springer

Poranen M. *et al.* (2002) Common principles in viral entry. *Annual Review of Microbiology*, **65**, 521–538

Young J. A. T. (2001) Virus entry and uncoating, Chapter 4 in Knipe D. M. and Howley P. M., editors-in-chief, *Fields Virology*, 4th edition, Lippincott, Williams and Wilkins

Journals

Döhner K., Nagel C.-H. and Sodeik B. (2005) Viral stop-and-go along microtubules: taking a ride with dynein and kinesins *Trends in Microbiology*, **13**, 320–327

Kemp P., Garcia L. R. and Molineux I. J. (2005) Changes in bacteriophage T7 virion structure at the initiation of infection *Virology*, **340**, 307–317

Kielian M. and Rey F. A. (2006) Virus membrane-fusion proteins: more than one way to make a hairpin *Nature Reviews Microbiology*, **4**, 67–76

Molineux I. J. (2006) Fifty-three years since Hershey and Chase; much ado about pressure but which pressure is it? *Virology*, **344**, 221–229

Pelkmans L. (2005) Viruses as probes for systems analysis of cellular signalling, cytoskeleton reorganization and endocytosis *Current Opinion in Microbiology*, **8**, 331–337

Smith A. E. and Helenius A. (2004) How viruses enter animal cells *Science*, **304**, 237–242

Whittaker G. R. (2003) Virus nuclear import *Advanced Drug Delivery Reviews*, **55**, 733–747

Wu E. and Nemerow G. R. (2004) Virus yoga: the role of flexibility in virus host cell recognition *Trends in Microbiology*, **12**, 162–169

6

Transcription, translation and transport

At a glance

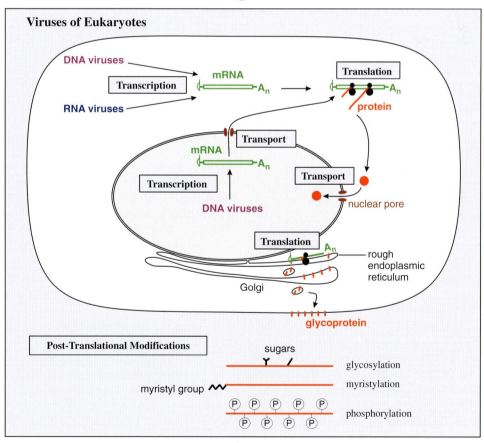

Viruses of Eukaryotes

Virology: Principles and Applications   John B. Carter and Venetia A. Saunders
© 2007 John Wiley & Sons, Ltd   ISBNs: 978-0-470-02386-0 (HB); 978-0-470-02387-7 (PB)

At a glance (continued)

Viruses of Bacteria

6.1   Introduction to transcription, translation and transport

You can probably think of several words beginning with 'trans-' and it is probable that most of them are derived from the Latin *trans*, meaning across. There are many 'trans' words used in virology and more broadly in biology, including transmission (Chapter 4), transposon (Chapter 20) and transformation (Chapter 22). The definitions of these words involve something going across.

In this chapter we deal with

- transcription = writing across

- translation = bearing across

- transport = carrying across.

For our purposes, transcription refers to the writing across of genetic information from a sequence of bases in a nucleic acid to the complementary sequence in messenger RNA (mRNA), while translation converts the genetic information from the language of bases in nucleic acids to the language of amino acids in proteins. Transcription and translation are steps 3 and 4 of our generalized replication cycle.

We also discuss in this chapter the transport of virus proteins and RNAs to particular locations in infected

cells. We start with an overview of virus transcription, and then we discuss these three 'trans' processes in eukaryotic cells. At the end of the chapter we point out some aspects of the processes that are different in bacterial cells.

6.2   Transcription of virus genomes

We have seen how there are four main categories of virus genome: dsDNA, ssDNA, dsRNA and ssRNA (Section 3.2). Because of distinct modes of transcription within the dsDNA and ssRNA categories a total of seven classes of viruses can be recognized (Figure 6.1).

This division of the viruses into classes based on genome type and mode of transcription was first suggested by David Baltimore and this scheme of virus classification is named after him. He initially proposed six classes.

In the summary of the scheme depicted in Figure 6.1 most of the nucleic acid strands are labelled (+) or (−). This labelling is relative to the virus mRNA, which is always designated (+). A nucleic acid strand that has the same sequence as mRNA is labelled (+) and a nucleic acid strand that has the sequence complementary to the mRNA is labelled (−).

The viruses with (+) RNA genomes (Classes IV and VI) have the same sequence as the virus mRNA. When these viruses infect cells, however, only the Class IV genomes can function as mRNA. These

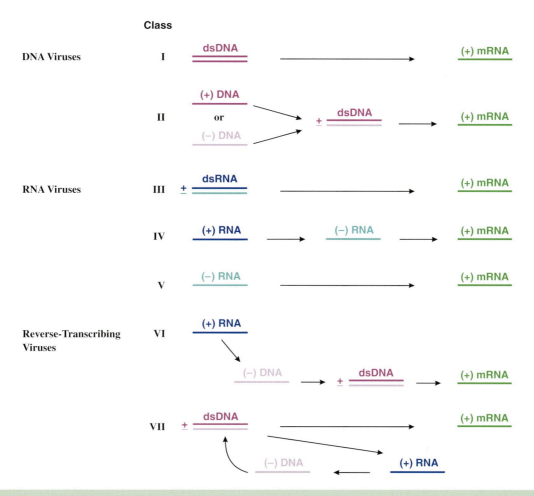

Figure 6.1 *Transcription of virus genomes.* (+) RNA and (+) DNA have the same sequence as the mRNA (except that in DNA thymine replaces uracil). (−) RNA and (−) DNA have the sequence complementary to the mRNA (except that in DNA thymine replaces uracil). (+) and (−) strands are not indicated for the dsDNA of the Class I viruses as the genomes of most of these viruses have open reading frames (ORFs) in both directions. (+) and (−) strands are indicated for the ssDNA of the Class II viruses. Most of these viruses have either a (+) or a (−) strand genome. A (+) RNA genome (dark blue) has the same sequence as the corresponding mRNA (green). The molecules are shown in different colours to indicate their different functions. In Class VII viruses the (+) RNA shown in blue (pregenome RNA) functions as a template for DNA synthesis (Section 18.8.6). Some of the DNA is used as a template for further transcription. Some ssDNA viruses and some ssRNA viruses have ambisense genomes. This means that that the polarity of the genome is part (+) and part (−).

viruses are commonly referred to as plus-strand (or positive-strand) RNA viruses. The Class V viruses are commonly referred to as minus-strand (or negative-strand) RNA viruses.

Class VI viruses must first reverse transcribe their ssRNA genomes to dsDNA before mRNA can be transcribed. Because they carry out transcription in reverse (RNA to DNA) Class VI viruses are known as retroviruses. The ability of some DNA viruses to carry out reverse transcription was discovered later; these viruses became known as pararetroviruses and Class VII was formed to accommodate them.

There are a few single-stranded nucleic acids of viruses where there is a mixture of (+) and (−) polarity within the strand, in other words there are open reading frames (ORFs) in both directions. Genomes of this type are known as ambisense, a word derived from the Latin *ambi*, meaning 'on both sides' (as in ambidextrous). Examples of ambisense genomes include the ssDNA genomes of the geminiviruses, which are plant viruses, and the ssRNA genomes of the arenaviruses, which are animal viruses and include the causative agent of Lassa fever.

6.2.1 Modifications to the central dogma

In 1958 Francis Crick proposed a 'central dogma of molecular biology'. James Watson, Crick's collaborator in deducing the structure of DNA, made significant contributions to the formulation of the dogma, which stated that the flow of genetic information is always from DNA to RNA and then to protein, with genetic information transmitted from one generation to the next through copying from DNA to DNA (Figure 6.2(a)). Increasing understanding of how viruses replicate their genomes necessitated some modifications to this dogma in 1970; many viruses have RNA genomes that are copied to RNA, and some viruses copy from RNA to DNA (Figure 6.2(b)).

6.3 Transcription in eukaryotes

We start this section with a brief summary of transcription from eukaryotic cell genes, as many viruses transcribe their genes by similar processes, some of them using parts of the cell transcription machinery (Figure 6.3). The expression of a gene is controlled by various sequences in the DNA:

- enhancers – sequences that contain binding sites for transcription factors, which affect the rate of transcription;

- a promoter – the 'on' switch;

- a terminator – the sequence that causes the enzyme to stop transcription.

6.3.1 Promoters and enhancers

The following consensus sequence is present in the promoters of many eukaryotic cell and virus genes:

T A T A A/T A A/T A/G

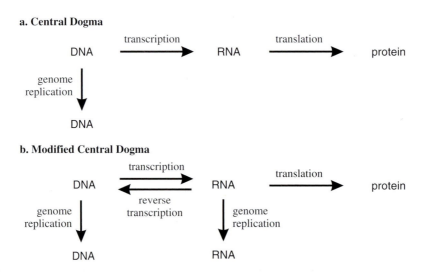

Figure 6.2 (a) Francis Crick's central dogma of molecular biology, which proposed that genetic information is transmitted from DNA to RNA to protein, and from DNA to DNA. (b) Modifications to the central dogma, required because of the various modes of virus transcription and genome replication.

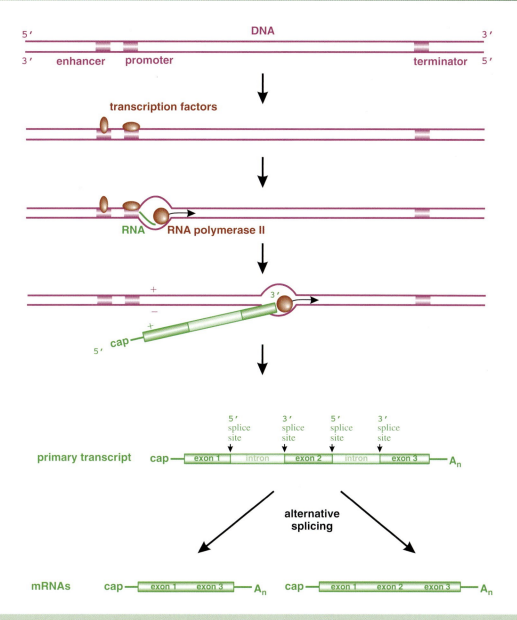

Figure 6.3 *Transcription from dsDNA in eukaryotes.* Transcription is initiated after transcription factors bind to sequences within promoters and enhancers. The primary transcript is capped at the 5′ end and polyadenylated at the 3′ end. The mRNAs are formed by removal of introns from the primary transcript.

The sequence is known as a TATA box and is usually located 25–30 bp upstream from the transcription start site. A TATA box is present, for example, in the single promoter of HIV-1 (Chapter 17), but in only one of the four promoters of hepatitis B virus (Section 18.8.3).

Enhancers contain sequences that bind transcription factors and these interactions may increase the rate of transcription starts by RNA polymerase II. Remarkably, some cell enhancers are up to 1 Mbp upstream or downstream from their promoters, though an enhancer and a promoter may come into close proximity as a result of DNA folding. Many enhancers are cell-type specific.

6.3.2 Transcription factors

Transcription factors are proteins that bind specifically to promoter and enhancer sequences to control gene expression. Some viruses produce their own transcription factors, such as herpes simplex virus VP16, which is a component of the virion (Section 11.5.2), and human T cell leukaemia virus I Tax protein, which is produced in the infected cell (Section 22.10.2).

Some cell transcription factors can activate or repress transcription of viral genes. Tissue-specific transcription factors are required by some viruses, which probably explains why some viruses are tissue specific.

Some cell transcription factors, known as general transcription factors, are involved in controlling the expression of genes in many cell types. An example is transcription factor IID (TFIID), which binds to the TATA box (Figure 6.4). TFIID is a complex of 13 polypeptides, one of which is the TATA box binding protein. After TFIID has bound to the TATA box other general transcription factors (TFIIA, IIB, IIE, IIF and IIH) and RNA polymerase II bind.

Among the cell transcription factors that bind to enhancers are

- AP-1 and AP-2 (activator proteins 1 and 2)

- Sp1 (stimulatory protein 1)

- NF-κB (nuclear factor κB).

Most of these transcription factors are involved in HIV-1 transcription (Section 17.4.3).

As well as activating gene expression, transcription factors are also involved in the repression of gene expression. All organisms regulate expression of their genes. A frog has different genes switched on depending on whether it is in the embryo, tadpole or adult stage. Similarly, a virus may have different genes switched on early and late in the replication cycle. For some viruses three phases (e.g. herpesviruses; Section 11.5.2) or four phases (e.g. baculoviruses) of gene expression can be distinguished.

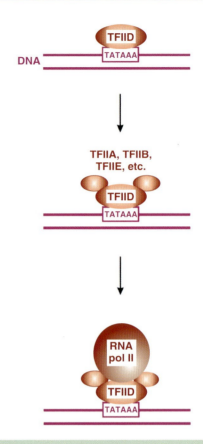

Figure 6.4 *Binding of transcription factors and RNA polymerase II at a TATA box.*

6.3.3 Transcriptases

Transcriptase is a general term for an enzyme that carries out transcription. Viruses that replicate in the nucleus generally use a cell enzyme, while viruses that replicate in the cytoplasm encode their own (Figure 6.5).

A DNA virus needs a DNA-dependent RNA polymerase to transcribe its genes into mRNA. Viruses that carry out transcription in the nucleus generally use the cell RNA polymerase II; these include the retroviruses, as well as many DNA viruses. DNA viruses that replicate in the cytoplasm use a virus-encoded enzyme because there is no appropriate cell enzyme in the cytoplasm.

An RNA virus (apart from the retroviruses) needs an RNA-dependent RNA polymerase to transcribe its

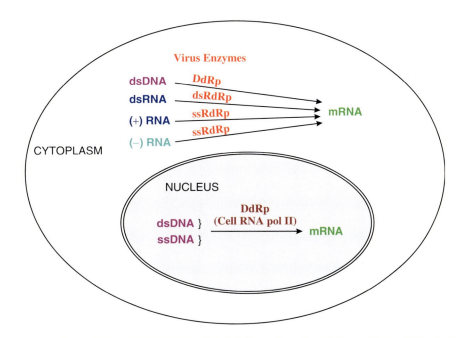

Figure 6.5 *Enzymes used by viruses to transcribe their genomes to mRNA.* A (+) RNA genome has the same sequence as the virus mRNA. A (−) RNA genome has the sequence complementary to that of the virus mRNA. Transcription from dsDNA in the nucleus applies not only to those dsDNA viruses that replicate in the nucleus, but also to the reverse transcribing viruses.

DdRp: DNA-dependent RNA polymerase

dsRdRp: double-stranded RNA-dependent RNA polymerase

ssRdRp: single-stranded RNA-dependent RNA polymerase

genes into mRNA. Each virus in Classes III, IV and V encodes its own enzyme, in spite of the fact that the cells of plants and some other eukaryotes encode ssRNA-dependent RNA polymerases.

All the viruses that carry out transcription in the cytoplasm, except the plus-strand RNA viruses, have the transcriptase in the virion so that the enzyme is immediately available to transcribe the virus genome when a cell is infected. Before the plus-strand RNA viruses can begin transcription they must translate copies of the enzyme from the genome RNA.

The retroviruses and the pararetroviruses perform reverse transcription (Section 6.2) using enzymes known as reverse transcriptases. These enzymes are RNA-dependent DNA polymerases, but they also have DNA-dependent DNA polymerase activity, as the process of reverse transcription involves synthesis of DNA using both RNA and DNA as the template.

6.3.4 Capping transcripts

Soon after RNA synthesis has begun, and while transcription is continuing, most transcripts are 'capped' at the 5′ end (Figure 6.3). The cap is a guanosine triphosphate joined to the end nucleotide by a 5′–5′ linkage, rather than the normal 5′–3′ linkage. A methyl group is added to the guanosine, and in some cases to one or both of the ribose residues on the first and second nucleotides (Figure 6.6). Throughout the book we shall use a cartoon cap to depict a cap on the 5′ end of an RNA molecule.

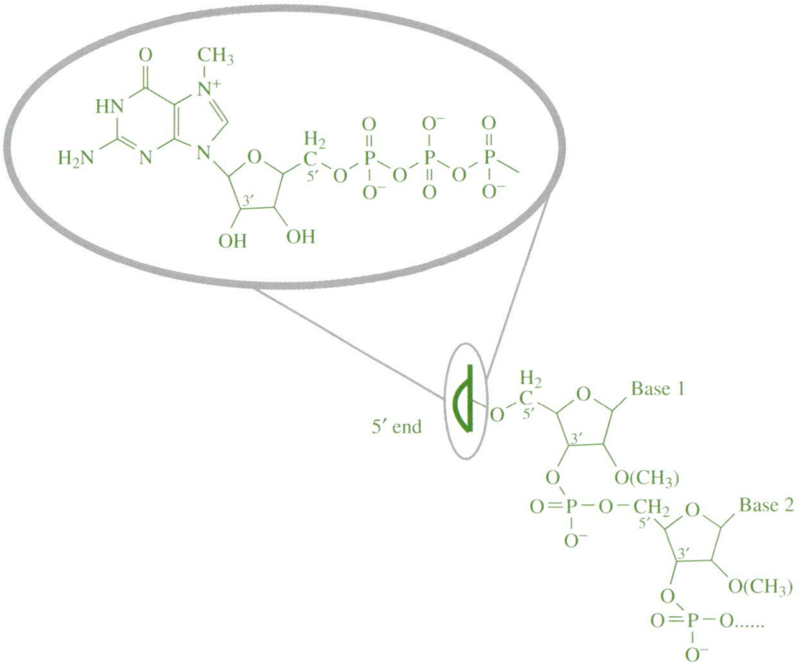

Figure 6.6 *Cap on 5′ end of messenger RNA.* The inset shows the structure of the cartoon cap, which depicts a methylated guanosine triphosphate linked to the first nucleotide by a 5′–5′ linkage.

Most eukaryotic cell and viral mRNAs have a cap at their 5′ end. The cap is thought to

- aid mRNA transport from the nucleus to the cytoplasm;

- protect the mRNA from degradation by exonucleases;

- be required for the initiation of translation.

The cell enzymes that carry out the capping activities are guanylyl transferases (they add the guanosine 5′-triphosphate) and methyl transferases (they add the methyl groups). These enzymes are located in the nucleus and most of the viruses that carry out transcription in the nucleus, like the retroviruses, use the cell enzymes. Many of the viruses that replicate in the cytoplasm, however, encode their own capping and methylating enzymes; these viruses include the poxviruses, the reoviruses and the coronaviruses.

Minus-strand RNA viruses with segmented genomes have evolved a mechanism to 'snatch' caps from cell mRNAs. These viruses include animal viruses, such as influenza viruses, and plant viruses, such as tomato spotted wilt virus. The complex of virus proteins making up the RNA polymerase binds to cellular capped mRNA, then an endonuclease activity associated with the complex cleaves the RNA, generally 10–20 nucleotides from the 5′ end. The capped oligonucleotides act as primers to initiate transcription of viral mRNA.

Not all mRNAs are capped. Picornaviruses, for example, do not cap their mRNAs; these viruses replicate in the cytoplasm, so their RNAs do not require transport from the nucleus, and translation is initiated by a mechanism that is not dependent upon a cap (Chapter 14).

6.3.5 Polyadenylation of transcripts

A series of adenosine residues (a polyadenylate tail; poly(A) tail) is added to the 3′ end of most primary transcripts of eukaryotes and their viruses. Polyadenylation probably increases the stability of mRNAs, and the poly(A) tail plays a role in the initiation of translation (Section 6.4.1). These functions can be provided in other ways, however, as some viruses, such as the reoviruses (Section 13.3.2), do not polyadenylate their mRNAs.

In most cases there is a polyadenylation signal about 10–30 bases upstream of the polyadenylation site. The polyadenylation signal AATAAA was first characterized in simian virus 40 in 1981. It has since been found that this sequence is used by many other animal viruses, such as HIV-1 (Section 17.4.3) and Rous sarcoma virus, as well as by animal cells. Some viruses use other sequences as polyadenylation signals; for example, the mammalian hepadnaviruses (Section 18.8.3) use TATAAA, a sequence that can function as a TATA box in other contexts!

In most cases the poly(A) tail is added by the following mechanism. During transcription the RNA polymerase proceeds along the template past the polyadenylation signal and the polyadenylation site. The newly synthesized RNA is then cleaved at the polyadenylation site and the poly(A) tail is added step by step by a complex of proteins, including a poly(A) polymerase. Some viruses have evolved alternative mechanisms to polyadenylate their mRNAs; these viruses include the picornaviruses (Section 14.4.4) and the rhabdoviruses (Section 15.4.2).

6.3.6 Splicing transcripts

Some primary transcripts are functional mRNAs, but most eukaryotic cell primary transcripts contain sequences (introns) that are removed. The remaining sequences (exons) are spliced at specific donor sites and acceptor sites to produce the mRNAs (Figure 6.3). A primary transcript may be cut and spliced in more than one way to produce two or more mRNA species. Some primary transcripts of viruses that replicate in the nucleus are processed in the same way to produce the virus mRNAs. The first evidence of split genes, as they are known, was reported in 1977 after studies with adenoviruses.

Further examples of viruses that have split genes are herpesviruses (Section 11.5.2), parvoviruses (Section 12.4.3) and retroviruses (Section 16.3.4). The simplest type of split gene consists of two exons separated by one intron, but some are much more complex; gene *K15* of Kaposi's-sarcoma-associated herpesvirus has eight exons and seven introns. The HIV-1 genome has a number of splice donor sites and acceptor sites; splicing of the primary transcript results in more than 30 different mRNA species (Section 17.4.3).

6.4 Translation in eukaryotes

A typical eukaryotic mRNA is monocistronic, i.e. it has one ORF from which one protein is translated (Figure 6.7). Sequences upstream and downstream of the ORF are not translated. Some large ORFs encode polyproteins, large proteins that are cleaved to form two or more functional proteins.

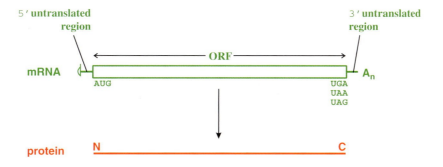

Figure 6.7 *Translation from a monocistronic mRNA.* There is one open reading frame (ORF), usually starting at the first AUG codon from the 5′ end of the mRNA, and ending at a stop codon (UGA, UAA or UAG). Translation occurs in the 5′ to 3′ direction, the N terminus of the protein being synthesized first.

6.4.1 Initiation of translation

As we have noted, most eukaryotic cell and virus mRNAs have a methylated nucleotide cap at the 5′ end and a poly(A) sequence at the 3′ end. These structures play key roles in the initiation of translation. The cap is especially important; it is the binding site for eukaryotic initiation factors (eIFs), a methionine tRNA charged with its amino acid, and a 40S ribosomal subunit (Figure 6.8). A poly(A)-binding protein binds to the poly(A) tail. The proteins bound at the ends of the RNA are able to interact, and it is thought that this interaction might allow circularization of the mRNA leading to stimulation of translation. Messenger RNAs that lack a cap and/or a poly(A) sequence might be circularized by other mechanisms.

The 40S ribosomal subunit is moved along the RNA in the 5′ to 3′ direction, scanning until an initiation codon is encountered in an appropriate sequence context. The initiation codon is normally AUG, and is normally the first AUG from the 5′ end. Some viruses, however, use other initiation codons; Sendai virus uses ACG as the initiation codon for one of its genes.

Some mRNAs are not capped and initiation of translation occurs by a different mechanism. A reduced set of eIFs binds not at the at the 5′ end, but at an

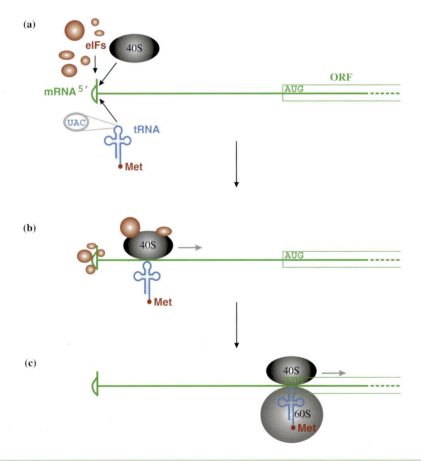

Figure 6.8 *Initiation of translation on a capped mRNA.* (a) Eukaryotic initiation factors (eIFs), a 40S ribosomal subunit and a methionine linked to its tRNA bind at the 5′ end of the mRNA. (b) The complex scans from the 5′ end of the mRNA. (c) When the first AUG codon is reached it is recognized by the anticodon UAC in the tRNA; a 60S ribosomal subunit is bound and initiation factors are released.

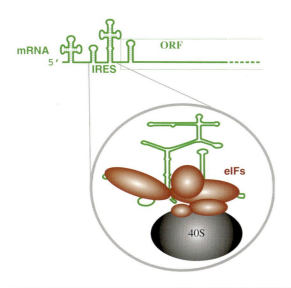

Figure 6.9 *Initiation of translation on an mRNA that is not capped.* A 40S ribosomal subunit and eIFs bind at an internal ribosome entry site (IRES) upstream of the ORF.

internal ribosome entry site (IRES), which has a high degree of secondary structure (Figure 6.9).

IRESs are present in a number of RNA viruses, including hepatitis C virus and the picornaviruses (Chapter 14). They have also been found in cell mRNAs and in one of the mRNAs of a DNA virus (Kaposi's-sarcoma-associated herpesvirus).

6.4.2 Translation from bicistronic mRNAs

Most eukaryotic cell and virus mRNAs have one ORF, but there are a number of virus mRNAs that have two or more ORFs. Some of these bicistronic and polycistronic mRNAs are functionally monocistronic, but some structurally bicistronic mRNAs are functionally bicistronic. A difference in the rate of translation of the two ORFs provides a mechanism for expressing two genes at different levels.

In many bicistronic mRNAs the ORFs overlap (Figure 6.10(a)); in others there is an ORF within an ORF (Figure 6.10(b)). One mechanism to read the second ORF involves leaky scanning; a 40S ribosomal subunit may overlook the ORF 1 start codon and initiate

translation at the start of ORF 2. The ORFs for the two proteins are in different reading frames, so the proteins that they encode are unrelated. Of course it is essential that the sequence 'makes sense' in both reading frames!

Another mechanism for reading a second ORF in an mRNA involves ribosomal frameshifting; a ribosome shifts into a different reading frame towards the end of ORF 1. It therefore does not recognize the ORF 1 stop codon, but continues along the mRNA, reading ORF 2 to produce an elongated version of the ORF 1 protein (Figure 6.10(c)). Frameshifting occurs when the ribosome moving along the RNA encounters a frameshift signal (a specific sequence) followed by a secondary structure, usually a pseudoknot (Section 3.2.2).

6.4.3 Co- and post-translational modification of proteins

During and after their translation proteins may undergo one or more modifications, including glycosylation, acylation and phosphorylation.

6.4.3.a Glycosylation

Glycosylation involves the addition of oligosaccharide groups to the polypeptide chain. When an oligosaccharide is linked through the –OH group of a serine or threonine residue the process is known as O-glycosylation; when it is linked through the $-NH_2$ group of an asparagine residue it is known as N-glycosylation.

Proteins destined for glycosylation are synthesized in the rough endoplasmic reticulum, where N-glycosylation commences. They are then transported to the Golgi complex (Section 6.5), where N-glycosylation is completed by enzymes such as α-mannosidases I and II and galactosyl transferase. O-glycosylation takes place in the Golgi complex.

Some glycoproteins have undergone only one type of glycosylation, such as the N-glycosylated gp120 of HIV-1, while many glycoproteins have undergone both O- and N-glycosylation, such as gC and gD of herpes simplex virus.

These glycoproteins of HIV and herpes simplex virus, like most virus glycoproteins, are integral membrane proteins that are components of the virion envelopes. Some glycoproteins, however, are not associated with virion envelopes; rotaviruses have

**Examples of Proteins Encoded
by Bicistronic mRNAs**

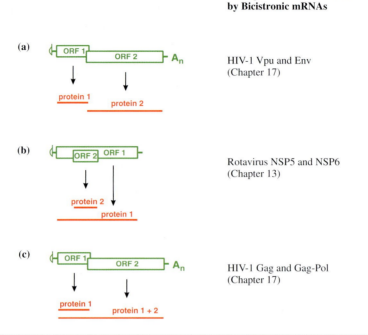

(a) HIV-1 Vpu and Env
(Chapter 17)

(b) Rotavirus NSP5 and NSP6
(Chapter 13)

(c) HIV-1 Gag and Gag-Pol
(Chapter 17)

Figure 6.10 *Translation from bicistronic mRNAs.* (a), (b) A ribosome may begin translation at the start of ORF 1, or scanning may be 'leaky' and translation may begin at the start of ORF 2. The two start codons are in different reading frames and the two proteins are unrelated. (c) ORF 2 is translated by a ribosomal frameshift, producing an extended version of protein 1.

naked virions but their surface protein (VP7) is glycosylated (Chapter 13). A few virus glycoproteins are non-structural proteins, such as the rotavirus protein NSP4.

6.4.3.b Acylation

Acylation is the addition of an acyl group (R–CO–) to a molecule. An acyl group that is commonly added to proteins is a myristyl group, where R is $CH_3–(CH_2)_{12}$. The myristyl group is linked to a glycine residue at the N terminus of the protein. Most viruses lack the enzyme N-myristyltransferase that is required for this modification; if one or more of their proteins is myristylated the process is carried out by a host enzyme.

Many myristylated proteins associate with membranes. This is true for the Gag proteins of most retroviruses (Chapter 16); if these proteins are not myristylated they do not associate with the plasma membrane and virion assembly does not take place.

Another example of a myristylated protein is the picornavirus capsid protein VP4 (Chapter 14).

6.4.3.c Phosphorylation

Phosphorylation involves the transfer of a phosphate group from a nucleotide, usually ATP, to the O of an –OH group of a serine, threonine or tyrosine residue. The transfer is carried out by protein kinases, which may be of cell and/or viral origin. The enzymes recognize short sequences of amino acids that bracket the residue to be phosphorylated.

Phosphorylation can alter the conformation, activity, localization and/or the stability of a protein, and many cell and viral processes involve protein phosphorylation. Many structural and non-structural virus proteins become phosphorylated, for example the phosphoproteins of rhabdoviruses (Chapter 15); one-sixth of the amino acid residues in the phosphoprotein of vesicular stomatitis virus are serine and threonine, and many of these are phosphorylated.

6.5 Transport in eukaryotic cells

We have already discussed how nucleocapsids and other virus structures are transported via microtubules and nuclear pores after entry into a cell (Section 5.2.5). Virus molecules synthesized in the infected cell must also be transported to particular sites. Virus mRNAs are transported from the nucleus to the cytoplasm, and virus proteins may be transported to various locations, including the nucleus (Figure 6.11).

Many proteins have a sequence of amino acids (a 'post code') that specifies their destination. Proteins destined to be incorporated into membranes have a signal sequence, which is a series of hydrophobic amino acid residues, either at the N terminus or internally. Protein synthesis begins on a free ribosome, but when the signal sequence has been synthesized it directs the polypeptide–ribosome complex to the endoplasmic reticulum, where protein synthesis continues. Regions of the endoplasmic reticulum with ribosomes associated are known as rough endoplasmic reticulum (Figure 6.11).

Each integral membrane protein has one or more membrane anchor sequences, which are rich in hydrophobic amino acid residues. For some of these proteins the signal sequence acts as a membrane anchor. Other integral membrane proteins, like the HIV-1 envelope protein, are moved through the membrane until the anchor sequence is reached, and then the signal sequence is removed by a host enzyme.

Many of the proteins synthesized in the rough endoplasmic reticulum are transported via vesicles to the Golgi complex, and most integral membrane

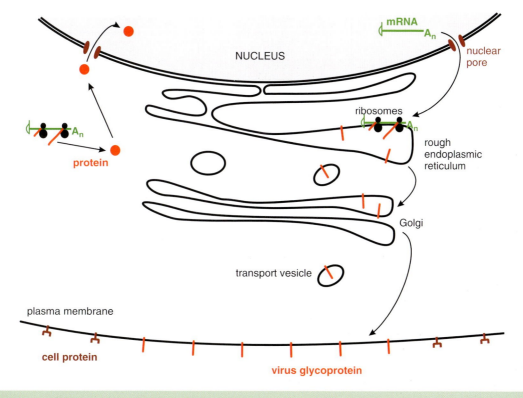

Figure 6.11 *Synthesis and intracellular transport of molecules.* Some proteins translated on free ribosomes are transported via nuclear pores to the nucleus. Some proteins translated in the rough endoplasmic reticulum are transported via the Golgi complex to the plasma membrane.

proteins become glycosylated in these membrane compartments. From here the glycoproteins may be transported to other membranes, such as the plasma membrane or the nuclear envelope. Progeny virions may bud from these membranes (Section 8.3.1).

Epithelial cells have apical (outer) and basolateral (inner) surfaces, which are composed of different lipids and proteins and are separated by 'tight junctions'. During infections of epithelial cells with enveloped viruses budding of virions from the plasma membrane may be restricted to either the apical surface or the basolateral surface.

Enrique Rodriguez-Boulan and David Sabatini used electron microscopy to examine thin sections of virus-infected epithelial cell cultures that retained their differentiated surfaces. They reported in 1978 that vesicular stomatitis virus buds from the basolateral surface, while influenza virus buds from the apical surface. This probably explains why most influenza virus infections of mammals are localized to the respiratory tract. It can be demonstrated that the glycoproteins of each virus are targeted to the appropriate surface (Figure 6.12). The targeting signal sequences and their locations within the proteins have been determined.

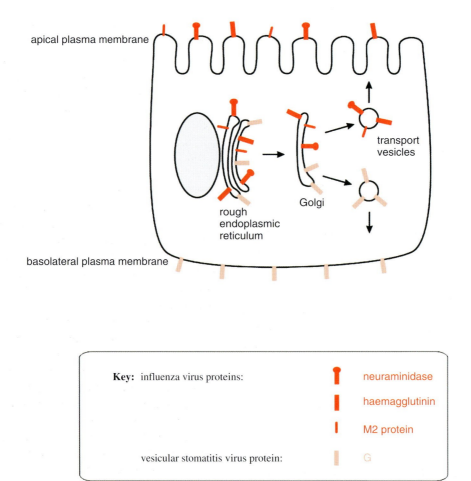

Key: influenza virus proteins:

neuraminidase

haemagglutinin

M2 protein

vesicular stomatitis virus protein:

G

Figure 6.12 *Targeting of virus envelope proteins to apical and basolateral surfaces of an epithelial cell.* Influenza A virus and vesicular stomatitis virus envelope proteins are transported to the apical and basolateral surfaces, respectively.

nuclear localization signal
TRQARRNRRRRWRERQR

nuclear export signal
LQLPPLERLTL

Figure 6.13 *HIV-1 Rev protein.* The nuclear localization signal is rich in arginine (R) residues. The nuclear export signal is rich in leucine (L) residues.

If the virus replicates in the nucleus then most, if not all, of the virus proteins must be transported into the nucleus. These proteins, like cell proteins that are transported into the nucleus, have a nuclear localization signal, which is rich in one or both of the basic amino acids lysine (K) and arginine (R). A common nuclear localization signal is

PKKKRKV.

This signal was first identified in a simian virus 40 protein known as large T antigen. A nuclear localization signal allows a protein to bind to cell proteins (importins) and subsequently to nuclear pore filaments (Figure 5.6), from where it is transported into the nucleus.

RNAs are also transported within the cell; for example, mRNAs synthesized in the nucleus must be exported through nuclear pores to the cytoplasm. The RNAs are taken to their destinations by proteins. The Rev protein of HIV-1 has both a nuclear localization signal and a nuclear export signal (Figure 6.13). The nuclear localization signal ensures that Rev is transported into the nucleus, where it binds specifically to HIV-1 RNA. The nuclear export signal ensures that Rev and its RNA cargo are transported from the nucleus to the cytoplasm via a nuclear pore.

6.6 Transcription and translation in bacteria

This section starts with a brief summary of transcription from prokaryotic genes. There is one type of DNA-dependent RNA polymerase for transcription in prokaryotes, and the *E. coli* holoenzyme comprises the catalytic core enzyme containing one β, one β′ and two α subunits, and a regulatory protein, the sigma

(σ) factor. The σ factor determines promoter specificity, enabling the polymerase to recognize and bind to specific promoters in the correct orientation to initiate transcription. Bacterial promoters contain characteristic sequences: the −10 region (Pribnow box) and the −35 region, centred 10 and 35 bp upstream of the transcription start site respectively, and recognized by the σ factor. Upstream of the −35 region some promoters additionally contain the UP element, which is recognized by the RNA polymerase α subunit and enhances binding of the enzyme.

Once transcription initiation is completed, the σ factor dissociates to be re-used in the process. Different σ factors recognize different promoters. There are two main sigma families: σ^{70} and σ^{54}. The σ^{70} family has many different σ factors, including the primary (vegetative) σ factors, directing transcription of genes for bacterial growth and metabolism, and alternative factors, e.g. σ^{32} for transcription of heat shock genes.

Transcription terminates after a terminator is transcribed. Termination may be rho independent (intrinsic) or rho dependent. Rho-independent termination depends on the template DNA sequence, which contains an inverted repeat and a string of adenine residues, located downstream of the rho-independent genes. The inverted repeat forms a stem-loop structure in the RNA transcript and adenine–uracil bonds form in the DNA–RNA hybrid, such that the RNA transcript can dissociate from the template. Rho-dependent termination involves a rho factor, which has helicase and ATPase activity, to unwind DNA–RNA hybrids when the polymerase is stalled at an inverted repeat. This leads to release of the transcript.

Bacterial cells and the viruses that infect them operate transcription and translation mechanisms that differ in a number of respects from those in eukaryotes. Introns are rarely present in the genes of bacteria and their viruses. Some phages use the host DNA-dependent RNA polymerase for their transcription, while others encode their own.

Whereas typical mRNAs in eukaryotes are monocistronic, in bacteria they are polycistronic, i.e. each mRNA has several ORFs (Figure 6.14). The mRNAs of bacteria and their viruses are never capped at the 5′ end, and are rarely polyadenylated at the 3′ end.

Figure 6.14 *Characteristics of bacterial and phage mRNA.* There are typically several ORFs, all of which may be translated at the same time. The 5′ end is not capped and the 3′ end is rarely polyadenylated.

Bacterial translation differs from eukaryotic translation in a number of features.

- Translation may start before transcription is complete. The lack of a nucleus allows transcription and translation to be coupled.

- The ribosomes are smaller (the ribosomal subunits have sedimentation coefficients of 30S and 50S).

- The 30S ribosomal subunit binds directly to a translation initiation region on the mRNA. Initiation generally involves interaction of the Shine-Dalgarno (S-D) sequence at the ribosome binding site (RBS) on mRNA and the anti-S-D sequence at the 3′ end of 16S rRNA in the 30S subunit. The RBS is located just upstream of the AUG start codon.

- The methionine of the initiator methionyl tRNA is generally formylated.

- A much smaller number of initiation factors is involved.

- All ORFs within an mRNA are translated and several may be translated concurrently.

A few phages have overlapping genes, which may be translated by reading through a stop codon or by ribosomal frameshifting (Section 6.4.2).

Learning outcomes

By the end of this chapter you should be able to

- explain how virus genes are transcribed and translated;

- describe the post-translational modifications that some virus proteins undergo;

- highlight differences in transcription and translation between prokaryotic and eukaryotic cells;

- discuss the transport of virus proteins and RNA within cells.

Sources of further information

Books

Brown T. A. (2006) *Genomes 3*, 3rd edition, Garland Science

Brown W. M. and Brown P. M. (2002) *Transcription*, Taylor and Francis

Lee D., Kapp L. D. and Lorsch J. R. (2004) The molecular mechanics of eukaryotic translation. *Annual Review of Biochemistry*, **73**, 657–704

Smith G. L., Murphy B. J. and Law M. (2003) Vaccinia virus motility *Annual Review of Microbiology*, **57**, 323–342

Historical papers

Crick F. (1970) Central dogma of molecular biology *Nature*, **227**, 561–563

Rodriguez-Boulan E. and Sabatini D. D. (1978) Asymmetric budding of viruses in epithelial monolayers: a model system for study of epithelial polarity *Proceedings of the National Academy of Sciences USA*, **75**, 5071–5075

Recent papers

Ahlquist P. (2006) Parallels among positive-strand RNA viruses, reverse-transcribing viruses and double-stranded RNA viruses *Nature Reviews Microbiology*, **4**, 371–382

Fechter P. and Brownlee G. G. (2005) Recognition of mRNA cap structures by viral and cellular proteins *Journal of General Virology*, **86**, 1239–1249

Kozak M. (2004) How strong is the case for regulation of the initiation step of translation by elements at the 3′ end of eukaryotic mRNAs? *Gene*, **343**, 41–54

7

Virus genome replication

At a glance

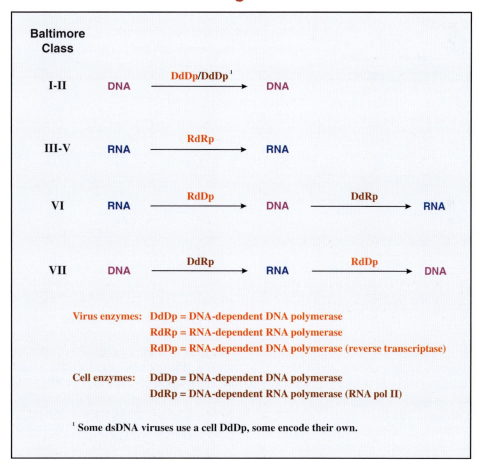

Virology: Principles and Applications   John B. Carter and Venetia A. Saunders
© 2007 John Wiley & Sons, Ltd   ISBNs: 978-0-470-02386-0 (HB); 978-0-470-02387-7 (PB)

At a glance (continued)

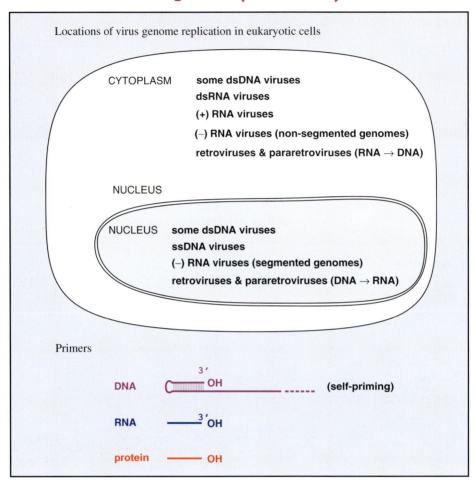

Locations of virus genome replication in eukaryotic cells

CYTOPLASM
- some dsDNA viruses
- dsRNA viruses
- (+) RNA viruses
- (−) RNA viruses (non-segmented genomes)
- retroviruses & pararetroviruses (RNA → DNA)

NUCLEUS

NUCLEUS
- some dsDNA viruses
- ssDNA viruses
- (−) RNA viruses (segmented genomes)
- retroviruses & pararetroviruses (DNA → RNA)

Primers

DNA — 3′ OH — — — — (self-priming)

RNA — 3′ OH

protein — OH

7.1 Overview of virus genome replication

In this chapter we consider the fifth step of our generalized replication cycle: genome replication. The genome of the infecting virus is replicated so that viral transcription can be amplified and to provide copies of the genome for progeny virions.

Generally, DNA viruses copy their genomes directly to DNA and RNA viruses copy their genomes directly to RNA. There are, however, some DNA viruses that replicate their genomes via an RNA intermediate and some RNA viruses that replicate their genomes via a DNA intermediate. The various replication modes of virus genomes are summarized in Figure 7.1.

Single-stranded genomes are designated as plus or minus depending on their relationship to the virus mRNA. Plus strand genomes have the same sequence as the mRNA (except that in DNA thymine replaces uracil), while minus-strand genomes have the sequence complementary to the mRNA. Single-stranded DNA is converted to dsDNA prior to copying.

There are two classes of viruses with (+) RNA genomes (Figure 7.1). Class IV viruses copy their (+) RNA genomes via a (−) RNA intermediate, while Class VI viruses replicate via a DNA intermediate. The

Class

DNA Viruses

RNA Viruses

Reverse-Transcribing Viruses

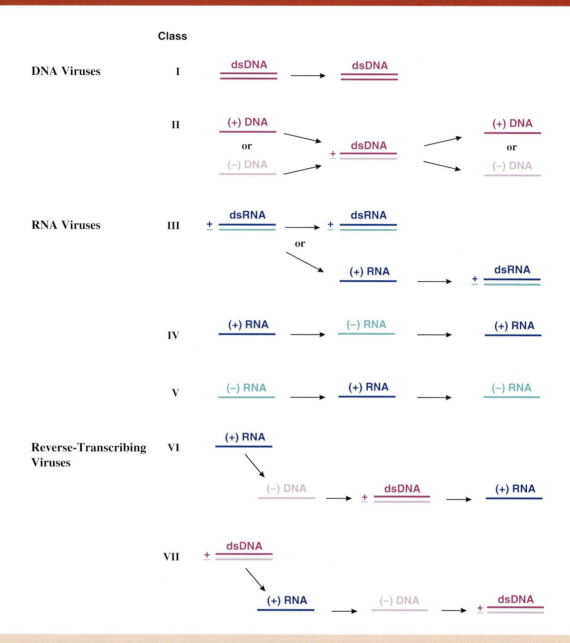

Figure 7.1 *Replication of virus genomes in the seven Baltimore Classes.* (+) RNA and (+) DNA have the same sequence as the mRNA (except that in DNA thymine replaces uracil). (−) RNA and (−) DNA have the sequence complementary to the mRNA (except that in DNA thymine replaces uracil). (+) and (−) strands are not indicated for the dsDNA of the Class I viruses as the genomes of most of these viruses have ORFs in both directions. (+) and (−) strands are indicated for the ssDNA of the Class II viruses. Most of these viruses have either a (+) or a (−) strand genome. Some ssDNA viruses and some ssRNA viruses have ambisense genomes.

synthesis of DNA from an RNA template (reverse transcription) is also a characteristic of Class VII viruses.

In this chapter we shall look at some general aspects of virus genome replication, and then we shall give individual attention to replication of the genomes of the DNA viruses, the RNA viruses and the reverse transcribing viruses.

7.2 Locations of virus genome replication in eukaryotic cells

As we saw in Chapter 5, when viruses infect eukaryotic cells the genomes of some are delivered to the cytoplasm and some are conveyed to the nucleus. The destination of a virus genome, and hence the location in which it is replicated, varies with the type of genome (Table 7.1).

The genomes of most DNA viruses are replicated in the nucleus, but those of some dsDNA viruses are replicated in the cytoplasm. The genomes of most RNA viruses are replicated in the cytoplasm, but those of the minus-strand RNA viruses with segmented genomes are replicated in the nucleus. The retroviruses and pararetroviruses are special cases: each replicates RNA to DNA in the cytoplasm and DNA to RNA in the nucleus.

7.3 Initiation of genome replication

Each virus genome has a specific sequence where nucleic acid replication is initiated. This sequence is recognized by the proteins that initiate replication.

Figure 7.2 3′ end of a ssDNA where self-priming of DNA synthesis can occur.

Nucleic acid replication requires priming, which is the first reaction of a nucleotide with an –OH group on a molecule at the initiation site. Replication of the genomes of many RNA viruses (including rotaviruses, and rhabdoviruses) initiates when the first nucleotide of the new strand base pairs with a nucleotide in the viral RNA. The initial nucleotide effectively acts as a primer for RNA replication when its 3′ –OH group becomes linked to the second nucleotide.

Some ssDNA viruses, such as parvoviruses, use self-priming. At the 3′ end of the DNA there are regions with complementary sequences that can base pair (Figure 7.2). The –OH group of the nucleotide at the 3′ end forms a linkage with the first nucleotide, then DNA synthesis proceeds by a rather complex process to ensure that the whole genome is copied.

In order to initiate the replication of many DNA genomes, and some RNA genomes, a molecule of RNA or protein is required to act as a primer.

7.3.1 RNA and protein primers

Synthesis of cell DNA commences after a region of the double helix has been unwound by a helicase and

Table 7.1 Locations of virus genome replication in eukaryotic cells

Virus genome	Cytoplasm	Nucleus
dsDNA	Some	Some
ssDNA		All
dsRNA	All	
(+) RNA	All	
(−) RNA (non-segmented genome)	All	
(−) RNA (segmented genome)		All
Retroviruses [(+) RNA]	ssRNA → dsDNA	dsDNA → ssRNA
Pararetroviruses [dsDNA]	ssRNA → dsDNA	dsDNA → ssRNA

after a primase has synthesized short sequences of RNA complementary to regions of the DNA. These RNAs act as primers; one is required for the leading strand, while multiple primers must be synthesized for the Okazaki fragments of the lagging strand. The first nucleotide of a new sequence of DNA is linked to the 3′ –OH group of the primer RNA.

Some DNA viruses also use RNA primers during the replication of their genomes. Some viruses, such as polyomaviruses, use the cell primase to synthesize their RNA primers, while others, such as herpesviruses and phage T7, encode their own primases.

During their replication cycle the retroviruses synthesize DNA from a (+) RNA template (Section 16.3.2). They use a cell transfer RNA to prime (−) DNA synthesis, then they use the 3′ –OH group in a polypurine tract of the partly degraded (+) RNA template to prime (+) DNA synthesis. The retrovirus DNA becomes integrated into a cell chromosome. If the infection is latent and the cell subsequently divides (Section 9.3.1), then the virus DNA is copied along with the cell DNA, using RNA primers synthesized by the cell primase.

For some viruses the primer for initiation of nucleic acid replication is the –OH group on a serine or tyrosine residue in a protein. DNA viruses that use protein primers include some animal viruses (e.g. adenoviruses) and some phages (e.g. tectiviruses). RNA viruses that use protein primers include some animal viruses (e.g. picornaviruses) and some plant viruses (e.g. luteoviruses).

Hepadnaviruses are DNA viruses that use a protein primer to initiate (−) DNA synthesis and an RNA primer to initiate (+) DNA synthesis (Section 18.8.6).

Protein primers (and the RNA primers of hepadnaviruses) are not removed once their role is performed and they are found linked to the 5′ ends of the genomes in virions (Section 3.2.3).

7.4 Polymerases

The key enzymes involved in virus genome replication are DNA polymerases and RNA polymerases. Many viruses encode their own polymerase, but some use a host cell enzyme (Figure 7.3).

A DNA virus requires a DNA-dependent DNA polymerase. Amongst the DNA viruses that replicate

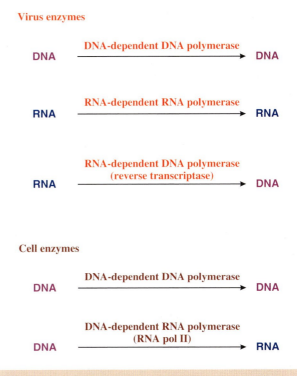

Figure 7.3 *Enzymes used by viruses to replicate their genomes.* Many viruses encode a polymerase to replicate their genome, but some use a cell enzyme.

in the nuclei of eukaryotic cells, viruses with small genomes (e.g. papillomaviruses) use the cell enzyme, while viruses with large genomes (e.g. herpesviruses) encode their own enzyme. Those DNA viruses that replicate in the cytoplasm must encode their own enzyme.

The enzyme that replicates the genome of an RNA virus is often referred to as a replicase; for many RNA viruses this is the same enzyme as that used for transcription (Section 6.3.3). The retroviruses and the pararetroviruses encode reverse transcriptases to transcribe from RNA to DNA, and use the host cell RNA polymerase II to transcribe from DNA to RNA.

Many viral polymerases form complexes with other viral and/or cell proteins to produce the active enzyme. Some of these additional proteins are processivity factors, for example an *Escherichia coli* thioredoxin molecule functions as a processivity factor for the DNA polymerase of phage T7.

7.5   DNA replication

The viruses of Class I (dsDNA) and Class II (ssDNA) replicate their genomes via dsDNA. The ssDNA viruses first synthesize a complementary strand to convert the genome into dsDNA.

Each viral DNA has at least one specific sequence (*ori*; replication origin) where replication is initiated. The proteins that initiate DNA replication bind to this site, and amongst these proteins are

- a helicase (unwinds the double helix at that site);

- a ssDNA binding protein (keeps the two strands apart);

- a DNA polymerase.

Viral dsDNA is generally replicated by a process similar to that used by cells to copy their genomes. The basic process and the enzymes involved are outlined in Figure 7.4. Fewer proteins are involved in bacterial systems than in eukaryotic systems; for example, the helicase–primase of phage T7 is a single protein molecule, while that of herpes simplex virus is a complex of three protein species.

DNA synthesis takes place near a replication fork. One of the daughter strands is the leading strand and the other is the lagging strand, synthesized as Okazaki fragments, which become joined by a DNA ligase. After a dsDNA molecule has been copied each of the daughter molecules contains a strand of the original molecule. This mode of replication is known as semi-conservative, in contrast to the conservative replication of some dsRNA viruses (Section 7.6).

Some DNA genomes are linear molecules, while some are covalently closed circles (Section 3.2). Some of the linear molecules are circularized prior to DNA replication, hence many DNA genomes are replicated as circular molecules, for which there are two modes of replication, known as theta and sigma (Figure 7.5). These terms refer to the shapes depicted in diagrams

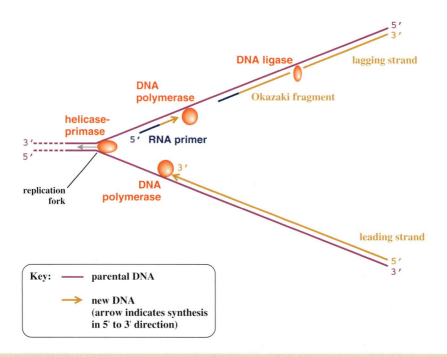

Key: —— parental DNA

→ new DNA
(arrow indicates synthesis
in 5' to 3' direction)

Figure 7.4 *DNA replication.* A helicase–primase unwinds the dsDNA and synthesizes RNA primers that are used by the DNA polymerase to initiate DNA synthesis. The leading strand is synthesized continuously, while the lagging strand is synthesized as Okazaki fragments that are joined together by a DNA ligase.

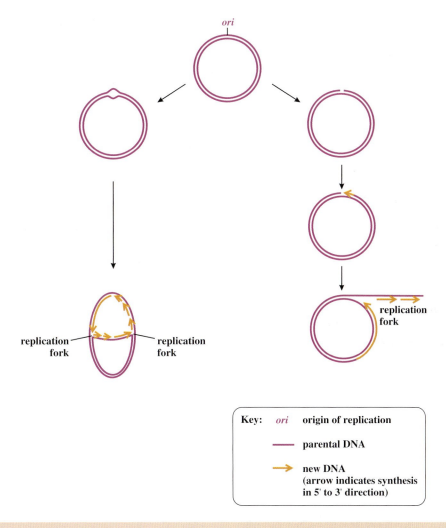

Figure 7.5 *Theta and sigma modes of DNA replication.* The theta structure is shown with two replication forks as a result of bidirectional replication from *ori*; unidirectional replication can also give rise to a theta structure.

of the replicating molecules, which resemble the Greek letters θ (theta) and σ (sigma). The sigma mode of replication is also known as a rolling circle mode. The genomes of some DNA viruses may be replicated by the theta mode of replication early in infection and the sigma mode late in infection.

Replication of the DNA of some viruses, such as herpesviruses (Section 11.5.3) and phage T4, results in the formation of very large DNA molecules called concatemers. Each concatemer is composed of multiple copies of the virus genome and the concatemers of some viruses are branched. When DNA is packaged during the assembly of a virion an endonuclease cuts a genome length from a concatemer.

7.6 Double-stranded RNA replication

Double-stranded RNA, like dsDNA, must be unwound with a helicase in order for the molecule to be replicated.

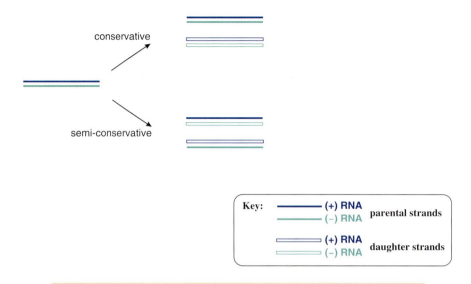

Figure 7.6 *Conservative and semi-conservative replication of dsRNA.*

Some dsRNA viruses, e.g. *Pseudomonas* phage φ6 (φ = Greek letter phi), replicate their genomes by a semi-conservative mechanism, similar to dsDNA replication (Section 7.5); each of the double-stranded progeny molecules is made up of a parental strand and a daughter strand. Other dsRNA viruses, including members of the family *Reoviridae* (Chapter 13), replicate by a mechanism designated as conservative because the double-stranded molecule of the infecting genome is conserved (Figure 7.6).

7.7   Single-stranded RNA replication

The ssRNA genomes of viruses in Classes IV and V are replicated by synthesis of complementary strands of RNA that are then used as templates for synthesis of new copies of the genome (Figure 7.1). The synthesis of each RNA molecule requires the recruitment of an RNA-dependent RNA polymerase to the 3′ end of the template, therefore both plus- and minus-strand RNA must have a binding site for the enzyme at the 3′ end.

An interesting point to note here is that all class IV viruses of eukaryotes replicate their RNA in association with cytoplasmic membranes. For many groups of viruses, including picornaviruses (Section 14.4.4), these membranes are derived mainly from the endoplasmic reticulum, but other membranous structures are used, including endosomes (by togaviruses) and chloroplasts (by tombusviruses). Viral proteins, including the RNA polymerases, are bound to the membranes.

During the replication of ssRNA both (+) and (−) strands of RNA accumulate in the infected cell, but not in equal amounts. Plus-strand RNA viruses accumulate an excess of (+) RNA over (−) RNA, and for minus-strand RNA viruses the reverse is true.

7.8   Reverse transcription

Some RNA viruses replicate their genomes via a DNA intermediate, while some DNA viruses replicate their genomes via an RNA intermediate (Figure 7.1). Both of these modes of genome replication involve reverse transcription, which has two major steps: synthesis of (−) DNA from a (+) RNA template followed by synthesis of a second DNA strand (Figure 7.7). Both steps are catalysed by a reverse transcriptase that is encoded by the virus.

Reverse transcription takes place within a viral structure in the cytoplasm of the infected cell. In later chapters the process is considered in more detail for the retroviruses (Section 16.3.2) and for hepatitis B virus (Section 18.8.6). No viruses of prokaryotes are known to carry out reverse transcription.

Figure 7.7 *Reverse transcription.* ssDNA is synthesized from an RNA template then the ssDNA is converted to dsDNA. A reverse transcriptase (RT) carries out both steps.

Learning outcomes

By the end of this chapter you should be able to

- state the locations within eukaryotic cells where different categories of virus genome are replicated;

- explain the role of primers in virus nucleic acid synthesis;

- discuss the roles of virus and host proteins in virus genome replication;

- outline the replication mechanisms of virus DNAs and RNAs;

- explain the term 'reverse transcription'.

Sources of further information

Books

Brown T. A. (2002) Chapter 13 in *Genomes*, 2nd edition, BIOS

Cann A. J., editor (2000) *DNA Virus Replication*, Oxford University Press

Journal

van Dijk A. A. *et al.* (2004) Initiation of viral RNA-dependent RNA polymerization *Journal of General Virology*, **85**, 1077–1093

8

Assembly and exit of virions from cells

At a glance

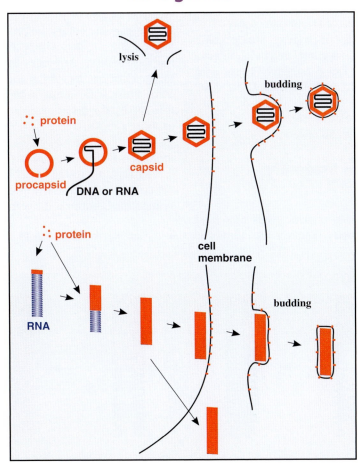

Virology: Principles and Applications   John B. Carter and Venetia A. Saunders
© 2007 John Wiley & Sons, Ltd   ISBNs: 978-0-470-02386-0 (HB); 978-0-470-02387-7 (PB)

8.1 Introduction to assembly and exit of virions from cells

In this chapter we deal with the final two stages of our generalized replication cycle: assembly of virions and their exit from the cell.

There is a requirement for a virion to be a stable structure that will survive in the environment as an infectious entity (though there is huge variation amongst viruses in the degree of stability, as discussed in Chapter 23). There is, however, also a requirement for a virion to become unstable during infection of a host cell so that the viral genome can be released. The virion must therefore have a built-in 'switch' that can initiate the change from stability to instability when the appropriate environment is encountered. The 'trigger' that activates the 'switch' might be binding to a receptor and/or a change in the concentration of H^+ or other ions.

Once threshold quantities of progeny virus genomes and structural proteins have accumulated in the infected cell, assembly of virions can commence. These components are assembled into nucleocapsids. If the virion also contains lipid then the assembly process also includes the acquisition of this component, either as an internal membrane or as an envelope.

8.2 Nucleocapsid assembly

8.2.1 Helical viruses

The assembly of virions and nucleocapsids of ssRNA viruses with helical symmetry (Section 3.4.1) involves coating the genome with multiple copies of a protein (Figure 8.1).

8.2.2 Icosahedral viruses

The assembly of virions and nucleocapsids of many viruses with icosahedral symmetry (Section 3.4.2) involves the construction of an empty protein shell, known as a procapsid, or a prohead in the case of a tailed bacteriophage. The procapsid is filled with a copy of the virus genome (Figure 8.2); during or after this process it may undergo modification to form the mature capsid. Modification of the procapsid may result in a change from a spherical to an icosahedral shape. For some viruses, including adenoviruses and picornaviruses, modification of the procapsid involves cleavage of one or more of the structural proteins.

The genome enters the procapsid through a channel located at a site that will become one of the vertices of the icosahedron. Any enzymes involved in packaging the genome are located at this site. The procapsids of some viruses, including herpesviruses, are modified at this site by the insertion of a 'portal' composed of one or more protein species.

The prohead of many tailed phages also has a portal through which a copy of the virus genome enters. A further function of the portal is as a 'connector' to join the tail to the head. Each of these three components is symmetrical, but in a different way:

- the connector has 12-fold rotational symmetry;

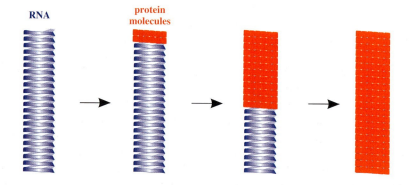

RNA   **protein molecules**

Figure 8.1 *Assembly of a helical nucleocapsid.* A few copies of a protein species bind to a helical ssRNA molecule, then more copies bind until the RNA is completely coated.

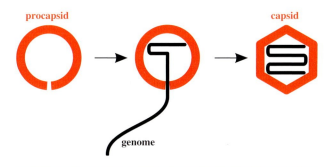

procapsid   capsid

genome

Figure 8.2 *Assembly of an icosahedral nucleocapsid.* A protein shell (procapsid) is constructed. A copy of the virus genome enters the procapsid, which undergoes modifications to form the capsid.

- the icosahedron vertex attached to one end of the connector has five-fold rotational symmetry (Section 3.4.2);

- the tail attached to the other end of the connector has six-fold rotational symmetry.

8.2.3 Genome packaging

In the context of packaging virus genomes during virion assembly the question 'How are virus genomes selected from all the cell and virus nucleic acids?' is an intriguing one. In a retrovirus-infected cell, for example, less than one per cent of the RNA is virus genome. In fact retroviruses do package a variety of cell RNAs, including tRNA, which plays a key role when the next host cell is infected (Chapter 16). Some other viruses also package cell nucleic acids; for example, tailed phages sometimes package host DNA.

For the most part, however, cell nucleic acids are not packaged. For some viruses it has been shown that this is achieved through the recognition by a virus protein of a specific virus genome sequence, known as a packaging signal. In single-stranded genomes the packaging signal is within a region of secondary structure. Most viruses with single-stranded genomes package either the plus strand or the minus strand, so the packaging signal must be present only in the strand to be packaged.

It is important that some copies of the virus genome are not packaged: those that are serving as templates

for synthesis of DNA, RNA or protein. The packaging signal must be masked in these copies of the genome.

Virus genomes are packaged into small volumes, which means that repulsion between the negative charges on their phosphate groups must be overcome. This may be aided by packaging basic proteins, which are positively charged, along with the genome. Some ssRNA viruses (e.g. rhabdoviruses, influenza viruses and retroviruses) coat their genomes with basic proteins, while some dsDNA viruses (e.g. adenoviruses and baculoviruses) have a basic protein closely associated with the genome. All of these basic proteins are virus encoded, but papillomaviruses and polyomaviruses coat their DNA with host cell histones, the basic proteins associated with the DNA in eukaryotic cells. Some viruses package other positively charged materials, such as polyamines and cations.

The dsDNA of large icosahedral viruses such as herpesviruses and the tailed phages is packed so tightly that the pressure within the capsid is about ten times greater than the pressure inside a champagne bottle.

8.2.4 Assembly mechanisms

In the laboratory, virions can be disassembled into their component molecules. For some viruses infectious virions can be reassembled from the purified components (protein and nucleic acid), under appropriate conditions of pH and in the presence of certain ions. The viruses that can self-assemble in this way are those with a relatively simple virion composed of a nucleic acid and one or a small number of protein species. Viruses that can self-assemble in a test tube are assumed to undergo self-assembly in the infected cell. Examples are tobacco mosaic virus (helical symmetry) and the ssRNA phages (icosahedral symmetry). Self-assembly is economical because no additional genetic information is needed for the assembly process.

The virions of more complex viruses, such as herpesviruses and the tailed phages, do not reassemble from their components in a test tube. The environment within the infected cell is required and the virions are constructed by a process of directed assembly.

Directed assembly of icosahedral viruses may involve proteins that are temporarily present while the virion is under construction, but are not present in the mature virion. These proteins are known as scaffolding proteins. Once their job is completed they are removed from the procapsid. Some are removed by proteolysis, while others remain intact and can be recycled. Scaffolding proteins of the tailed phages play roles in determining the size and shape of the phage head.

Some viruses synthesize polyproteins from which the individual virus proteins are cleaved by proteases; many of the cleavages take place during the assembly process. Picornaviruses and retroviruses are two groups of viruses that utilize this strategy. The final cleavage of a picornavirus polyprotein does not occur until a copy of the genome enters the procapsid (Section 14.4.5), while the internal structures of the retrovirus virion are formed from polyprotein cleavage products late in the assembly process (Chapter 16).

8.3 Formation of virion membranes

Enveloped virions acquire their membrane envelopes by one of two mechanisms; either they modify a host cell membrane and then nucleocapsids bud through it, or the virus directs synthesis of new membrane, which forms around the nucleocapsids.

8.3.1 Budding through cell membranes

Most enveloped viruses acquire their envelopes by budding through a membrane of the host cell (Figure 8.3). For viruses with eukaryotic hosts this membrane is often the plasma membrane; the virions of most retroviruses and rhabdoviruses acquire their envelopes in this way. Regions of membrane through which budding

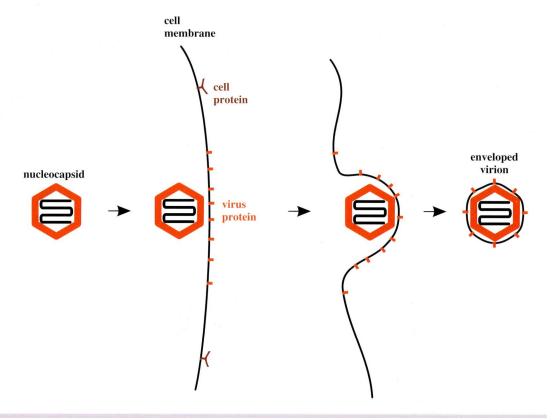

Figure 8.3 *Acquisition of a virion envelope by budding.* Virus membrane proteins, usually glycoproteins, become incorporated into regions of a cell membrane, often the plasma membrane. A nucleocapsid buds through the membrane, which pinches off to form the virion envelope.

will occur become modified by the insertion of one or more species of virus protein, the vast majority of which are glycoproteins (Section 6.4.3.a).

Integral proteins in a lipid bilayer have a degree of mobility as they are 'floating in a sea' of lipid. Virus proteins accumulate in regions of membrane from which cell proteins, to a large extent, become excluded; either the virus proteins repel the cell proteins, and/or the virus protein molecules have affinities for each other. Cell proteins may not be totally excluded from these regions and may become incorporated into virus envelopes; for example, HIV envelopes contain major histocompatibility complex class II proteins. Nucleocapsids accumulate adjacent to the regions of membrane containing virus protein.

Budding of virions involves interaction between the cytoplasmic tail of a virus glycoprotein in the membrane and another virus protein. In a number of virus groups, including paramyxoviruses and rhabdoviruses, this protein is the M (membrane, matrix) protein (Figure 8.4(a)). M proteins have an affinity for membranes, and bind to nucleocapsids as well as to the virus glycoproteins, 'stitching' the two together during budding.

The vital role played by the M protein in the budding process has been demonstrated using mutants of measles virus (a paramyxovirus) and rabies virus (a rhabdovirus). Mutants defective in the M protein gene either do not bud at all or produce severely reduced amounts of budded virus from infected cells. Roles similar to that of the M protein are played by the M1 protein of influenza A virus (Section 3.5.1) and the MA (matrix) domain of the retrovirus Gag protein (Section 16.3.6).

Not all enveloped viruses have a layer of protein between the envelope and the nucleocapsid (Figure 8.4(b)). In the virions of alphaviruses (e.g. yellow fever virus) the surface of the nucleocapsid interacts directly with the cytoplasmic tails of the glycoproteins in the membrane.

Viruses that bud from the cell do so from particular regions of the plasma membrane, and if the cell is polarized then budding may take place primarily from one surface. Body surfaces, such as the respiratory tract, are lined with epithelial cells that are polarized. Each cell has an apical (outer) and a basolateral (inner) surface and many viruses bud preferentially from one of these. Influenza A virus buds almost exclusively from the apical surface, while vesicular stomatitis virus buds almost exclusively from the basolateral surface. Experiments in which the envelope protein genes of these viruses were expressed in cells indicated that each protein has a signal that targets it to the surface from which the virus buds (Section 6.5).

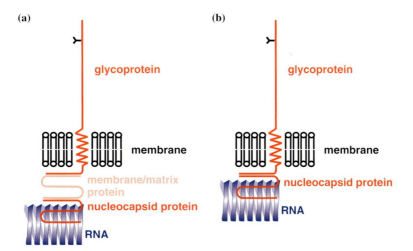

Figure 8.4 *Interactions between proteins of enveloped viruses.* Viruses (a) with and (b) without a membrane/matrix protein layer between the envelope and the nucleocapsid.

The late stage of budding involves the membrane pinching off and the release of the newly formed virion. There is evidence that host cell proteins bind to virus proteins and play vital roles in this process. Binding sites for these cell proteins have been identified in the Gag proteins of most retroviruses and in the matrix proteins of rhabdoviruses, paramyxoviruses and filoviruses. Because they are involved in the late stage of budding the protein binding sites are known as late (L) domains. Mutations in these domains can result in failure of virions to be released from the cell surface.

The release of influenza virus particles from cells requires the activity of a virion enzyme: a neuraminidase activity associated with an envelope glycoprotein. At budding sites cleavage of neuraminic acid, the substrate for the enzyme, is essential for the release of virions. Some anti-influenza virus drugs have been developed that act by inhibiting the neuraminidase (Section 25.3.5).

Some enveloped viruses bud from cell membranes other than the plasma membrane. Herpesvirus nucleocapsids are constructed in the nucleus and begin their journey to the cytoplasm by budding through the inner membrane of the nuclear envelope; the envelope of the mature virion is acquired in the cytoplasm by budding into vesicles derived from the Golgi complex (Section 11.5.4). Hepadnaviruses bud into a membrane compartment between the endoplasmic reticulum and the Golgi complex (Section 18.8.7).

One of the rare enveloped phages, *Acholeplasma laidlawii* virus L2, acquires its envelope by budding through the cell membrane of its mycoplasma host.

8.3.2 De novo *synthesis of viral membranes*

A minority of viruses direct the synthesis of lipid membrane late in the replication cycle. In some cases the membrane forms a virion envelope (e.g. poxviruses); in other cases the membrane forms a layer below the surface of the capsid (e.g. iridoviruses; Section 3.5.2).

Baculoviruses produce two types of enveloped virion during their replication. One type of virion has the function of spreading the infection to other cells within the host, and this virion acquires its envelope by budding from the plasma membrane. The other type of virion has the function of infecting new

Table 8.1   Summary of virion membrane acquisition

Site of virion assembly	Origin of virion membrane	Examples	
		Virion envelope	Internal virion membrane
Eukaryotic cell cytoplasm	Plasma membrane	Most retroviruses Most rhabdoviruses	
	Post-RER membrane	Hepadnaviruses	
	De novo synthesis	Poxviruses	Iridoviruses
Eukaryotic cell nucleus	Inner nuclear membrane	Nucleorhabdoviruses	
	De novo synthesis	Baculovirus virions that will be occluded	
Prokaryotic cell	Cell membrane	*Acholeplasma laidlawii* virus L2	
	De novo synthesis	*Pseudomonas* phage φ6	*Alteromonas* phage PM2

host individuals. Its envelope is laid down around nucleocapsids within the nucleus and the resulting virions become incorporated into occlusion bodies (Section 3.6).

Among the viruses of prokaryotes, the enveloped *Pseudomonas* phage φ6 acquires its envelope in the cytoplasm. The lipids are derived from the host cell membrane and the virion is released by lysis of the cell. The virion of *Alteromonas* phage PM2 is not enveloped, but it has a membrane between outer and inner protein shells. The membrane is synthesized in the infected cell.

The various origins of virion membranes are summarized in Table 8.1.

8.4 Virion exit from the infected cell

The virions of many viruses are released from the infected cell when it bursts (lyses), a process that may be initiated by the virus. Many phages produce enzymes (lysins, such as lysozymes) that break bonds in the peptidoglycan of the host bacterial cell walls. Other phages synthesize proteins that inhibit host enzymes with roles in cell wall synthesis; this leads to weakening of the cell wall and ultimately to lysis.

Average yields of infectious virions per cell (burst sizes; Section 2.8.3) vary considerably. A typical burst size for phage T4 (a large virus in a small host cell) is 200, while a typical burst size for a picornavirus (a small virus in a large host cell) is 100 000.

Many viruses do not lyse their host cells; instead, progeny virions are released from the cells over a period of time. Obviously the concept of burst size is not relevant for this situation. We have already seen how virions that acquire an envelope from a cell surface membrane leave the cell by budding. Virions that acquire envelopes from internal membranes of the cell exit the cell in other ways. Some are transported to the cell surface in vesicles, which fuse with the plasma membrane to release the virions. Others, such as vaccinia virus, attach via motor proteins to microtubules (Section 5.2.5) and use this transport system to reach the cell surface where they are released.

The filamentous phages are naked viruses with a unique mechanism for exiting from their bacterial hosts: extrusion through the cell surface. The ssDNA genomes of the phages are coated with protein within the cell; as they are extruded from the cell they lose this protein and acquire their capsids (Section 19.4.2.c).

After leaving the host cell most virions remain inert unless they encounter a new host cell. There are some cases, however, where virions undergo morphological changes because maturation continues after they have left the cell. In the retrovirus virion, for example, polyprotein cleavages commence during budding and continue after the virion has been released from the cell (Section 16.3.6). In another example an unusual lemon-shaped phage of an archaeon (*Acidianus convivator*) develops two long tails *after* it has left its host.

Because of their modes of transmission, most plant viruses leave their host cells in ways that differ from those of animal and bacterial viruses. Plant cells are separated from each other by thick cell walls, but in many of them there are channels, called plasmodesmata, through which the plant transports materials. Viruses are able to spread within the host by passing from cell to cell through plasmodesmata (Section 9.4.1). For spread to new hosts many plant viruses leave the host cell as a component of the meal of a vector (e.g. an aphid or a nematode) that feeds by ingesting the contents of cells (Section 4.2).

Learning outcomes

By the end of this chapter you should be able to

- describe the assembly mechanisms for nucleocapsids with (a) helical symmetry and (b) icosahedral symmetry;

- discuss the origins of internal virion membranes and of virion envelopes;

- explain the roles played by membrane/matrix proteins in the budding of some enveloped viruses;

- describe mechanisms used by viruses to exit from cells.

Sources of further information

Books

Harrison S. C. (2005) Mechanism of membrane fusion by viral envelope proteins, pp 231–261 in Roy P., editor *Advances in Virus Research*, Vol. 64, Elsevier

Hunter E. (2001) Virus assembly, Chapter 8 in Knipe D. M. and Howley P. M., editors-in-chief, *Fields Virology*, 4th edition, Lippincott, Williams and Wilkins

Schmitt A. P. and Lamb R. A. (2004) Escaping from the cell: assembly and budding of negative-strand RNA viruses, pp 145–196 in Yawaoka Y., editor, *Biology of Negative Strand RNA Viruses: the Power of Reverse Genetics*, Springer

Journals

Bieniasz P. D. (2006) Late budding domains and host proteins in enveloped virus release *Virology*, **344**, 55–63

Chazal N. and Gerlier D. (2003) Virus entry, assembly, budding, and membrane rafts *Microbiology and Molecular Biology Reviews*, **67**, 226–237

Herrero-Martínez E. *et al.* (2005) Vaccinia virus intracellular enveloped virions move to the cell periphery on microtubules in the absence of the A36R protein *Journal of General Virology*, **86**, 2961–2968

Loessner M. J. (2005) Bacteriophage endolysins – current state of research and applications *Current Opinion in Microbiology*, **8**, 480–487

Steven A. C. *et al.* (2005) Virus maturation: dynamics and mechanism of a stabilizing structural transition that leads to infectivity *Current Opinion in Structural Biology*, **15**, 227–236

9

Outcomes of infection for the host

At a glance

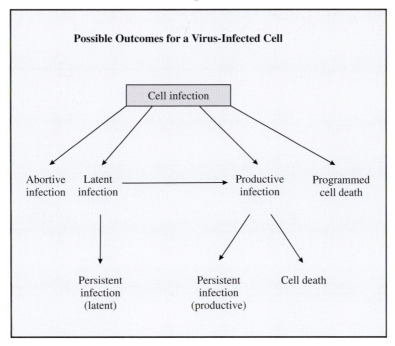

Possible Outcomes for a Virus-Infected Cell

Virology: Principles and Applications John B. Carter and Venetia A. Saunders
© 2007 John Wiley & Sons, Ltd ISBNs: 978-0-470-02386-0 (HB); 978-0-470-02387-7 (PB)

At a glance (continued)

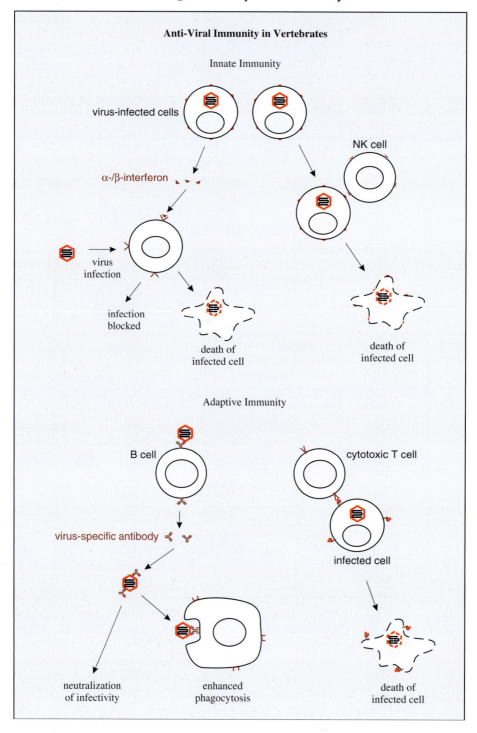

9.1 Introduction to outcomes of infection for the host

In the previous few chapters we have looked at aspects of the virus replication cycle that culminate in the exit of infective progeny virions from an infected cell. When this is the outcome the infection is said to be productive. Virions may be released when the host cell lyses, or the cell may survive releasing virions for a period, which may be short, as in the case of HIV infection, or it may be long, as in the case of hepatitis B virus infection.

Some virus infections, however, are not productive, for a variety of reasons.

- An infection may become latent with the virus genome persisting, perhaps for the lifetime of the cell, and perhaps in the daughter cells if the cell divides.

- An infection may be abortive, in which case neither a latent infection nor a productive infection is established. One cause of an abortive infection is a virus with a mutated genome; the virus is said to be defective. It is unable to undergo a complete replication cycle, unless the cell is also infected with a virus that can provide the missing function(s). A virus that is able to provide functions for a defective virus is known as a helper virus.

Some virus infections persist in their hosts for long periods, sometimes for life. In some cases persistent infections are productive, e.g. HIV; in other cases persistence may take the form of periods of latency alternating with periods of productive infection, e.g. many herpesviruses. Long-term infection with some viruses may lead to the development of cancer in the infected host (Chapter 22).

Some phages, including the filamentous phages (Section 19.4.2), initiate persistent productive infections in cells of their bacterial hosts. Under appropriate conditions the cells can survive for long periods releasing progeny virions.

The outcome of a virus infection from the point of view of the host may vary from harmless at one end of the scale, through deleterious effects of varying severity, to death at the other end of the scale. The outcome may hinge on complex interplay between a variety of host, viral and environmental factors.

Over time the organisms that are hosts to viruses have evolved anti-viral defences, but the viruses have not sat idly by. The relationship between a virus and its host usually involves an arms race and viruses have developed various countermeasures against host defences. Some of these countermeasures are indicated in boxes throughout this chapter.

In this chapter we concentrate mainly on outcomes of virus infections in vertebrate hosts, and we start by considering some aspects of host immunity that can affect outcomes of infection.

9.2 Factors affecting outcomes of infection

A major factor affecting the outcome of infection is the efficiency of the immune systems of the host. In animals there are both innate and adaptive immune systems.

9.2.1 Innate immunity in vertebrates

Some aspects of innate immunity that viruses might encounter prior to infecting cells were discussed in Section 4.3.3. Further components of innate immunity that might be encountered after cells have become infected are interferons and natural killer (NK) cells.

9.2.1.a Interferons

Interferons are proteins synthesized and secreted by cells in response to virus infection. A potent trigger for interferon production is dsRNA, which is produced, not only by dsRNA viruses, but also by ssRNA viruses as they replicate.

The roles of interferons are to protect adjacent cells from infection and to activate T cell-mediated immunity (Section 9.2.2.b). There are a number of types of interferon. Alpha- and beta- (α- and β-) interferons are produced by most cell types when they become infected with viruses. After secretion, the interferon molecules diffuse to nearby cells, where they can trigger various anti-viral activities by binding to interferon receptors (Figure 9.1).

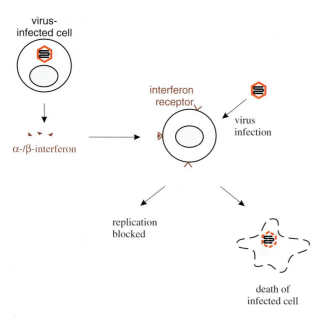

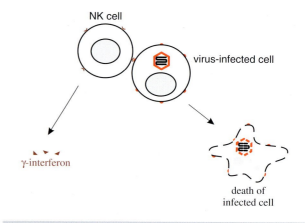

Figure 9.2 *Activities of natural killer (NK) cells.*

a number of effects, including stimulation of antigen presentation and activation of phagocytes and NK cells.

Figure 9.1 *Activities of α- and β-interferons.* Interferons released from virus-infected cells bind to interferon receptors on other cells. This interaction induces processes that, if the cells become infected, either block virus replication or kill the infected cells.

Anti-viral activities triggered by α- and β-interferons include the following.

- Activation of genes that encode antiviral proteins, such as dsRNA-dependent protein kinase R and RNase L.

- Stimulation of production of major histocompatibility (MHC) class I molecules and proteasome proteins; these molecules enhance the presentation of viral peptides on the infected cell surface to T cells.

- Activation of NK cells (Section 9.2.1.b).

- Induction of apoptosis (Section 9.2.4).

Another type of interferon, gamma- (γ-) interferon, is produced mainly by T cells and NK cells when triggered by certain molecules (e.g. interleukin-2) released during immune responses. γ-interferon has

---

Viral countermeasures against interferons

Many viruses produce proteins that inhibit either the production of interferons or their activities. The NS1 protein of influenza A virus and the NS3-4A protein of hepatitis C virus block pathways involved in interferon production. Some viruses, such as poliovirus, prevent the synthesis of interferons as a result of a general inhibition of cell gene expression.

9.2.1.b Natural killer (NK) cells

NK cells are present throughout the body, but mainly in the blood. They recognize changes in the surface molecules of virus-infected cells as a result of infection, though they do not recognize specific antigens, unlike B cells and T cells (Section 9.2.2). After recognizing virus-infected cells as target cells, NK cells are able to bind to them and kill them (Figure 9.2).

NK cells kill their target cells either by releasing perforins, which are proteins that are inserted into the plasma membrane of the virus-infected cell, or by inducing apoptosis (Section 9.2.4). Also, on binding to infected cells, NK cells release γ-interferon.

Viral countermeasures against NK cells

The presence of HIV particles in the blood alters the expression of a number of molecules on the surface of NK cells. This reduces the efficiency of NK cell activities, including the ability to kill virus-infected cells and to secrete γ-interferon.

9.2.1.c APOBEC3 proteins

There are enzymes in the cells of humans and animals that can interfere with the replication of retroviruses. These enzymes can induce lethal mutations by deaminating deoxycytidine to deoxyuridine during reverse transcription. The enzymes delight in the name apolipoprotein B mRNA-editing enzyme, catalytic polypeptide-like 3 proteins; the abbreviation APOBEC3 proteins is easier to cope with!

Several of these proteins in human cells can interfere with the replication of HIV. Two of them (APOBEC3F and APOBEC3G) can be incorporated into HIV virions and taken into the next cell, where they can wreak havoc during reverse transcription.

Viral countermeasures against APOBEC3 proteins

In the cytoplasm of an HIV-infected cell the virus protein Vif can bind APOBEC3G, triggering its degradation and hence preventing its incorporation into virions.

9.2.2 Adaptive immunity in vertebrates

An important outcome of virus infection in a vertebrate host is the development of a virus-specific immune response triggered by the virus antigens. Regions of antigens known as epitopes bind to specific receptors on lymphocytes, activating cascades of events that result in the immune response.

Lymphocytes are the key cells in specific immune responses. There are two classes of lymphocyte: B lymphocytes (B cells), which develop in the Bursa of Fabricius in birds and in the bone marrow in mammals, and T lymphocytes (T cells), which develop in the thymus. Each lymphocyte is specific for a particular epitope as a result of the presence of epitope-specific receptors on the cell surface. Naïve lymphocytes are those that have not encountered their specific epitopes; these cells have surface molecules and circulation patterns in the body distinct from lymphocytes that have previously encountered their epitopes.

9.2.2.a Antibodies

Antibodies are glycoproteins of a type known as immunoglobulins. The basic structure of an antibody molecule is shown in Figure 9.3. The molecule is constructed from two 'heavy' and two 'light' polypeptide chains and contains two antigen-binding sites and a region known as the Fc (Fragment crystallizable) region.

There are several classes of immunoglobulin (Ig), the most significant from the point of view of antiviral immunity being IgG and IgM in the blood, and IgA at mucosal surfaces. IgG molecules are monomers with the structure depicted in Figure 9.3, while IgA and IgM molecules are normally dimers and pentamers, respectively.

Antigen-specific antibodies are synthesized by plasma cells, which develop from a B cell after it has been stimulated by interaction between the antigen and a specific immunoglobulin receptor at the cell surface. Antibodies play important roles in several aspects of anti-viral immunity, some of which are summarized in Figure 9.4.

Virus-specific antibody can coat both virions and virus-infected cells, and this may lead to their destruction by a variety of mechanisms. A number of cell types

Figure 9.3 *An antibody molecule.* The molecule consists of four polypeptide chains held together by disulphide bonds. The chains are arranged to form two sites that can bind to a specific antigen.

Fc: Fragment crystallizable

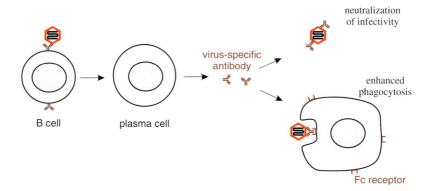

Figure 9.4 *Production of virus-specific antibody and some of its roles in immunity.* Virus-specific antibody is secreted by plasma cells derived from B cells that have recognized virus antigens. The antibody may bind to virions and neutralize their infectivity. Antibody-coated virions may be phagocytosed after attachment of antibody Fc regions to receptors on phagocytes.

in the immune system have receptors for the Fc region of IgG (Figure 9.3), allowing these cells to attach to antibody-coated virions and cells. Cell types that have IgG Fc receptors include

- neutrophils and macrophages (these cell types are phagocytes and may phagocytose the antibody-coated materials; they may also kill cells without phagocytosing them)

- NK cells (these cells may kill virus-infected cells by insertion of perforins into their membranes).

Another outcome of antibody binding to virus antigens is the activation of complement, which was briefly introduced in Section 4.3.3. This can have a number of anti-viral effects, one of which involves insertion of complement protein complexes into membranes of virus-infected cells and enveloped virions, leading to destruction of these cells and virions. Another anti-viral effect of complement occurs when virions become coated with complement proteins. There are receptors for some of these proteins on neutrophils and macrophages, so phagocytosis of virions is enhanced.

A further effect of antibody binding to virions is neutralization of infectivity, which may occur by a variety of mechanisms.

- *Release of nucleic acid from virions.* In studies with several viruses, including poliovirus, it was found

that antibodies can attach to virions, and then detach leaving empty capsids devoid of their genomes.

- *Prevention of virion attachment to cell receptors.* Antibody bound to a virion may mask virus attachment sites. Not all virus attachment sites, however, are accessible to antibodies; those of most picornaviruses are in deep canyons (Section 14.3.1).

- *Release of virions that have attached to cell receptors.*

- *Inhibition of entry into the cell.* Antibody coating fusion proteins on an enveloped virion may inhibit fusion of the envelope with a cell membrane.

- *Inhibition of genome uncoating.*

9.2.2.b T cells

Several days after antigenic stimulation naïve or memory T cells develop into effector T cells, of which there are two classes.

- Helper T cells secrete specific cytokines and are characterized by the presence of CD4 molecules at the cell surface. Helper T cells play essential roles in the initiation of immune responses, for example in triggering B cells to develop into antibody-secreting cells and in the maturation of cytotoxic T cells.

- Cytotoxic T cells kill virus-infected cells and are characterized by the presence of CD8 molecules at

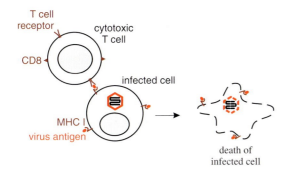

Figure 9.5 *Killing of a virus-infected cell by a cytotoxic T cell.* A cytotoxic T cell kills a virus-infected cell after the T cell receptors have recognized fragments of virus antigens displayed in association with MHC class I molecules at the cell surface.

the cell surface (Figure 9.5). There is a requirement for viral antigens to be expressed at the surface of target cells. The antigens may be virion surface proteins (e.g. envelope glycoproteins) though often the target antigens are internal virion components, or even non-structural proteins. Cytotoxic T cells specific for early virus proteins may destroy virus-infected cells long before any infectious virus is produced.

The viral antigens are displayed on the surface of infected cells in association with MHC class I molecules, flagging infected cells for destruction by cytotoxic T cells. Cytotoxic T cells can kill target cells by insertion of proteins (perforins) into their membranes or by inducing apoptosis.

Viral countermeasures against cytotoxic T cells

Some viruses, such as herpesviruses, reduce the level of expression of MHC class I molecules at the surface of infected cells, thereby making it more difficult for cytotoxic T cells to recognize infected cells.

9.2.2.c Immunological memory

The quantity and quality of the adaptive immune response depends on whether or not the host is encountering the virus for the first time. Some B cells

and T cells can survive as memory cells long after the first or subsequent encounters. Memory cells have returned to a resting state, from which they can be reactivated if they encounter the same antigen again. These cells are the basis of immunological memory, which can be formed as a result of a natural infection, but also as a result of encountering antigens in vaccines.

The outcome of infection of a vertebrate animal with a virus may depend on whether or not the host has immunological memory of the virus antigens. If immunological memory is present then signs and symptoms of disease are likely to be less severe, or totally absent.

9.2.3 RNA silencing

RNA silencing, also known as post-transcriptional gene silencing or RNA interference (RNAi), is an intracellular process that is induced by dsRNA. The process results in the destruction of mRNAs that have the same sequence as the inducing dsRNA; both cellular and viral mRNAs can be destroyed.

The process, which is outlined in Figure 9.6, involves cleavage of the dsRNA into small interfering dsRNAs (siRNAs), 21–25 bp long. The cleavages are performed by a protein complex containing an enzyme called Dicer, which belongs to the RNase III family. These enzymes are unusual in that they show specificity for dsRNAs and cleave them to fragments with 3′ overhangs of two or three nucleotides. Each of the siRNAs joins a complex of proteins to form a RISC (RNA-induced silencing complex). The double-stranded siRNA is unwound and the minus-strand RNA remains associated with the complex, which is now an activated RISC. The minus-strand RNA selects the target mRNA by complementary base-pairing, then the mRNA is degraded in that region.

RNA silencing has been found in plants and fungi, and in invertebrate and vertebrate animals. In plants it is considered to be an important antiviral defence mechanism, and this may be true for fungi and animals too. RNA silencing can also be initiated by the introduction of synthetic dsRNA (Section 2.9.3).

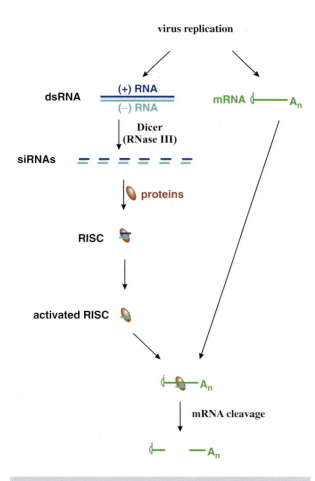

Figure 9.6 *RNA silencing*. dsRNA is processed by Dicer to siRNAs, which associate with RISC proteins. An activated RISC contains minus-strand RNA, which targets the complex to the specific mRNA. The mRNA is cleaved by a RISC protein.

RISC: RNA-induced silencing complex

siRNA: small interfering RNA

Viral countermeasures against RNA silencing

Some plant viruses encode proteins that can suppress RNA silencing, for example the 'helper-component proteinase' of potyviruses and the P19 protein of tombusviruses are strong suppressors of RNA silencing.

9.2.4 Programmed cell death

Virus infection of a cell may initiate a process that causes the death of the cell before progeny virus has been produced, hence preventing the spread of infection to other cells. In animal cells this suicide mechanism is known as apoptosis. It is triggered, not only by virus infection, but also when the life-span of cells, such as epithelial cells, is complete.

Bacteria have developed similar mechanisms to protect the species from phage infection. The death of a host bacterium before any progeny phage has been produced protects other susceptible cells from infection. These mechanisms have been found in *Escherichia coli* and in many other species.

If a virus-infected cell successfully completes the process of programmed cell death then it altruistically commits suicide for the benefit of either a multicellular host or a population of unicellular hosts.

Viral countermeasures against apoptosis

Many viruses have evolved mechanisms that can suppress apoptosis at a variety of points in the process, for example several DNA viruses encode proteins related to the cell BCL-2 proteins that control apoptosis. These viral proteins block apoptosis, resulting in the survival of host cells and the completion of virus replication cycles.

9.3   Non-productive infections

Under some circumstances a virus infects a cell, but the replication cycle is not completed. If the virus genome persists in the cell the infection is said to be latent; otherwise, it is an abortive infection. These situations are discussed in the next two sections.

9.3.1 Latent infections

When a latent infection is initiated the virus genome is maintained in the infected cell either as a sequence of DNA integrated into the cell genome, or as multiple copies of covalently closed circular DNA (Figure 9.7). In eukaryotic cells the virus DNA is associated with

(a)

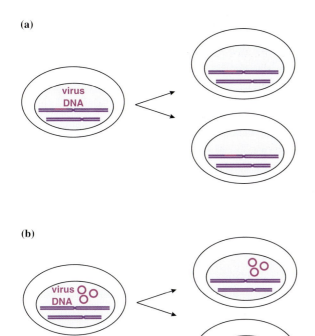

(b)

Figure 9.7 *Maintenance of virus genomes in cells with latent infections.* (a) Virus DNA integrated into a cell chromosome. After infection of the cell the virus genome is integrated into the genome of the host cell, e.g. retroviruses. (b) Virus DNA present as multiple copies of circular molecules. After infection of the cell the virus genome is circularized and replicated, e.g. herpesviruses. If a latently infected cell divides the virus genome is normally replicated along with the cell genome so that each daughter cell contains a copy/copies of the virus genome.

host cell histones, which play roles in the maintenance of latency.

In the case of retroviruses, the integration of a copy of the virus genome into the cell genome is an early stage in the virus replication cycle (Section 16.3.3). When the infection is latent events do not progress beyond this stage, but if the intracellular environment changes to become favourable for replication of the virus then the latent infection may progress to a productive infection.

The phenomenon of latent phage infection of a bacterium is known as lysogeny and the phage is said to be temperate (Section 19.1). The phage genome (the prophage) can persist in the cell in ways similar to the persistence of latent animal virus infections (Figure 9.7). In many cases the prophage is integrated into the bacterial genome, but some prophages persist as non-integrated circular DNA. A temperate phage genome may include a gene that can confer a selective advantage on the bacterial host. Some virulence factors of parasitic bacteria are encoded by temperate phages, an example being the Shiga toxin produced by some strains of *E. coli*.

During a latent infection the virus genome may be totally shut down, or a few virus genes may be expressed as proteins and/or non-coding RNAs. If the latently infected cell divides then the virus DNA is replicated and each daughter cell has one or more copies of the virus genome (Figure 9.7). When this happens the virus has achieved its objective of replicating its genes, and it has done so without destroying the host cell!

Many latent infections may be activated to become productive infections, a process known as induction. This may occur if the following happen.

- A eukaryotic host cell moves into another phase of the cell cycle (Figure 4.5).

- The host cell is irradiated with ultra-violet light. This might trigger, for example, a latent phage infection to replicate and lyse the host cell, or a latent herpes simplex virus infection to replicate and cause a cold sore.

- A host organism becomes immunocompromised. This is another trigger that can reactivate a latent herpes simplex virus infection.

- The host cell becomes infected with a second virus that provides a function that the first virus lacks. Terms used for the first and second viruses in this relationship are satellite virus and helper virus, respectively. Examples are given in Table 9.1.

9.3.2 Abortive infections

An abortive infection is one where the infection is non-productive and the virus genome does not persist in the

Table 9.1 Examples of satellite/helper viruses

Satellite virus	Helper virus
Hepatitis delta virus	Hepatitis B virus
Adeno-associated viruses	Adenovirus

Table 9.2 Involvement of plant virus movement proteins (MPs) in the spread of infection via plasmodesmata

Mode of transport through plasmodesmata	Virus examples
Virus RNA–MP complexes transported	Tobacco mosaic virus
	Cowpea chlorotic mottle virus
Virus RNA–coat protein–MP complexes transported	Cucumber mosaic virus
Virions transported through tubules composed of MP	Cowpea mosaic virus

cell as a latent infection. An infection might be abortive for reasons concerning the cell, the environmental conditions and/or the virus. A virus may initiate productive infections in some cell types (permissive cells) and abortive infections in other cell types (non-permissive cells). Some abortive infections can kill the host cell.

Some virions may contain mutated genomes that may be able to initiate the replication cycle, but may not have the full set of functional genes necessary to complete it. Such virions are said to be defective, and if they infect cells without the means of complementing the mutated or missing genes the result is an abortive infection.

There are a number of types of defective virus. One type is known as a defective interfering particle (DIP). DIPs often arise in the laboratory after animal viruses have been passaged several times in cell culture at high multiplicities of infection. DIPs have also been reported after passage in chick embryos and in mice, and DIPs of plant viruses have been observed.

Most DIPs contain less nucleic acid than the standard virus from which they were derived. By themselves DIPs are either non-infectious or they can initiate only abortive infections. The standard virus can act as helper virus to a DIP, so in cells that are co-infected with a DIP and the standard virus the DIP is able to replicate its genome and produce progeny DIPs. When this happens the replication of the DIP interferes with the replication of the standard virus and reduces its yield, hence 'interfering' particle.

9.4 Productive infections

9.4.1 Spread of infections within multicellular hosts

Progeny virions from the first-infected cell of a multicellular host may infect cells nearby, for example common cold viruses and rotaviruses might infect other epithelial cells lining the respiratory tract and the intestinal tract, respectively. Most adjacent animal cells are separated from each other only by their plasma membranes, providing opportunities for direct cell-to-cell spread for some viruses.

Plant viruses are able to spread from cell to cell by passing through plasmodesmata, and each plant virus encodes between one and four specialized proteins that enable them to do this. These proteins are known as movement proteins (MPs) and they act in a variety of ways. Some MPs form complexes either with virus RNA or with virus RNA plus coat protein molecules; other MPs form tubular structures through which fully encapsidated virus RNA is transported (Table 9.2). MPs are multifunctional proteins with other roles in the virus replication cycles.

In some circumstances progeny virions from infected cells may be transported to distant sites in the host, where susceptible cells may become infected. In the animal body the blood and the nerves may act as transport vehicles, while in plants transport may occur via the phloem.

9.4.2 Disease

Many virus infections result in no disease in the host, while at the other end of the scale a virus infection may result in fatal disease, such as rabies or AIDS. Between these extremes is a huge range of severity in diseases caused by viruses.

Disease may be manifest as symptoms and/or signs. In medical terminology symptoms are subjective features, such as abdominal pain and fatigue, whereas signs are objective features, such as blood in the faeces and skin rashes. Symptoms can be recognized only by the infected individual, whereas signs can be recognized by others. Plant virologists and insect virologists sometimes use the word symptom in connection with objective features, such as leaf lesions in a plant (Figure 2.17) and iridescence in an insect (Chapter 1, at a glance).

Infection with some viruses, such as dependoviruses, some herpesviruses and some reoviruses, apparently never causes disease, while infection with other viruses, such as human parvovirus B19, poliovirus and hepatitis B virus, may or may not result in disease. Infections that do not result in disease are said to be subclinical or asymptomatic.

The outcome of a complex interplay of factors determines whether or not a virus infection results in disease, and if so the severity of the disease. These factors may include virus factors, host factors and human intervention.

Not all viruses are pathogens (disease-causing agents) and some are pathogenic only under certain circumstances. For pathogenic viruses factors that can affect the outcome of infection include the following.

- *The virulence of the virus strain.* The virulence of a virus (or any micro-organism) is a measure of the severity of disease it is capable of causing. Influenza A type H5N1 can be said to be more virulent in humans than types H1N1 and H3N2 (Section 21.5.2), as type H5N1 causes more severe disease than the other two types.

- *The dose of virus.* A larger dose of virus may result in a shorter incubation period (the time between infection and the first appearance of signs and/or symptoms).

Host factors that can affect the outcome of infection include the effectiveness of immune systems (discussed in Chapter 4 and earlier in this chapter), and these in turn vary with age and nutritional status of the host. It should be noted that a strong immune response by the host does not guarantee the elimination of the virus. HIV continues to replicate in the presence of high levels of HIV-specific antibody and T cells. In some cases signs and/or symptoms may result from the host's immune response against the virus. The measles rash and herpes simplex lesions are clinical manifestations of attempts by the human body to destroy virus-infected cells.

Human interventions that could influence the outcome of a virus infection in a human or an animal include the administration of anti-viral antibodies or anti-viral drugs (Chapter 25).

Recovery of the host from a virus disease may be accompanied by elimination of the virus, as normally occurs after infection with common cold viruses in the respiratory tract and rotavirus infections in the intestine. Alternatively, recovery may be accompanied by establishment of a long-term infection, perhaps for life.Such persistent infections, whether they are productive or latent, may have no further consequences for the host, or they may cause disease when the host is older. Childhood infection with varicella-zoster virus may result in a lifetime latent infection, which may be reactivated, causing shingles (Section 11.2.2). Long-term persistent infection of humans and animals with certain viruses may lead to the development of cancer (Chapter 22).

Learning outcomes

By the end of this chapter you should be able to

- describe the major components of innate and adaptive immunity in vertebrates;

- outline the process of RNA silencing;

- explain programmed cell death;

- explain the terms
  - productive infection
  - non-productive infection
  - latent infection
  - abortive infection
  - defective virus;

- discuss the spread of virus infections within animal bodies and within plants;

- discuss the factors that determine whether virus infection results in disease.

Sources of further information

Books (general immunology)

Mims C. A. *et al.* (2004) Sections 2 and 3 in *Medical Microbiology*, 3rd edition, Mosby

Playfair J. H. L. and Bancroft G. J. (2004) *Infection and Immunity*, 2nd edition, Oxford University Press

Rabson A., Roitt I. and Delves P. (2005) *Really Essential Medical Immunology*, 2nd edition, Blackwell

Books (immune responses to virus infections)

Digard P., Nash A. A. and Randall R. E., editors (2005) *Molecular Pathogenesis of Virus Infections*, 64th Symposium of the Society for General Microbiology, Cambridge University Press

Chen Y.-B., Fannjang Y. and Hardwick J. M. (2004) Cell death in viral infections, Chapter 17 in *When Cells Die II*, editors Lockshin R. A. and Zakeri Z., Wiley

Whitton J. L. and Oldstone M. B. A. (2001) The immune response to viruses, Chapter 11 in Knipe D. M. and Howley P. M., editors-in-chief, *Fields Virology*, 4th edition, Lippincott, Williams and Wilkins

Historical paper

Isaacs A. and Lindenmann J. (1957) Virus interference. 1. The interferon *Proceedings of the Royal Society of London B*, **147**, 258–267

Recent papers

Cuconati A. and White E. (2002) Viral homologs of BCL-2: role of apoptosis in the regulation of virus infection *Genes and Development*, **16**, 2465–2478

Doehle B. P., Schäfer A. and Cullen B. R. (2005) Human APOBEC3B is a potent inhibitor of HIV-1 infectivity and is resistant to HIV-1 Vif *Virology*, **339**, 281–288

Fauci A. S., Mavilio D. and Kottilil S. (2005) NK cells in HIV infection: paradigm for protection or targets for ambush *Nature Reviews Immunology*, **5**, 835–843

Haller O., Kochs G. and Weber F. (2006) The interferon response circuit: induction and suppression by pathogenic viruses *Virology*, **344**, 119–130

Iannello A. *et al.* (2006) Viral strategies for evading antiviral cellular immune responses of the host *Journal of Leukocyte Biology*, **79**, 16–35

Lieberman P. M. (2006) Chromatin regulation of virus infection *Trends In Microbiology*, **14**, 132–140

Scholthof H. B. (2005) Plant virus transport: motions of functional equivalence *Trends in Plant Science*, **10**, 376–382

Soosaar J. L. M. *et al.* (2005) Mechanisms of plant resistance to viruses *Nature Reviews Microbiology*, **3**, 789–798

Stram Y. and Kuzntzova L. (2006) Inhibition of viruses by RNA interference *Virus Genes*, **32**, 299–306

10

Classification and nomenclature of viruses

At a glance

Authority: International Committee on Taxonomy of Viruses

http://www.ncbi.nlm.nih.gov/ICTVdb/index.htm

Taxonomic groups: Order

Family

Subfamily

Genus

Species

Phylogenetic trees - based on genome sequences

- indicate relationships between viruses:

Baltimore Classification: - seven classes of viruses

- based on genome type and transcription.

Virology: Principles and Applications   John B. Carter and Venetia A. Saunders
© 2007 John Wiley & Sons, Ltd   ISBNs: 978-0-470-02386-0 (HB); 978-0-470-02387-7 (PB)

10.1 History of virus classification and nomenclature

Virologists are no different to other scientists in that they find it useful to classify the objects of their study into groups and sub-groups. In the early days, when little was known about viruses, they were loosely grouped on the basis of criteria such as the type of host, the type of disease caused by infection and whether the virus is transmitted by an arthropod vector.

As more was learnt about the characteristics of virus particles some of these began to be used for the purposes of classification, for example

- whether the nucleic acid is DNA or RNA

- whether the nucleic acid is single stranded or double stranded

- whether or not the genome is segmented

- the size of the virion

- whether the capsid has helical symmetry or icosahedral symmetry

- whether the virion is naked or enveloped.

Various combinations of these criteria produced some useful virus groups, but there was no single approach to the naming of the groups, and names were derived in a variety of ways:

- small, icosahedral, single-stranded DNA viruses of animals were called parvoviruses (Latin *parvus* = small);

- *ne*matode-transmitted *po*lyhedral (icosahedral) viruses of plants were called nepoviruses;

- phages T2, T4 and T6 were called T even phages.

Serological relationships between viruses were investigated, and distinct strains (serotypes) could be distinguished in serological tests using antisera against purified virions. Serotypes reflect differences in virus proteins and have been found for many types of virus, including rotaviruses and foot and mouth disease virus.

10.1.1 International Committee on Taxonomy of Viruses

By 1966 it was decided that some order had to be brought to the business of naming viruses and classifying them into groups, and the International Committee on Taxonomy of Viruses (ICTV) was formed. The committee now has many working groups and is advised by virologists around the world.

The ICTV lays down the rules for the nomenclature and classification of viruses, and it considers proposals for new taxonomic groups and virus names. Those that are approved are published in book form (Please see *Sources of further information* at the end of this chapter.) and on the web; these sources should be consulted for definitive information. The web site for this book (www.wiley.com/go/carter) has links to relevant web sites.

10.2 Modern virus classification and nomenclature

For a long time virologists were reluctant to use the taxonomic groups such as family, subfamily, genus

Table 10.1 Taxonomic groups of viruses

Taxonomic group	Suffix	Example 1	Example 2	Example 3
Order	-virales	*Caudovirales*	*Mononegavirales*	*Nidovirales*
Family	-viridae	*Myoviridae*	*Paramyxoviridae*	*Coronaviridae*
Subfamily	-virinae	–	*Paramyxovirinae*	–
Genus	-virus	*T4-like viruses*	*Morbillivirus*	*Coronavirus*
Species	–	*Enterobacteria phage T4*	*Measles virus*	*Severe acute respiratory syndrome virus*

and species that have long been used to classify living organisms, but taxonomic groups of viruses have gradually been accepted and are now established (Table 10.1). Some virus families have been grouped into orders, but higher taxonomic groupings, such as class and phylum, are not used. Only some virus families are divided into subfamilies.

Each order, family, subfamily and genus is defined by viral characteristics that are necessary for membership of that group, whereas members of a species have characteristics in common but no one characteristic is essential for membership of the species. Many species contain variants known as virus strains, serotypes (differences are detected by differences in antigens) or genotypes (differences are detected by differences in genome sequence).

Many of the early names of virus groups were used to form the names of families and genera, e.g. the picornaviruses became the family *Picornaviridae*. Each taxonomic group has its own suffix and the formal names are printed in italic with the first letter in upper case (Table 10.1), e.g. the genus *Morbillivirus*. When common names are used, however, they are not in italic and the first letter is in lower case (unless it is the first word of a sentence), e.g. the morbilliviruses.

10.2.1 Classification based on genome sequences

Now that technologies for sequencing virus genomes and for determining genome organization are readily available, the modern approach to virus classification is based on comparisons of genome sequences and organizations. The degree of similarity between virus genomes can be assessed using computer programs, and can be represented in diagrams known as phylogenetic trees because they show the likely phylogeny (evolutionary development) of the viruses. Phylogenetic trees may be of various types (Figure 10.1).

- Rooted – the tree begins at a root which is assumed to be the ancestor of the viruses in the tree.

- Unrooted – no assumption is made about the ancestor of the viruses in the tree.

The branches of a phylogenetic tree indicate how sequences are related. The branches may be scaled

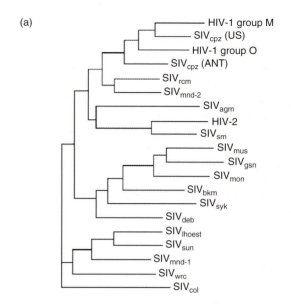

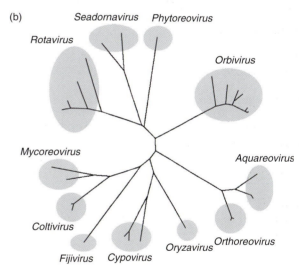

Figure 10.1 *Phylogenetic trees.* (a) Rooted tree of human immunodeficiency viruses (HIVs) and simian immunodeficiency viruses (SIVs) based on the RT-IN region of the *pol* gene. Data from Takemura *et al.* (2005) *Journal of General Virology*, 86, 1967. (b) Unrooted tree derived from sequences of the VP1 proteins of viruses in the family *Reoviridae*. The named clusters are the genera of the family (Chapter 13). From Attoui *et al.* (2005) *Journal of General Virology*, 86, 3409. Both figures redrawn with the permission of the authors and the Society for General Microbiology.

or unscaled; if they are scaled, their lengths represent genetic distances between sequences.

In many cases analysis of the sequence and organization of virus genomes has supported earlier classifications of viruses, e.g. the genera of the family *Reoviridae* (Figure 10.1(b)). Another example is the rhabdoviruses (Chapter 15), which were originally grouped together because of their bullet-shaped morphology, but it turns out that they are also related genetically.

10.2.2 Nomenclature of viruses and taxonomic groups

The naming of individual viruses has been a rather haphazard business, with somewhat different approaches taken for viruses of different host types. Bacterial viruses were simply allotted codes, such as T1, T2 and φX174. Viruses of humans and other vertebrates were commonly named after the diseases that they cause, e.g. measles virus, smallpox virus, foot and mouth disease virus, though some were named after the city, town or river where the disease was first reported, e.g. Newcastle disease virus, Norwalk virus, Ebola virus. Some of these original names have been adopted as the formal names of the viruses.

Some of the place names where viruses were first found have become incorporated into the names of virus families and genera (Table 10.2).

Table 10.2 Names of virus families and genera derived from place names

Place name	Family/genus name
Bunyamwera (Uganda)	Family *Bunyaviridae*
Ebola (river in Zaire)	Genus *Ebolavirus*
Hantaan (river in South Korea)	Genus *Hantavirus*
Hendra (Australia) and Nipah (Malaysia)	Genus *Henipavirus*
Norwalk (United States)	Genus *Norovirus*

Table 10.3 Names of families and genera of plant viruses based on the host and signs of disease

Host and disease signs	Family/genus name
Brome mosaic	Family *Bromoviridae*
Cauliflower mosaic	Family *Caulimoviridae*
Cowpea mosaic	Family *Comoviridae*
Tobacco mosaic	Genus *Tobamovirus*
Tobacco rattle	Genus *Tobravirus*
Tomato bushy stunt	Family *Tombusviridae*

Many insect viruses were named after the insect, with an indication of the effect of infection on the host. A virus was isolated from *Tipula paludosa* larvae that were iridescent as a result of the large quantities of virions in their tissues (see photograph in Chapter 1, at a glance). Another virus was isolated from *Autographa californica* larvae that had large polyhedral structures in the nuclei of infected cells. These viruses were named *Tipula* iridescent virus and *Autographa californica* nuclear polyhedrosis virus.

Most plant viruses were given names with two components: the host and signs of disease, e.g. potato yellow dwarf virus, tobacco rattle virus. Some of these names have been used as the bases for family and genus names (Table 10.3).

As in other areas of biology, many names of virus taxonomic groups are based on Latin words, while some have Greek origins; a sample is given in Table 10.4.

The student of virology thus gains some grounding in the classical languages! We can note that both Latin and Greek translations of 'thread' have been used to name the filoviruses and the closteroviruses, respectively. Both of these families have thread-shaped virions. Similarly, Latin and Greek translations of 'small' have been used to name the parvoviruses (animal viruses) and the microviruses (phages). The word for 'small' from a third language was used when devising a name for small RNA viruses; the Spanish '*pico*' was linked to 'RNA' to form 'picornaviruses'.

Table 10.4 Names of virus families and genera based on Latin and Greek words. Note that there are two Latin words meaning yellow. One was used to name the flaviviruses (animal viruses) and the other was used to name the luteoviruses (plant viruses)

		Translation	Reason for name	Family/genus name
Latin	*Arena*	Sand	Ribosomes in virions resemble sand grains in thin section	Family *Arenaviridae*
	Baculum	Stick	Capsid shape	Family *Baculoviridae*
	Filum	Thread	Virion shape	Family *Filoviridae*
	Flavus	Yellow	Yellow fever virus	Family *Flaviviridae*
	Luteus	Yellow	Barley yellow dwarf virus	Family *Luteoviridae*
	Parvus	Small	Virion size	Family *Parvoviridae*
	Tenuis	Thin, fine	Virion shape	Genus *Tenuivirus*
	Toga	Cloak	Virion is enveloped	Family *Togaviridae*
Greek	*Kloster*	Thread	Virion shape	Family *Closteroviridae*
	Kystis	Bladder, sack	Virion is enveloped	Family *Cystoviridae*
	Mikros	Small	Virion size	Family *Microviridae*
	Pous	Foot	Phages with short tails	Family *Podoviridae*

10.3 Baltimore classification of viruses

We have seen how viruses can be grouped into seven classes on the basis of the type of genome and the way in which the genome is transcribed and replicated (Sections 6.2 and 7.1). This approach to virus classification was first suggested by David Baltimore, after whom the scheme is named. An advantage of the Baltimore classification is its differentiation between plus-strand RNA viruses that do (class VI) and do not (class IV) carry out reverse transcription, and between dsDNA viruses that do (class VII) and do not (class I) carry out reverse transcription.

In the chapters that follow we shall examine in depth a representative family of viruses from each of the seven Baltimore classes.

Learning outcomes

By the end of this chapter you should be able to

- evaluate the traditional criteria used to classify viruses into families and genera;

- write family and genus names in the correct format;

- explain how genome sequence data are used to classify viruses;

- evaluate phylogenetic trees;

- explain the basis of the Baltimore classification of viruses.

Sources of further information

Book

Fauquet C. M. *et al.*, editors (2005) *Virus Taxonomy: Eighth Report of the International Committee on Taxonomy of Viruses*, Elsevier

Journals

Bao Y. *et al.* (2004) National Center for Biotechnology Information Viral Genomes Project *Journal of Virology*, **78**, 7291–7298

van Regenmortel M. H. V. & Mahy B. W. J. (2004) Emerging issues in virus taxonomy *Emerging Infectious Diseases*, **10**(1) http://www.cdc.gov/ncidod/eid/vol10no1/03-0279.htm

11

Herpesviruses (and other dsDNA viruses)

At a glance

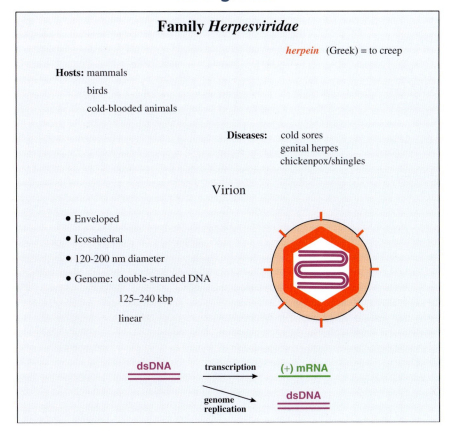

Family *Herpesviridae*

herpein (Greek) = to creep

Hosts: mammals

birds

cold-blooded animals

Diseases: cold sores
genital herpes
chickenpox/shingles

Virion

- Enveloped
- Icosahedral
- 120-200 nm diameter
- Genome: double-stranded DNA

125–240 kbp

linear

dsDNA → transcription → (+) mRNA

genome replication → dsDNA

Virology: Principles and Applications   John B. Carter and Venetia A. Saunders
© 2007 John Wiley & Sons, Ltd   ISBNs: 978-0-470-02386-0 (HB); 978-0-470-02387-7 (PB)

11.1   Introduction to herpesviruses

The herpesviruses derive their name from the Greek word *herpein*, meaning to creep. More than 100 herpesviruses have been isolated from a range of hosts that includes mammals, birds, fish, reptiles, amphibians and molluscs. Eight of these viruses are human viruses (Section 11.2).

A notable characteristic of herpesviruses is that, once they have infected a host, they often remain as persistent infections for the lifetime of the host. These infections are often latent infections, which can be reactivated from time to time, especially if the host becomes immunocompromised. Both primary and reactivated herpesvirus infections can either be asymptomatic or can result in disease of varying severity. The outcome depends on the interplay between the particular virus and its host, and especially on the immune status of the host.

11.2   The human herpesviruses

There are eight herpesviruses known in man, and most adults in the world are persistently infected with most of them.

11.2.1 Herpes simplex viruses 1 and 2

Herpes simplex viruses 1 and 2 (HSV-1 and HSV-2) initially infect epithelial cells of the oral or genital mucosa, the skin or the cornea. The virus may enter neurones and may be transported to their nuclei, where they may establish latent infections.

HSV-1 commonly infects via the lips or the nose between the ages of 6 and 18 months. A latent infection may be reactivated if, for example, the host becomes stressed or immunosuppressed. Reactivation results in the production of virions, which in about 20–40 per cent of cases are transported within the neurone to the initial site of infection, where they cause productive infection in epithelial cells, resulting in a cold sore. Occasionally there may be serious complications such as encephalitis, especially in immunocompromised hosts.

HSV-2 is the usual causative agent of genital herpes, which is a sexually transmitted disease. In newborn babies infection can result in serious disease, with a mortality rate of about 54 per cent.

Although the face and the genitals are the normal sites of infection for HSV-1 and HSV-2, respectively there are increasing numbers of cases where HSV-1 infects the genitals and HSV-2 infects the face (Figure 11.1).

11.2.2 Varicella-zoster virus

Infection with varicella-zoster virus usually occurs in childhood and causes varicella (chickenpox), when the virus spreads through the blood to the skin, causing a rash. It may also spread to nerve cells, where it may establish a latent infection. The nerves

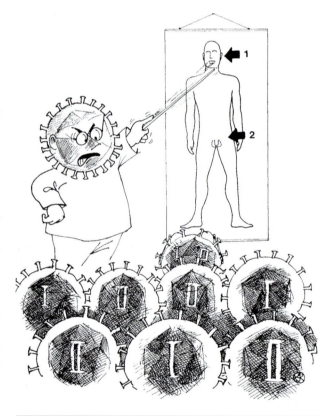

Figure 11.1 *'Get your priorities right'.* From: Haaheim, Pattison and Whiteley (2002) *A Practical Guide to Clinical Virology*, 2nd edition. Reproduced by permission of John Wiley and Sons.

most often affected are those in the face or the trunk, and these are the areas most commonly affected in zoster (shingles) when a latent infection is reactivated.

11.2.3 Epstein-Barr virus

Epstein-Barr virus (EBV) is transmitted in saliva. Epithelial cells are infected first then the infection spreads to B cells, which are the main host cell type for this virus. More than 90 per cent of people become infected with EBV, usually during the first years of life, when infection results in few or no symptoms. In developed countries some individuals do not become infected until adolescence or adulthood. A proportion of these individuals develop infectious mononucleosis (glandular fever), commonly called 'the kissing disease' by doctors. EBV is associated with a number of tumours in man (Chapter 22).

11.2.4 Human cytomegalovirus

In the vast majority of infections with human cytomegalovirus symptoms are either absent or they are mild. In a pregnant woman, however, the virus can infect the placenta and then the foetus, for whom the consequences may be serious. In the US about one per cent of babies are born infected with the virus (about 40 000 per year). In about seven per cent of these there is evidence of virus-induced damage at birth, including small brain size and enlargement of the liver and spleen. In other individuals damage develops at a later stage; the damage may be manifest in a number of ways, such as hearing loss and mental retardation.

Human cytomegalovirus can also cause severe disease (e.g. pneumonitis, hepatitis) in immunocompromised patients such as those with AIDS, those who have received treatment for cancer and those who are immunosuppressed because they have received an organ transplant.

11.2.5 Human herpesvirus 6

There are two types of human herpesvirus 6, known as HHV-6A and HHV-6B. Infection of a child with the latter can cause a fever and the sudden appearance of a rash known as exanthem subitum.

11.2.6 Human herpesvirus 7

Human herpesvirus 7 was first isolated from a culture of CD4 T cells that developed a cytopathic effect; the cells were from a healthy person. The virus has been associated with some cases of exanthem subitum.

11.2.7 Kaposi's sarcoma-associated herpesvirus

Kaposi's sarcoma-associated herpesvirus was discovered in 1994 and is named after the tumour with which the virus is associated (Chapter 22).

11.3 The herpesvirus virion

Herpesviruses have relatively complex virions composed of a large number of protein species organized into three distinct structures: capsid, tegument and envelope (Figure 11.2). The virus genome is a linear dsDNA molecule, which varies in size within the herpesvirus family from 125 to 240 kbp. The DNA is housed in the capsid, which is icosahedral, and the capsid is surrounded by the tegument. The HSV-1 tegument contains at least 15 protein species and some virus mRNA molecules. The envelope contains a large number of spikes (600–750 in HSV-1) composed of ten or more glycoprotein species. There are several different sizes of spike.

A number of schemes have evolved for the nomenclature of herpesvirus proteins, with the result that an individual protein may be referred to in the literature by two or more different names. Most of the structural proteins are commonly named VP (virus protein). In HSV-1 the most abundant proteins in the capsid and the tegument are VP5 and VP16, respectively. In the envelope there are at least 12 species of glycoprotein, each of which is prefixed 'g', for example gB, gC and gD.

The capsid is constructed from 162 capsomeres, 12 of which are pentons and the remainder of which are

(a) Virion components

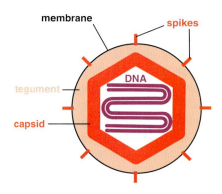

Electron cryo-tomographic visualizations

(b) Central slice through   (c) Cut-away view
reconstructed volume

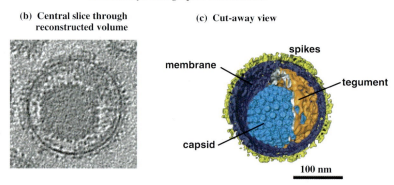

Figure 11.2 *The herpesvirus virion.* (b), (c) Images of herpes simplex virus from Grünewald and Cyrklaff (2006) *Current Opinion in Microbiology*, **9**, 437. Reproduced by permission of the authors and Elsevier Limited.

hexons (Figure 11.3). In HSV-1 the capsomeres are constructed from VP5: a penton is made from five molecules of VP5 and a hexon from six molecules. Other proteins make up structures called triplexes, which connect the capsomeres.

11.4 HSV-1 genome organization organization

HSV-1 is one of the most-studied herpesviruses; we shall look at its genome organization and then at its replication. The genome consists of two unique sequences each flanked by repeat sequences (Figure 11.4(a)). The unique sequences are not of equal

length: the longer is designated U_L and the shorter is designated U_S. The HSV-1 genome encodes at least 74 proteins plus some RNAs that are not translated (Figure 11.4(b)). Both strands of the DNA are used for coding. The inverted repeats contain some genes, so the genome contains two copies of these genes, one in each strand.

11.5 HSV-1 replication

Although HSV-1 infects only humans in nature, a variety of animal species and cell cultures can be infected in the laboratory. The replication cycle of the

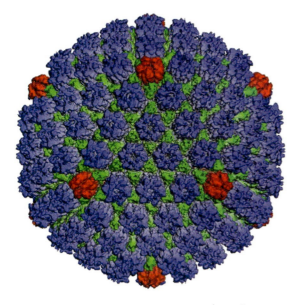

red : pentons

blue : hexons

green : triplexes

Figure 11.3 *The HSV-1 capsid.* Reconstructed image from cryo-electron microscopy, courtesy of Professor Wah Chiu, Baylor College of Medicine, Houston, TX. Reinterpretation of data from Zhou *et al.* (2000) *Science*, **288**, 877, with permission of the American Association for the Advancement of Science.

virus has been studied in cell cultures from a number of species, including humans, monkeys, mice and dogs.

11.5.1 Attachment and entry

The sequence of events at the cell surface usually involves the HSV-1 virion binding initially to heparan sulphate, and then to the main receptor. The latter can be one of several types of cell surface molecule including some nectins, which are cell adhesion molecules. The virion envelope then fuses with the plasma membrane (Figure 11.5). Infection may also occur by endocytosis, followed by fusion between the virion envelope and the endosome membrane.

At least five of the glycoprotein species in the envelope are involved in these processes (Table 11.1).

The nucleocapsid and the tegument proteins are released into the cytoplasm and the nucleocapsid must then be transported to the nucleus, where virus replication takes place. When the host cell is a neurone this journey is a long one; it has been estimated that it would take at least 200 years for the nucleocapsid to reach the nucleus by passive diffusion. In fact, the nucleocapsid is rapidly transported along microtubules to the vicinity of a nuclear pore (Figure 11.6). The virus DNA is released into the nucleus, where the linear molecule is converted into a covalently closed circular molecule, which becomes associated with cell histones.

(a) The major regions of the HSV-1 genome

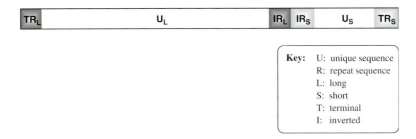

Key: U: unique sequence
      R: repeat sequence
      L: long
      S: short
      T: terminal
      I: inverted

Figure 11.4 *The HSV-1 genome.* Most of the genes are labelled according to their location in the genome:
UL in the unique long region
US in the unique short region
RL in the repeats flanking the unique long region.
Some genes are labelled α or γ. This refers to the phase of gene expression (see Section 11.5.2). LAT: latency-associated transcript (see Section 11.6).
Modified with permission of the authors Gerald Myers and Thomas Brettin, from data on Los Alamos National Laboratory website(http://hpv-web.lanl.gov/).

(b) HSV-1 genes

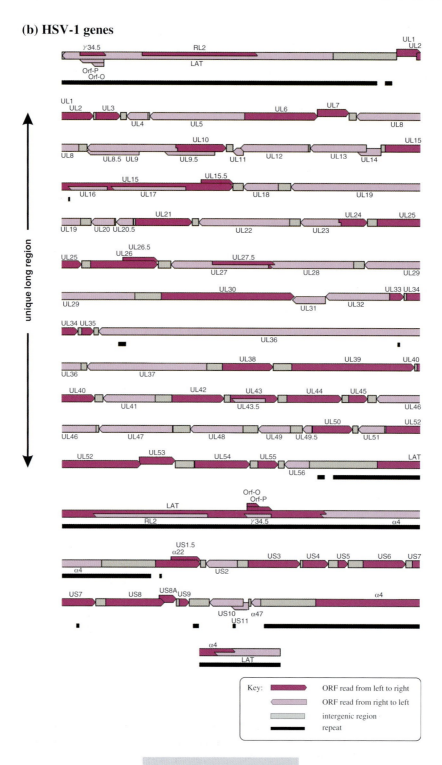

Figure 11.4   *(continued)*

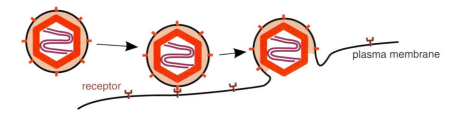

Figure 11.5 *HSV-1 attachment and entry into the cell.* Virus glycoproteins bind to receptors on the cell then the virion envelope fuses with the plasma membrane.

Table 11.1 HSV-1 glycoproteins involved in attachment and entry into the host cell

Process	HSV-1 glycoprotein(s) involved
Attachment to heparan sulphate	gB or gC
Attachment to main cell receptor	gD
Fusion of virion envelope and plasma membrane	gB + gH/gL complex

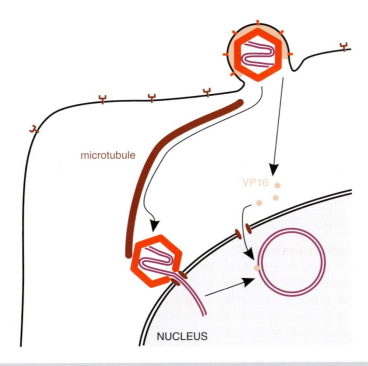

Figure 11.6 *Transport of HSV-1 DNA and VP16 into the nucleus.* The nucleocapsid is transported along a microtubule to a site close to the nucleus. Docking of the nucleocapsid at a nuclear pore is followed by release of the genome into the nucleus. Molecules of VP16 released from the tegument are also transported into the nucleus.

Studies with antibodies specific for tegument proteins have provided evidence that these proteins are transported to several sites in the cell. At these sites tegument proteins play a variety of roles, including the down-regulation of host DNA, RNA and protein synthesis. One tegument protein known as virion host shutoff (vhs) protein degrades cell mRNA. Other tegument proteins are involved in the activation of virus genes, in particular the major tegument protein, VP16, which is transported to the nucleus, where it becomes associated with the virus DNA.

11.5.2 Transcription and translation

Herpesvirus genes are expressed in three phases: immediate early (IE), early (E) and late (L) (Figure 11.7).

Some authors refer to these phases by the Greek letters α, β and γ. There are introns in a few of the HSV-1 genes, mainly in the IE genes.

The IE genes are activated by VP16. It was noted above that VP16 from infecting virions associates with the virus DNA. It does this by binding to a complex of cell proteins including Oct-1, which binds to the sequence TAATGARAT present in the promoter of each of the IE genes (Figure 11.8). VP16 then acts as a transcription factor to recruit the host RNA polymerase II and associated initiation components to each IE gene.

There are five IE proteins and all are transcription factors with roles in switching on E and L genes and in down-regulating the expression of some of these genes. At least some of the IE proteins have more than one role.

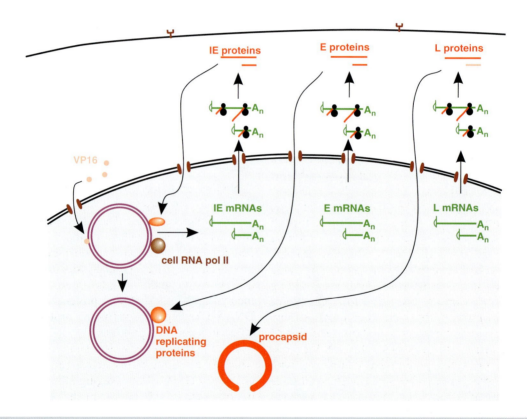

Figure 11.7 *HSV-1 transcription and translation.* There are three phases of transcription and translation:
IE: immediate early
E: early
L: late.

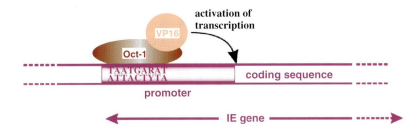

Figure 11.8 *Activation of transcription of HSV-1 IE genes by VP16.*
R: purine
Y: pyrimidine

Some of the E proteins have roles in virus DNA replication (next section), which takes place in discrete regions of the nucleus known as replication compartments. Virus DNA and proteins accumulate in these compartments along with cell RNA polymerase II, which transcribes the L genes. Most of the L proteins are the virus structural proteins.

11.5.3 Genome replication

The virus DNA is replicated by E proteins, seven of which are essential for the process (Figure 11.9). Copies of an origin-binding protein bind at one of three *ori* sites in the virus DNA. The protein has a helicase activity, causing the double helix to unwind at that site, and it has an affinity for the resulting single strands of DNA. The double helix is prevented from re-forming by the binding of copies of a ssDNA-binding protein.

The *ori* site is then bound by a complex of three proteins, which act as a helicase, further unwinding the double helix to form a replication fork. On one of the strands a complex of the same three proteins (now acting as a primase) synthesizes a short sequence of RNA complementary to the DNA. This RNA acts as a primer, and the DNA polymerase complexed with a polymerase processivity factor starts to synthesize the leading strand of DNA. On the other genome strand primases synthesize short RNAs, which are the primers for the synthesis of the Okazaki fragments of the lagging strand of DNA.

In addition to the activities of the seven proteins just described, other virus proteins, such as a thymidine kinase, may also be involved in DNA replication.

It is thought that the circular DNA is first amplified by θ replication, and that later the replication mode switches to σ, also known as rolling circle (Section 7.5). The latter is the predominant mode of herpesvirus DNA replication, and the products are long DNA molecules known as concatemers (Figure 11.10), each of which consists of multiple copies of the virus genome.

11.5.4 Assembly and exit of virions from the cell

The virus envelope glycoproteins are synthesized in the rough endoplasmic reticulum and are transported to the Golgi complex. The other structural proteins, such as VP5, accumulate in the nuclear replication compartments, where procapsids are constructed. A procapsid is more rounded than a mature capsid and its structural integrity is maintained by the incorporation of scaffolding proteins. Before or during DNA packaging the scaffolding proteins are removed by a virus-encoded protease.

Each procapsid acquires a genome-length of DNA, which is cut from a concatemer (Figure 11.11). Each cleavage occurs at a packaging signal at the junction of two copies of the genome. The DNA enters the procapsid via a portal at one of the vertices of the icosahedron. There is evidence from electron microscopy of capsids treated with antibody specific for pUL6 (the protein encoded by the *UL6* gene) that this protein forms the portal (Figure 11.12). It is likely that the DNA passes through the portal both when entering

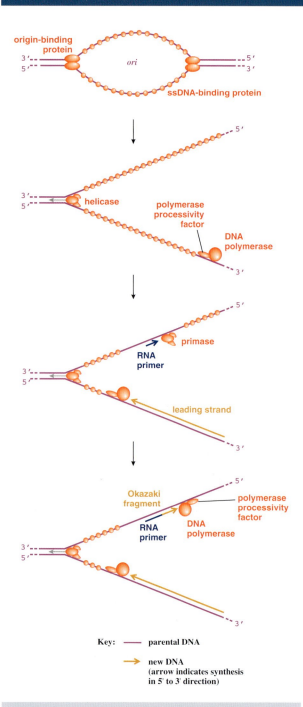

Key: —— parental DNA

→ new DNA
(arrow indicates synthesis
in 5' to 3' direction)

Figure 11.9 *Roles of HSV-1 proteins in DNA replication.* The roles of the seven virus proteins essential for DNA replication are shown. The helicase and the primase are complexes of the same three proteins.

the procapsid during packaging and on leaving the capsid during the infection process, the portal playing a similar role to the head–tail connector in a tailed phage particle (Section 3.4.2.d).

Once the nucleocapsid has been constructed it must acquire its tegument and envelope and then the resulting virion must be released from the cell; these are complex processes and many of the details are still unclear. It is thought that the sequence of events is as follows (Figure 11.13):

- budding through the inner membrane of the nuclear envelope, giving the nucleocapsid a temporary envelope (Figure 11.13(b));

- fusion of the temporary envelope with the outer membrane of the nuclear envelope, releasing the nucleocapsid into the cytoplasm;

- acquisition of VP16 and other components of the tegument;

- acquisition of the virion envelope by budding into a vesicle derived from the Golgi complex (Figure 11.13(c), (d));

- fusion of the vesicle membrane with the plasma membrane, releasing the virion from the cell (Figure 11.13(e)).

Although the virion envelopes are derived from membranes within the cell, virus glycoproteins are also expressed at the cell surface. These glycoproteins can cause fusion between infected cells and non-infected cells, resulting in the formation of giant cells known as syncytia, which can be observed both in infected cell cultures (Figure 2.11) and in HSV lesions. Infected cell cultures are normally destroyed 18–24 hours after inoculation.

11.5.5 Overview of HSV-1 replication

The HSV-1 replication cycle is summarized in Figure 11.14.

11.6 Latent herpesvirus infection

When infection of a cell with a herpesvirus results in latency rather than a productive infection, multiple

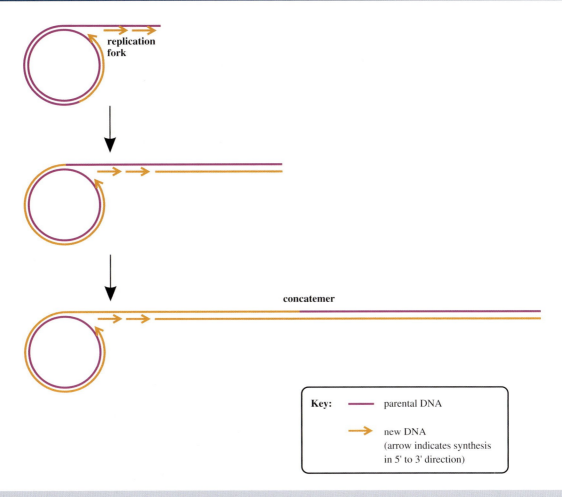

Key: ——— parental DNA

——→ new DNA
(arrow indicates synthesis
in 5' to 3' direction)

Figure 11.10 *Formation of a concatemer.* See Figure 7.5 for earlier stages of rolling circle replication of DNA. Colour coding indicates the fates of the two parental strands of DNA.

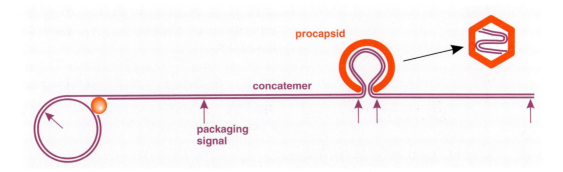

Figure 11.11 *Packaging HSV-1 DNA into a procapsid.* A genome-length of DNA enters a procapsid and is cleaved at packaging signals.

genome is switched off during latency, but a few regions are transcribed and a few RNAs are synthesized; some viruses also synthesize a few proteins.

No virus proteins are required to maintain latency in cells that do not divide, so none are produced in neurones latently infected with HSV-1. Virus RNAs, however, are synthesized; these are known as latency-associated transcripts (LATs). Primary transcripts are synthesized from the *LAT* gene, which is located in the terminal repeats of the genome (Figure 11.4). The LATs undergo splicing, and at least one of them plays a role in inhibiting apoptosis, thereby ensuring the survival of the neurone with its latent HSV-1 infection.

EBV, in contrast, becomes latent in memory B cells, which divide from time to time; the virus therefore synthesizes proteins needed to maintain the copy number of its genome when the host cell divides.

The likelihood of a latent herpesvirus infection becoming reactivated is increased if the host becomes immunocompromised; the greater the degree to which the host is immunocompromised, the greater the likelihood of reactivation.

11.7 Other dsDNA viruses

Some further examples of viruses with dsDNA genomes in Baltimore class I are given in Table 11.2. Baculoviruses infect insects and other invertebrates. Some of them are used as insecticides and some are used as gene vectors in protein expression systems.

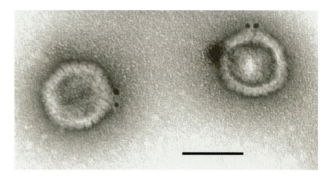

Figure 11.12 *Evidence for pUL6 at one vertex of the HSV-1 capsid.* The preparation was treated with antibody specific for pUL6 followed by an anti-antibody conjugated to gold beads. The bar represents 100 nm. From Newcomb *et al.* (2001) *Journal of Virology,* **75**, 10923. Reproduced by permission of the American Society for Microbiology and the author.

Papillomaviruses are the causative agents of warts (papillomas) and some of them cause human cancers. The dsDNA phages are considered further in Section 19.5.

Most dsDNA viruses, like the herpesviruses, encode genes in both strands of the DNA, so the strands cannot be designated as plus and minus. There are some dsDNA viruses, however, such as the papillomaviruses and phage T7, that encode all of their genes in one strand of the DNA.

Table 11.2   Examples of dsDNA viruses

	Family	Example
Animal viruses	*Baculoviridae*	*Autographa californica* nucleopolyhedrovirus
	Papillomaviridae	Human papillomaviruses (Section 22.2)
	Polyomaviridae	Simian virus 40 (Section 22.3)
	Poxviridae	Smallpox virus
Bacterial viruses	*Myoviridae*	Phage T4 (Section 19.5.1)

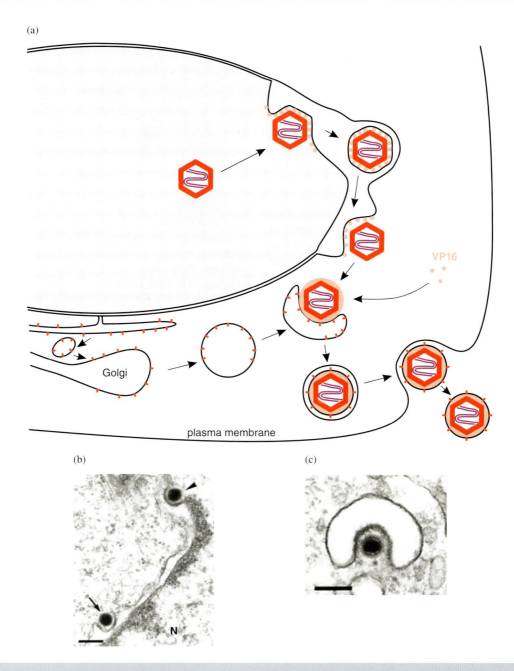

(a)

VP16

Golgi

plasma membrane

(b)

(c)

N

Figure 11.13 *Final stages of HSV-1 assembly.* (a) Transport of nucleocapsids to the cytoplasm, envelopment, and exit from the cell. (b) Nucleocapsids leaving the nucleus (N) through envelopment (arrowhead) and loss of the temporary envelope (arrow). The bar represents 200 nm. (c) Nucleocapsid budding into a vesicle in the cytoplasm. The bar represents 150 nm. (d) Virions inside vesicles. The bar represents 150 nm. (e) Release of virions at the plasma membrane. The bar represents 150 nm. (b) From Granzow *et al.* (2001) *Journal of Virology*, **75**, 3675. (c)–(e) From Mettenleiter (2002) *Journal of Virology*, **76**, 1537. Electron micrographs reproduced by permission of the authors and the American Society for Microbiology.

(d)

(e)

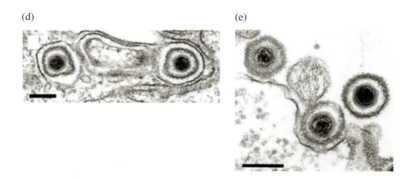

Figure 11.13   (*continued*)

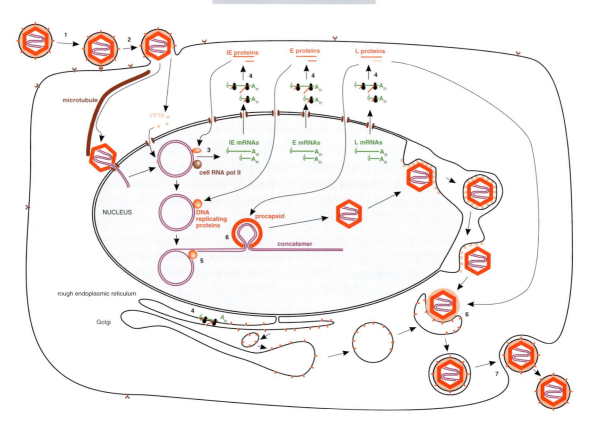

1. **A**ttachment
2. **E**ntry
3. **T**ranscription
4. **T**ranslation
5. **G**enome replication
6. **A**ssembly
7. **E**xit

Figure 11.14   *The HSV-1 replication cycle.* Stages in virion assembly include procapsid construction, packaging a copy of the genome in the procapsid and acquisition of the envelope by budding into a vesicle within the cytoplasm.

Learning outcomes

By the end of this chapter you should be able to

- list the human herpesviruses and explain their importance;

- describe the phenomenon of herpesvirus persistence and explain its importance;

- describe the HSV-1 virion;

- outline the main features of the HSV-1 genome;

- describe the replication cycle of HSV-1.

Sources of further information

Books

Freeman M. L., Decman V. and Hendricks R. L. (2005) Neurons and host immunity conspire to maintain herpes simplex virus in a latent state, pp. 203–213 in *Molecular Pathogenesis of Virus Infections*, editors Digard P., Nash A. A. and Randall R. E., 64th Symposium of the Society for General Microbiology, Cambridge University Press

Knipe D. M. and Howley P. M., editors-in-chief (2001) *Fields Virology*, 4th edition, Chapters 71–83, Lippincott, Williams and Wilkins

Sandri-Goldin R. M., editor (2006) *Alpha Herpesviruses: Molecular and Cellular Biology*, Caister

Journals

Babb R. (2001) DNA recognition by the herpes simplex virus transactivator VP16: a novel DNA-binding structure *Molecular and Cellular Biology*, **21**, 4700–4712

De Bolle L., Naesens L. and De Clercq E. (2005) Update on human herpesvirus 6 biology, clinical features, and therapy *Clinical Microbiology Reviews*, **18**, 217–245

Efstathioua S. and Preston C. M. (2005) Towards an understanding of the molecular basis of herpes simplex virus latency *Virus Research*, **111**, 108–119

Garner J. A. (2003) Herpes simplex virion entry into and intracellular transport within mammalian cells *Advanced Drug Delivery Reviews*, **55**, 1497–1513

Gupta A. *et al*. (2006) Anti-apoptotic function of a microRNA encoded by the HSV-1 latency-associated transcript *Nature*, **442**, 82–85

Hagglund R. and Roizman B. (2004) Role of ICP0 in the strategy of conquest of the host cell by herpes simplex virus 1 *Journal of Virology*, **78**, 2169–2178

Kimberlin D. W. (2004) Neonatal herpes simplex infection *Clinical Microbiology Reviews*, **17**, 1–13

McGeoch D. J., Rixon F. J. and Davison A. J. (2006) Topics in herpesvirus genomics and evolution *Virus Research*, **117**, 90–104

Mettenleiter T. C., Klupp B. G. and Granzow H. (2006) Herpesvirus assembly: a tale of two membranes *Current Opinion in Microbiology*, **9**, 423–429

Nicola A. V. and Straus S. E. (2004) Cellular and viral requirements for rapid endocytic entry of herpes simplex virus *Journal of Virology*, **78**, 7508–7517

Nishiyama Y. (2004) Herpes simplex virus gene products: the accessories reflect her lifestyle well *Reviews in Medical Virology*, **14**, 33–46

Sinclair J. and Sissons P. (2006) Latency and reactivation of human cytomegalovirus *Journal of General Virology*, **87**, 1763–1779

12

Parvoviruses (and other ssDNA viruses)

At a glance

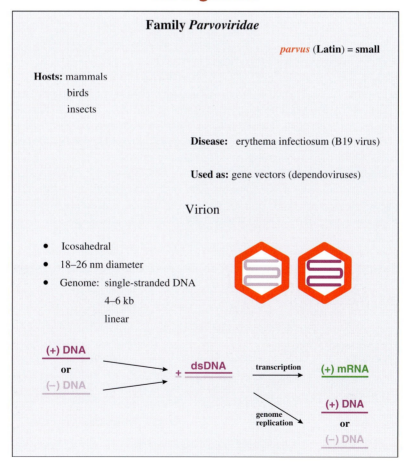

Family *Parvoviridae*

parvus (**Latin**) = small

Hosts: mammals
birds
insects

Disease: erythema infectiosum (B19 virus)

Used as: gene vectors (dependoviruses)

Virion

- Icosahedral
- 18–26 nm diameter
- Genome: single-stranded DNA
4–6 kb
linear

(+) DNA
or → ± **dsDNA** —transcription→ **(+) mRNA**
(−) DNA

genome replication →
(+) DNA
or
(−) DNA

Virology: Principles and Applications   John B. Carter and Venetia A. Saunders
© 2007 John Wiley & Sons, Ltd   ISBNs: 978-0-470-02386-0 (HB); 978-0-470-02387-7 (PB)

12.1 Introduction to parvoviruses

Parvoviruses are amongst the smallest known viruses, with virions in the range 18–26 nm in diameter. They derive their name from the Latin *parvus* (= small). The family *Parvoviridae* has been divided into two subfamilies: the *Parvovirinae* (vertebrate viruses) and the *Densovirinae* (invertebrate viruses). Some of the genera and species of the two subfamilies are shown in Table 12.1.

The subfamily *Parvovirinae* includes the genus *Dependovirus*, the members of which are defective, normally replicating only when the cell is co-infected with a helper virus. Other parvoviruses that do not require helper viruses are known as autonomous parvoviruses.

12.2 Examples of parvoviruses

12.2.1 Dependoviruses

The first dependovirus to be discovered was observed in the electron microscope as a contaminant of an adenovirus preparation (Figure 12.1). The contaminant turned out to be a satellite virus (Section 9.3.1): a defective virus dependent on the help of the adenovirus for replication. The satellite virus was therefore called an adeno-associated virus. Other dependoviruses (distinct serotypes) have since been found in adenovirus preparations and in infected humans and other species. Results of surveys using serological methods and PCR detection of virus DNA indicate that dependovirus infections are widespread.

Not all dependoviruses have an absolute requirement for the help of an adenovirus. Other DNA viruses (e.g.

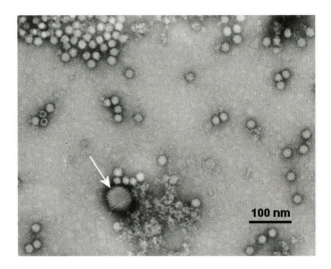

Figure 12.1 *Virions of adenovirus (arrowed) and dependovirus.* Reproduced with permission of Professor M. Stewart McNulty, The Queen's University of Belfast.

herpesviruses) may sometimes act as helpers, and some dependoviruses may replicate in the absence of a helper virus under certain circumstances.

Dependoviruses are valuable gene vectors. They are used to introduce genes into cell cultures for mass-production of the proteins encoded by those genes, and they are being investigated as possible vectors to introduce genes into the cells of patients for the treatment of various genetic diseases and cancers. One of the advantages of dependoviruses for such applications is the fact that they are not known to cause any disease, in contrast to other viruses under investigation, such as retroviruses (Section 16.5).

Table 12.1 Some of the viruses in the family *Parvoviridae*

Subfamily	Genus	Species
Parvovirinae	*Dependovirus*	Adeno-associated virus 2
	Parvovirus	Minute virus of mice
		Feline panleukopenia virus
	Erythrovirus	B19 virus
	Bocavirus	Human bocavirus
Densovirinae	*Iteravirus*	*Bombyx mori* densovirus

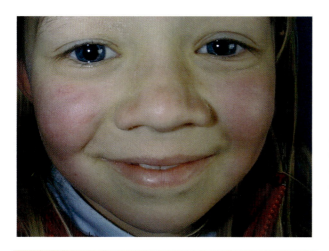

Figure 12.2 *Child with fifth disease.* Reproduced by permission of the New Zealand Dermatological Society.

12.2.2 Autonomous parvoviruses

A parvovirus that does not require a helper virus was discovered in serum from a healthy blood donor. The virus, named after a batch of blood labelled B19, infects red blood cell precursors. Many infections with B19 are without signs or symptoms, but some result in disease, such as fifth disease (erythema infectiosum), in which affected children develop a 'slapped-cheek' appearance (Figure 12.2).

Other diseases caused by B19 virus include

- acute arthritis

- aplastic anaemia in persons with chronic haemolytic anaemia

- hydrops foetalis (infection may be transmitted from a pregnant woman to the foetus and may kill the foetus).

In 2005 a new human parvovirus was discovered using a technique for molecular screening of nasopharyngeal aspirates from children with lower respiratory tract disease. The virus is related to known parvoviruses in the genus *Bocavirus*.

Viruses in the subfamily *Densovirinae* cause the formation of dense inclusions in the nucleus of the infected cell. Some of these viruses are pathogens of the silkworm (*Bombyx mori*), and can cause economic damage to the silk industry.

12.3 Parvovirus virion

Parvoviruses are small viruses of simple structure with the ssDNA genome enclosed within a capsid that has icosahedral symmetry (Figure 12.3).

12.3.1 Capsid

The parvovirus capsid has icosahedral symmetry and is built from 60 protein molecules. One protein species forms the majority of the capsid structure and there are small amounts of between one and three other protein species, depending on the virus. The proteins are numbered in order of size, with VP1 the largest;

(a) Virion components

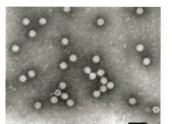

(b) Electron micrograph of negatively stained virions

Bar represents 50 nm.

(c) Reconstructed image from cryo-electron microscopy

Bar represents 10 nm.

Figure 12.3 *The parvovirus virion.* The virions in Figure 12.3(b), (c) are adeno-associated virus 5, from Walters *et al.* (2004) *Journal of Virology*, **78**, 3361. Reproduced by permission of the American Society for Microbiology and the authors.

the smaller proteins are shorter versions of VP1. Each protein species contains an eight-stranded β-barrel structure that is common to many viral capsid proteins, including those of the picornaviruses (Chapter 14).

The virion is roughly spherical, with surface protrusions and canyons (Figure 12.3(c)). At each of the vertices of the icosahedron there is a protrusion with a pore at the centre.

12.3.2 Genome

Parvoviruses have genomes composed of linear ssDNA in the size range 4–6 kb. At each end of a DNA molecule there are a number of short complementary sequences that can base pair to form a secondary structure (Figure 12.4). Some parvovirus genomes have sequences at their ends known as inverted terminal repeats (ITRs), where the sequence at one end is complementary to, and in the opposite orientation to, the sequence at the other end (Section 3.2.6). As the sequences are complementary, the ends have identical secondary structures (Figure 12.4(a)). Other parvoviruses have a unique sequence, and therefore a unique secondary structure, at each end of the DNA (Figure 12.4(b)).

During replication, parvoviruses with ITRs generate and package equal numbers of (+) and (−) strands of DNA, while most viruses with unique sequences at the termini do not. The percentages of virions containing (+) DNA and (−) DNA therefore vary with different viruses (Table 12.2).

In a (−) DNA the genes for non-structural proteins are towards the 3′ end and the structural protein genes are towards the 5′ end (Figure 12.5).

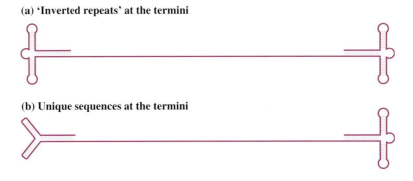

(a) 'Inverted repeats' at the termini

(b) Unique sequences at the termini

Figure 12.4   *Base pairing at the termini of parvovirus DNA.*

Table 12.2   Percentages of parvovirus virions containing (+) and (−) strand DNA

Virus	Sequences at the DNA termini[1]	% virions containing (+) DNA	% virions containing (−) DNA
B19 virus	ITRs	50	50
Dependoviruses	ITRs	50	50
Minute virus of mice	Unique	1	99

[1]The termini of a parvovirus DNA are either 'inverted repeats' (ITRs; Figure 12.4(a)) or unique (Figure 12.4(b)).

ssDNA (minus strand)

3′   5′

genes for   capsid protein genes
non-structural proteins

Figure 12.5 *Parvovirus genome organization.*

12.4 Parvovirus replication

The small genome of a parvovirus can encode only a few proteins, so the virus depends on its host cell (or another virus) to provide important proteins. Some of these cell proteins (a DNA polymerase and other proteins involved in DNA replication) are available only during the S phase of the cell cycle (Figure 4.5), when DNA synthesis takes place. This restricts the opportunity for parvovirus replication to the S phase. Contrast this situation with that of the large DNA viruses, such as the herpesviruses (Chapter 11), which encode their own DNA-replicating enzymes, allowing them to replicate in any phase of the cell cycle.

This account of parvovirus replication is based on studies with several parvoviruses. Some aspects specific to the dependoviruses are covered in Section 12.4.6.

12.4.1 Attachment and entry

A virion attaches to receptors on the surface of a potential host cell (Figure 12.6). In the case of B19 virus the host cell is a red blood cell precursor and the receptor is the blood group P antigen. The virion enters the cell by endocytosis and is released from the endosome into the cytoplasm, where it associates with microtubules and is transported to a nuclear pore. With a diameter of 18–26 nm, the parvovirus virion is small enough to pass through a nuclear pore, unlike the herpesvirus nucleocapsid (Section 11.5.1), though there is evidence that the virion must undergo some structural changes before it can be transported into the nucleus. Nuclear localization signals have been found in the capsid proteins of some parvoviruses.

12.4.2 Single-stranded DNA to double-stranded DNA

In the nucleus the single-stranded virus genome is converted to dsDNA by a cell DNA polymerase (Figure 12.7). The ends of the genome are double stranded as a result of base pairing (Figure 12.4), and at the 3′ end the –OH group acts as a primer to which the enzyme binds.

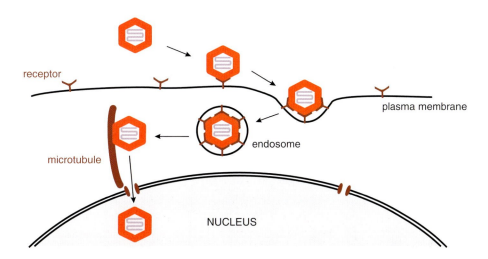

Figure 12.6 *Parvovirus attachment and entry.* A virion is taken into the cell by endocytosis. After release from the endosome it is transported on a microtubule to a site close to the nucleus. It is likely that the virion enters the nucleus through a nuclear pore.

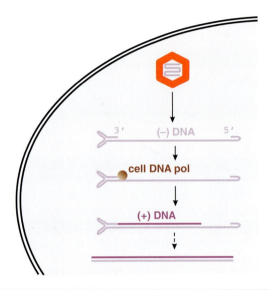

Figure 12.7 *Conversion of ssDNA to dsDNA by the cell DNA polymerase.* Not all steps are shown, as indicated by the dashed arrow.

12.4.3 Transcription and translation

The cell RNA polymerase II transcribes the virus genes and cell transcription factors play key roles. The primary transcript(s) undergo various splicing events to produce two size classes of mRNA (Figure 12.8). The larger mRNAs encode the non-structural proteins and the smaller mRNAs encode the structural proteins (see genome organization, Figure 12.5).

The non-structural proteins are phosphorylated and play roles in the control of gene expression and in DNA replication.

12.4.4 DNA replication

After and virion assembly conversion of the ssDNA genome to dsDNA (Figure 12.7), the DNA is replicated by a mechanism called rolling-hairpin replication. This is a leading strand mechanism and sets parvoviruses apart from other DNA viruses, which replicate their genomes through leading and lagging strand synthesis (Section 7.5).

Procapsids are constructed from the structural proteins and each is filled by a copy of the virus

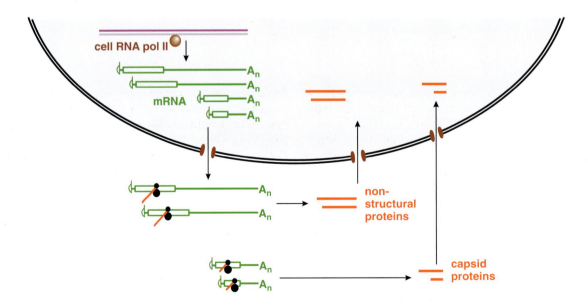

Figure 12.8 *Parvovirus transcription and translation.* The cell RNA polymerase II transcribes the virus genome. After translation much of the virus protein is transported to the nucleus.

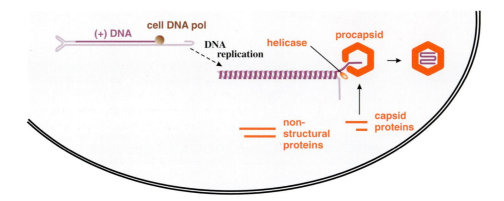

Figure 12.9 *Parvovirus virion assembly.* The virus DNA is replicated (details not shown) prior to encapsidation of one of the strands in a procapsid.

genome, either a (+) DNA or a (−) DNA as appropriate. One of the non-structural proteins functions as a helicase to unwind the dsDNA so that a single strand can enter the procapsid (Figure 12.9).

12.4.5 Overview of parvovirus replication

The parvovirus replication cycle is summarized in Figure 12.10.

12.4.6 Dependovirus replication

When a cell is co-infected with a dependovirus and an appropriate helper virus there is a productive infection with both viruses (Figure 12.11(a)).

In nature it is perhaps more common for a dependovirus to infect a cell in the absence of a helper virus, in which case the virus genome, after it has been converted to dsDNA, may be integrated into a cell chromosome (Figure 12.11(b)). Integration occurs as a result of recombination between the cell and viral DNAs and results in a latent infection. In human cells the virus DNA is integrated at a specific site in chromosome 19. Latent dependovirus infections have been found in a number of cell lines of human and monkey origin. If a cell with a latent dependoviral genome becomes infected with an appropriate helper virus then a productive infection with both viruses can ensue (Figure 12.11(c)).

12.5 Other ssDNA viruses

Some further examples of viruses with ssDNA genomes are given in Table 12.3. These viruses, and indeed the majority of the known ssDNA viruses, have circular genomes. The only viruses known with ssDNA linear genomes are the parvoviruses, the main subjects of this chapter. The ssDNA phages are considered further in Section 19.4.

Table 12.3 Examples of ssDNA viruses

	Family	Example
Animal viruses	*Circoviridae*	Porcine circovirus
Plant viruses	*Geminiviridae*	Maize streak virus
Bacterial viruses	*Microviridae*	Phage φX174

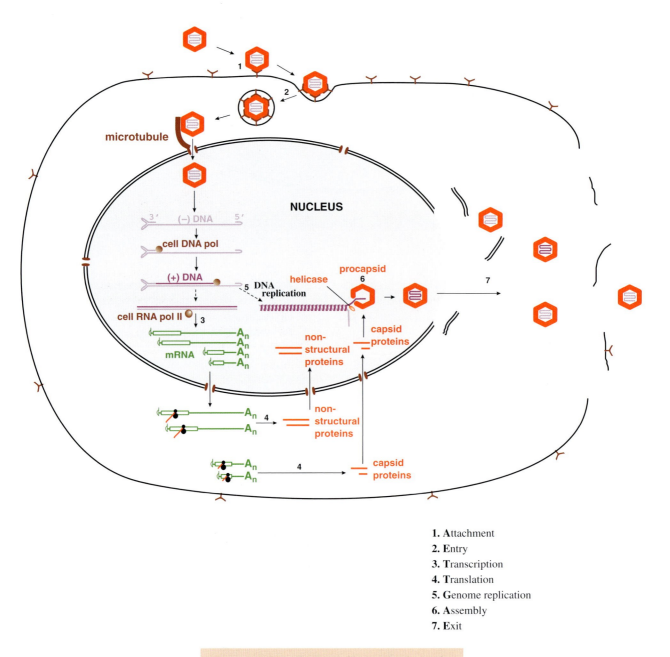

1. **A**ttachment
2. **E**ntry
3. **T**ranscription
4. **T**ranslation
5. **G**enome replication
6. **A**ssembly
7. **E**xit

Figure 12.10 *The parvovirus replication cycle.*

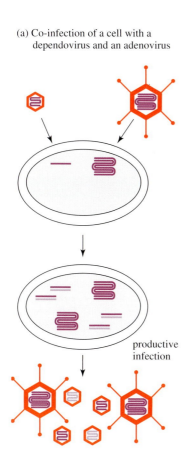

(a) Co-infection of a cell with a dependovirus and an adenovirus

productive infection

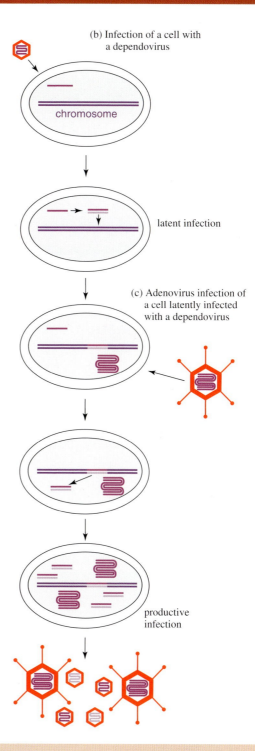

(b) Infection of a cell with a dependovirus

chromosome

latent infection

(c) Adenovirus infection of a cell latently infected with a dependovirus

productive infection

Figure 12.11 *Dependovirus replication*. There is a productive dependovirus infection only when the cell is co-infected with a helper virus.

Learning outcomes

By the end of this chapter you should be able to

- give examples of parvoviruses and explain their importance;

- describe the parvovirus virion;

- outline the main features of the parvovirus genome;

- describe the replication cycle of parvoviruses;

- explain the difference between autonomous and defective parvoviruses.

Sources of further information

Book

Knipe D. M. and Howley P. M., editors-in-chief (2001) *Fields Virology*, 4th edition, Chapters 69 and 70, Lippincott, Williams and Wilkins

Journals

Allander T. *et al.* (2005) Cloning of a human parvovirus by molecular screening of respiratory tract samples *Proceedings of the National Academy of Sciences of the USA*, **102**, 12891–12896

Carter B. J. (2004) Adeno-associated virus and the development of adeno-associated virus vectors: a historical perspective *Molecular Therapy*, **10**, 981–989

Corcoran A. and Doyle S. (2004) Advances in the biology, diagnosis and host–pathogen interactions of parvovirus B19 *Journal of Medical Microbiology*, **53**, 459–475

Flotte T. R. (2005) Adeno-associated virus-based gene therapy for inherited disorders *Pediatric Research*, **58**, 1143–1147

Gao G. *et al.* (2004) Clades of adeno-associated viruses are widely disseminated in human tissues *Journal of Virology*, **78**, 6381–6388

Heegaard E. D. and Brown K. E. (2002) Human parvovirus B19 *Clinical Microbiology Reviews*, **15**, 485–505

Hueffer K. and Parrish C. R. (2003) Parvovirus host range, cell tropism and evolution *Current Opinion in Microbiology*, **6**, 392–398

Vihinen-Ranta M. *et al.* (2004) Pathways of cell infection by parvoviruses and adeno-associated viruses *Journal of Virology*, **78**, 6709–6714

13

Reoviruses
(and other dsRNA viruses)

At a glance

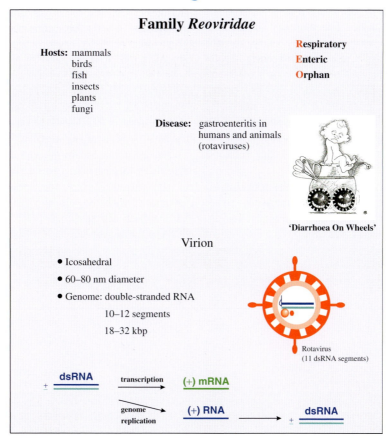

Family *Reoviridae*

Hosts: mammals
birds
fish
insects
plants
fungi

Respiratory
Enteric
Orphan

Disease: gastroenteritis in
humans and animals
(rotaviruses)

'Diarrhoea On Wheels'

Virion

- Icosahedral
- 60–80 nm diameter
- Genome: double-stranded RNA

    10–12 segments

    18–32 kbp

Rotavirus
(11 dsRNA segments)

$$\underset{+}{-}\ \text{dsRNA} \xrightarrow{\text{transcription}} \text{(+) mRNA}$$

$$\xrightarrow[\text{replication}]{\text{genome}} \text{(+) RNA} \longrightarrow \underset{+}{-}\ \text{dsRNA}$$

Virology: Principles and Applications   John B. Carter and Venetia A. Saunders
© 2007 John Wiley & Sons, Ltd   ISBNs: 978-0-470-02386-0 (HB); 978-0-470-02387-7 (PB)

'Diarrhoea On Wheels' from Haaheim *et al.* (2002) *A Practical Guide to Clinical Virology*, 2nd edition. Reproduced by permission of John Wiley & Sons

13.1 Introduction to reoviruses

Icosahedral viruses with dsRNA genomes isolated from the *r*espiratory tracts and *e*nteric tracts of humans and animals, and with which no disease could be associated (*o*rphan), became known as *reo*viruses. A large number of similar viruses have been found in mammals, birds, fish, invertebrates including insects, plants and fungi. Many of these viruses are causative agents of disease, but the original name has been preserved in the family name *Reoviridae*, and has been incorporated into the names of several of the genera within the family (Table 13.1).

The original reoviruses are incorporated into the genus *Orthoreovirus*. The avian viruses are important disease agents, but most orthoreovirus infections of mammals are asymptomatic. The majority of humans become infected with orthoreoviruses early in life and have specific serum antibodies by early adulthood.

It is interesting to note that most of the plant-infecting reoviruses are transmitted between plants by insect vectors (Chapter 4). The viruses replicate in both the plant and the insect, generally causing disease in the plant, but little or no harm to the infected insect.

The main focus of this chapter is on the rotaviruses, which have been the subjects of intensive study because they are amongst the most important agents of gastroenteritis in humans and animals.

13.2 Rotavirus virion

Rotaviruses were first described in 1963, when they were observed during electron microscopy of faecal samples from monkeys and mice. Spherical virions about 75 nm in diameter, with structures resembling the spokes of a wheel, were described (Figure 13.1), so the viruses were named after the Latin word *rota* (= wheel). Similar viruses were observed 10 years later during electron microscopy of faecal samples from children with diarrhoea.

The virion, which has icosahedral symmetry, is also known as a triple-layered particle as the capsid has three layers, each constructed from a distinct virus protein (VP). The inner and middle layers, constructed from VP2 and VP6 respectively, are perforated by channels. The middle layer contains the 'spokes' of the 'wheel' and is the major component of the virion. The outer layer is constructed from VP7, which is glycosylated. It is unusual to find a glycoprotein in a naked virion, but VP7 is associated with a membrane within the cell before it is incorporated into the virion (Section 13.3.3). Three other protein species are found in the virion: VP1 and VP3 in the core, and VP4, which forms 60 spikes at the surface.

The proteins are numbered in order of their sizes. The three largest proteins are found towards the centre of the virion; within the inner capsid layer (VP2) associated with the genome are 12 copies of VP1 and VP3, which are enzymes. VP1 is the RNA-dependent RNA polymerase while VP3 has guanylyl transferase

Table 13.1 Some of the genera in the family *Reoviridae*

Genus	Hosts	Number of dsRNA segments
Cypovirus	Insects	10
Orbivirus	Mammals, birds, invertebrates	10
Orthoreovirus	Mammals, birds	10
Phytoreovirus	Plants, insects	12
Rotavirus	Mammals, birds	11

(a) Virion components

Only four of the sixty VP4 spikes and one of the eleven dsRNA segments are shown.

(b) Cryo-electron micrograph

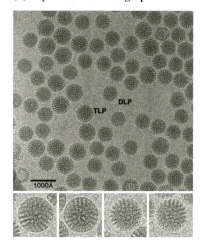

DLP: double-layered particle
TLP: triple-layered particle
Magnified views of two TLPs (left) and two DLPs (right) are shown.

(c) Electron micrograph of negatively stained virions

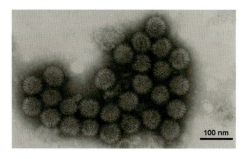

(d) Reconstructed image from cryo-electron microscopy

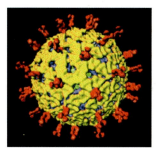

Red   : VP4
Yellow : VP7

Figure 13.1 *The rotavirus virion.* (b) From Baker *et al.* (1999) *Microbiology and Molecular Biology Reviews*, **63**, 862, by permission of the American Society for Microbiology and the author. (c) By permission of Professor M. Stewart McNulty, The Queen's University of Belfast. (d) From: López and Arias (2004) *Trends in Microbiology*, **12**, 271, courtesy of Dr. B.V.V. Prasad, Baylor College of Medicine, Houston, TX, US. Reproduced by permission of Elsevier Limited.

and methyl transferase activities. There is one copy each of VP1 and VP3 attached to the inner capsid layer at each of the 12 vertices of the icosahedron.

The 11 dsRNA segments that make up the rotavirus genome can be separated according to size by electrophoresis in a sodium dodecyl sulphate–polyacrylamide gel (Figure 13.2). Each RNA segment encodes one protein, with the exception of one segment, which encodes two proteins. Hence 12 proteins are encoded: six structural proteins (VP) and six non-structural proteins (NSP). In the diagrams the RNA strands are colour-coded to distinguish the plus strand (the coding strand) from the minus strand.

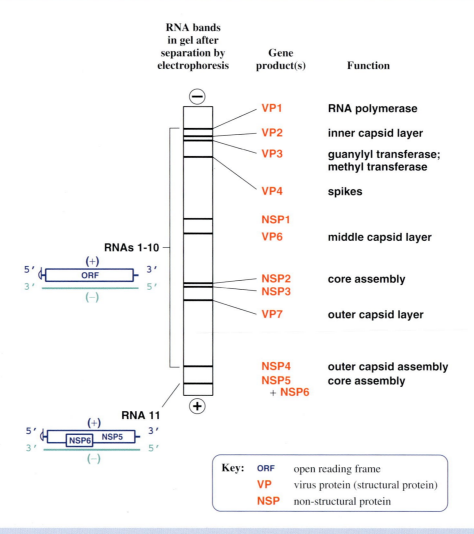

Figure 13.2 *The rotavirus genome and gene products.* The plus strand of each of the 11 RNA segments is capped. There is one ORF in each segment, except for the smallest (RNA 11), in which there are two ORFs in different reading frames. NSP5 and NSP6 therefore do not share a common sequence. The diagram shows a typical pattern of rotavirus RNA segments in a gel after electrophoretic separation, and the protein(s) encoded by each segment.

13.3 Rotavirus replication

Rotaviruses infect cells called enterocytes at the ends of the villi (finger-like extensions) in the small intestine (Figure 13.3).

13.3.1 Attachment and entry

The mechanisms by which rotaviruses attach to and enter their host cells are complex and many details remain uncertain. Cleavage of the spike protein VP4 to its products VP8* and VP5* by proteolytic enzymes such as trypsin (Figure 13.4) results in much more rapid entry into the cell. Binding of a virion to a cell is initially via sites on the spike proteins (VP4, VP8*, VP5*) and then via sites on the capsid surface (the glycoprotein VP7). It is likely that these proteins interact with a number of cell surface proteins; there is evidence that VP5* and VP7 bind to integrins.

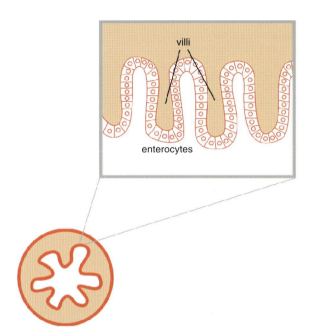

Figure 13.3 *Section of the small intestine.* Rotaviruses infect cells near the tips of the villi.

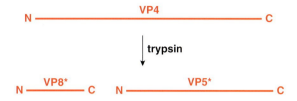

Figure 13.4 *Cleavage of the rotavirus spike protein VP4.*

hydrophobic region of VP5*. This region is hidden in the uncleaved VP4, so virions with spike proteins that have not been cleaved are unable to enter by this mechanism.

13.3.2 Early events

The outer layer of the virion is removed, leaving a double-layered particle (Figure 13.1(b)) in which transcription is activated (Figure 13.6). It is likely that each of the 11 genome segments is associated with a molecule of VP1, which synthesizes a new copy of the (+) RNA, and a molecule of VP3, which caps the 5′ end of the new RNA. The nucleotides for RNA synthesis enter the particles through the channels in the protein layers and the transcripts are extruded from the particles through the same channels. The transcripts are not polyadenylated.

There are two possible ways in which a virion can enter the cell: direct penetration of the virion across the plasma membrane and endocytosis (Figure 13.5). It is thought that direct penetration is mediated by a

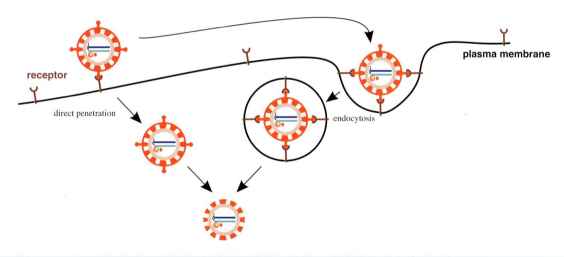

Figure 13.5 *Modes of rotavirus entry into the host cell.* A rotavirus virion may either penetrate the plasma membrane or it may be endocytosed.

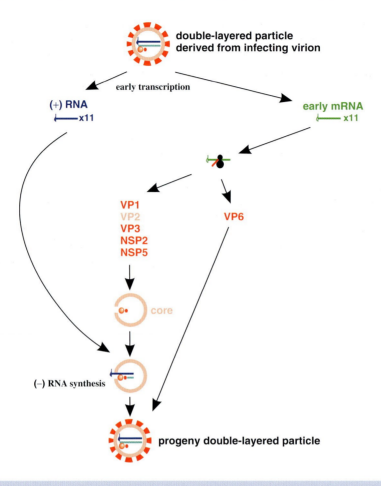

Figure 13.6 *Early transcription, translation, and assembly of double-layered particles.* Some of the (+) RNA that is synthesized functions as mRNA (green). NSP2 and NSP5 play roles in the assembly of viroplasms, where cores are assembled from VP1, VP2 and VP3. Some of the (+) RNA (dark blue) functions as templates for synthesis of (−) RNA (light blue). VP6 is added to the core to form a progeny double-layered particle.

Some of the virus proteins undergo co- and post-translational modifications: VP2 and VP3 are myristylated while NSP5 is phosphorylated and O-glycosylated. The virus proteins accumulate within the cytoplasm in discrete regions known as viroplasms. There is evidence that NSP2 and NSP5 play roles in the formation of viroplasms. Transient expression of the genes for these proteins resulted in the formation of viroplasm-like structures, while reduced expression of NSP5 by RNA interference (Section 2.9.3) resulted in fewer and smaller viroplasms. In the viroplasms virion cores are assembled from VP1, VP2 and VP3.

Newly synthesized (+) RNAs enter the cores, and a rigorous selection procedure ensures that each core receives one each of the 11 RNA species, i.e. a full genome complement. This procedure, which operates for all viruses with multipartite genomes, involves the recognition of a unique sequence in each genome segment.

Synthesis of (−) RNA takes place during the entry of the (+) strands into the core (Figure 13.6), VP1 again acting as the RNA polymerase. The dsRNA of the infecting virion therefore remains intact and the mode of replication is conservative, in contrast to the replication of the dsRNAs of some other viruses (Section 7.6). VP6 is added to the core, forming the second layer of the capsid. The resulting structure is a double-layered particle similar to that derived from the infecting virion.

13.3.3 Late events

A further round of transcription takes place within the double-layered particles. In contrast to the early transcripts, the late transcripts are not capped (Figure 13.7). The translation machinery of the cell undergoes changes that result in it selecting uncapped transcripts in preference to capped transcripts. Hence, translation of cell proteins is shut down while translation of virus proteins continues.

The virus requires different quantities of each of its 12 proteins; for example, a large amount of the main capsid protein (VP6) is required, but a relatively small amount of VP1. Although there are equimolar amounts of each of the 11 dsRNAs, the mRNA species are not made in equimolar amounts, and there are also controls at the level of translation. Hence, there are several mechanisms used by the virus to control the quantity of each protein produced.

The final stages of virion assembly involve the addition of the outer layer of the capsid and the spikes. VP7 and NSP4 are synthesized and N-glycosylated in the rough endoplasmic reticulum, where they remain localized in the membrane. NSP4 has binding sites for both VP4 and double-layered particles (Figure 13.7, inset). After binding these components the immature

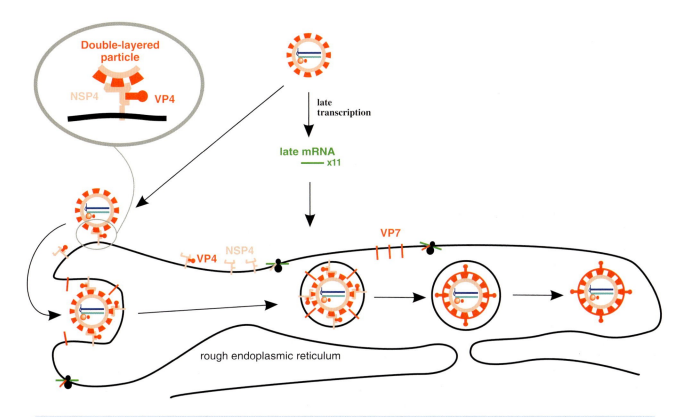

Figure 13.7 *Late transcription, translation and virion assembly.* NSP4 is synthesized in the endoplasmic reticulum, where it binds VP4 and a double-layered particle (inset). This complex buds into the endoplasmic reticulum, where the final stages of assembly occur. VP7, also synthesized in the endoplasmic reticulum, is added to form the outer layer of the capsid and the VP4 spikes are added.

virion buds through the membrane into a vesicle within the endoplasmic reticulum.

The vesicle membrane forms a temporary 'envelope' which contains VP7. Cleavage of VP7 molecules releases them from the membrane to build up the outer layer of the virion and VP4 is added to form the virion spikes. There is some uncertainty as to whether the spikes are added within the endoplasmic reticulum or at another location. Virions are released from the cell either by lysis or by exocytosis.

13.3.4 Overview of rotavirus replication

The rotavirus replication cycle is summarized in Figure 13.8.

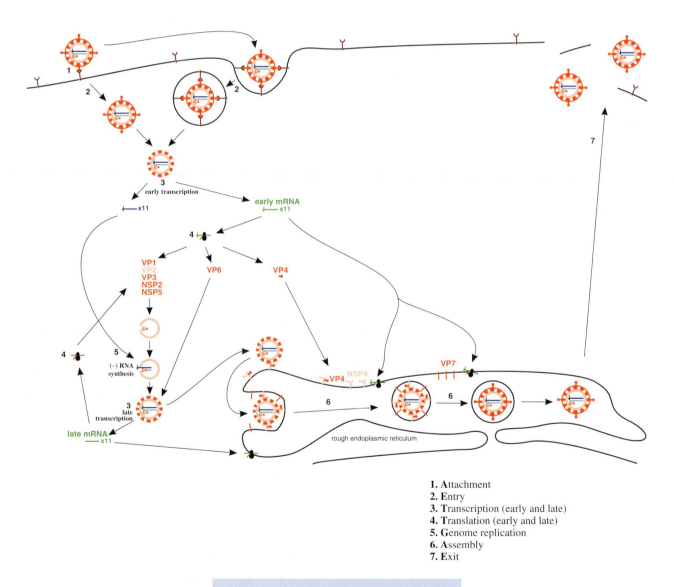

1. **A**ttachment
2. **E**ntry
3. **T**ranscription (early and late)
4. **T**ranslation (early and late)
5. **G**enome replication
6. **A**ssembly
7. **E**xit

Figure 13.8 *The rotavirus replication cycle.*

13.3.5 Rotavirus disease

Enterocytes on the villi are destroyed as a result of rotavirus infection and this leads to reduced absorption of water, salts and sugars from the gut. There is evidence that the tight junctions between cells are damaged by the non-structural protein NSP4, allowing leakage of fluid into the gut. These effects of virus infection, together with the secretion of water and solutes by secretory cells, result in diarrhoea and this can lead to dehydration. Treatment is a relatively simple matter of rehydrating the patient with a solution of salts and sugar, but the means for this are not available in some parts of the world. It estimated that each year there are around half a million deaths of infants and young children due to rotavirus infections.

Although most rotavirus infections are confined to the gut, there is evidence that on occasion the virus may cross the gut and infect other tissues.

13.4 Other dsRNA viruses

The virions of most dsRNA viruses have icosahedral symmetry and most of them are naked; an exception is the family *Cystoviridae*, the members of which are enveloped bacteriophages (Section 19.3). Some further examples of viruses with dsRNA genomes are given in Table 13.2.

A problem that dsRNA viruses must overcome is the fact that dsRNA is a potent inducer of a number of cell defence mechanisms including apoptosis, interferon production and RNA silencing (Chapter 9). Most of these viruses, including the rotaviruses, have solved this problem by ensuring that the viral dsRNA is always enclosed within virus protein structures, and is therefore never free in the cytoplasm to trigger these defences.

Learning outcomes

By the end of this chapter you should be able to

- explain how viruses in the family *Reoviridae* are classified into genera;

- describe the rotavirus virion;

- discuss the main events of the rotavirus replication cycle;

- explain how rotaviruses cause disease.

Sources of further information

Book

Roy P., editor (2006) *Reoviruses: Entry, Assembly and Morphogenesis*, Springer

Journals

Campagna M. *et al.* (2005) RNA interference of rotavirus segment 11 mRNA reveals the essential role of NSP5 in the virus replicative cycle *Journal of General Virology*, **86**, 1481–1487

Chen D. *et al.* (2001) Features of the 3′-consensus sequence of rotavirus mRNAs critical to minus strand synthesis *Virology* **282**, 221–229

Table 13.2 Examples of dsRNA viruses

	Family	Example
Animal viruses	*Birnaviridae*	Infectious bursal disease virus
Plant viruses	*Partitiviridae*	White clover cryptic virus 1
Fungal viruses	*Chrysoviridae*	*Penicillium chrysogenum* virus
Bacterial viruses	*Cystoviridae*	*Pseudomonas* phage φ6

López S. and Arias C. F. (2004) Multistep entry of rotavirus into cells: a Versaillesque dance *Trends in Microbiology*, **12**, 271–278

López T. *et al.* (2005) Silencing the morphogenesis of rotavirus *Journal of Virology*, **79**, 184–192

Mertens P. (2004) The dsRNA viruses *Virus Research*, **101**, 3–13

Pesavento J. B. *et al.* (2005) pH-induced conformational change of the rotavirus VP4 spike: implications for cell entry and antibody neutralization *Journal of Virology*, **79**, 8572–8580

Taraporewala Z. F. and Patton J. T. (2004) Nonstructural proteins involved in genome packaging and replication of rotaviruses and other members of the Reoviridae *Virus Research*, **101**, 57–66

Widdowson M.-A. *et al.* (2005) Rotavirus disease and its prevention *Current Opinion in Gastroenterology*, **21**, 26–31

14

Picornaviruses (and other plus-strand RNA viruses)

At a glance

Family *Picornaviridae*

Pico (= small) **RNA** viruses

Hosts: mammals
birds

Diseases: common cold
polio
hepatitis A
foot and mouth disease

Virion

- icosahedral
- 25–30 nm diameter
- genome: single-stranded RNA

  Plus polarity

  7–8 kb

  Covalently linked protein (VPg)

(+) RNA	→	(−) RNA	transcription	(+) mRNA
			genome replication	(+) RNA

The virus genome can function as mRNA.

Virology: Principles and Applications   John B. Carter and Venetia A. Saunders
© 2007 John Wiley & Sons, Ltd   ISBNs: 978-0-470-02386-0 (HB); 978-0-470-02387-7 (PB)

14.1   Introduction to picornaviruses

Members of the family *Picornaviridae* are found in mammals and birds; some of the genera in the family and some of the important viruses are listed in Table 14.1.

Poliovirus was one of the first viruses to be propagated in cell culture (Enders, Weller and Robbins, 1949) and was also one of the first to be plaque purified (Dulbecco and Vogt, 1954). Most picornaviruses grow readily in cell culture, are easy to purify and are stable, making them popular viruses for laboratory studies.

The picornaviruses are class IV viruses; their genome is a plus-strand RNA that functions as mRNA once it is released into a host cell. The first virus molecules to be synthesized in an infected cell are therefore proteins, in contrast to all other viruses, where transcription of virus genes must occur before virus protein synthesis can start.

14.2   Some important picornaviruses

14.2.1 Hepatitis A virus

Hepatitis A is especially prevalent in developing countries with poor sanitation. In most infants and young children infection is asymptomatic or mild, and leads to life-long immunity. When adults become infected about 75% develop jaundice; severe hepatitis is a rare complication, which can be fatal.

14.2.2 Poliovirus

Poliovirus has been the subject of much research effort because of the devastating paralysis that it can cause. In fact, the majority of poliovirus infections are relatively harmless infections of the oropharynx and the gut. Serious disease occurs only after other tissues become infected, resulting in viraemia (virus in the blood) and spread of infection to the central nervous system. This is a very rare event in babies, who still have anti-poliovirus antibodies acquired from their mothers. If, however, there is less poliovirus in the human environment, so that most infections occur after these antibodies have disappeared, then infection of the central nervous system is more likely. Ironically, polio is a disease primarily associated with improving standards of hygiene and sanitation!

Poliovirus infection of the central nervous system may result in meningitis (from which most patients recover completely), encephalitis and/or paralytic poliomyelitis. The latter is due to virus replication in motor neurones of the spinal cord or the brain stem, resulting in paralysis of limbs and/or breathing muscles (Figure 14.1).

The fear of paralysis was a major motivation for research programmes to develop polio vaccines in the mid-20th century, and these efforts led to the development of inactivated vaccines and live attenuated vaccines (Chapter 24). The use of both types of vaccine has proved very effective in preventing polio, so that by the start of the 21st century the annual number of world cases had fallen to 3500, which was 1% of the number of cases in 1988. Polio has been eradicated from many parts of the world and there is hope that the disease will soon be completely eradicated.

14.2.3 Coxsackieviruses

In 1948 investigators in the US town of Coxsackie injected faecal specimens from two suspected polio

Table 14.1   Examples of picornaviruses

Genus	Name derivation from Greek word	Example(s)
Hepatovirus	*Hepatos* = liver	Hepatitis A virus
Enterovirus	*Enteron* = intestine	Poliovirus
		Coxsackieviruses
Rhinovirus	*Rhinos* = nose	Common cold viruses
Aphthovirus	*Aphtha* = vesicles in the mouth	Foot and mouth disease virus

(a)

(b)

Figure 14.1 (a) *Children with paralysed limbs as a result of polio.* Reproduced by permission of the World Health Organisation. (b) Respirators ('iron lungs') used in the mid-20th century to aid the breathing of polio victims. Reproduced by permission of the US Centers for Disease Control and Prevention.

cases into suckling mice; lesions developed in the skeletal muscles of the mice and the first coxsackieviruses had been isolated. Coxsackieviruses cause a range of medical conditions, including myocarditis (heart disease), meningitis and rashes.

14.2.4 Rhinoviruses

Rhinoviruses are the most common agents causing upper respiratory tract infections in humans. Most children have had at least one rhinovirus infection by the age of 2 years, and in adults rhinoviruses account for about 50% of common colds. Rhinoviruses replicate in the epithelium of the upper respiratory tract, where the temperature is around 33–35 °C. For at least some human rhinoviruses the optimum temperature for replication is in this range, though many human rhinoviruses are able to replicate at 37 °C and some of these are probably responsible for disease in the lower respiratory tract.

14.2.5 Foot and mouth disease virus

Foot and mouth disease virus has a much wider host range than other picornaviruses; it infects mammals such as cattle, sheep, goats and pigs, causing lesions on the feet (Figure 14.2) and in the mouth (see Chapter 1 'At a glance').

   Outbreaks of foot and mouth disease can have serious economic consequences, as milk yields drop, young infected animals may die and in developing

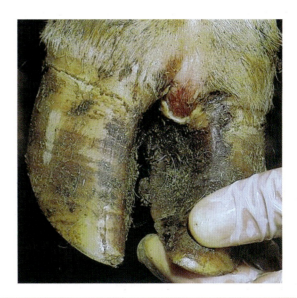

Figure 14.2 *Foot and mouth disease virus lesion.* Reproduced by permission of the Animal Sciences Group, Wageningen UR.

countries animals become unfit for their crucial roles in ploughing and transport. Restrictions on trade in susceptible animal species cause further economic damage. A massive outbreak in the UK in 2001 was brought under control through a series of measures that included the slaughter of more than 6 million animals (Figure 14.3). The estimated cost of the outbreak to the nation was over 8 billion pounds.

Figure 14.3 Foot and mouth disease pyre at a UK farm, 2001. Courtesy of Simon Ledingham.

14.3 Picornavirus Virion

Picornaviruses are small RNA viruses of relatively simple structure (Figure 14.4). The RNA is enclosed by a capsid, which is roughly spherical and has a diameter of about 25–30 nm.

14.3.1 Capsid

The picornavirus capsid has icosahedral symmetry and is made from 60 copies each of four virus proteins

(a) Virion components

(b) Electron micrograph of negatively stained virions of poliovirus

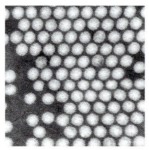

Figure 14.4 *The picornavirus virion.* VPg: virus protein, genome linked. Electron micrograph courtesy of J. Esposito (US Centers for Disease Control and Prevention) and Professor Frederick A. Murphy (University of Texas Medical Branch).

(VP1–4, numbered from the largest to the smallest). Each of the proteins VP1–3 contains an eight-stranded β-barrel, like many virus capsid proteins, including those of the parvoviruses (Chapter 12). The capsid of human rhinovirus 14 has been studied extensively and some of the results are summarized in Figure 14.5. Parts of VP1–3 are at the surface of the virion, while the N termini of VP1–3 and all 60 VP4 molecules are completely internal.

In many picornaviruses, including polioviruses and rhinoviruses, there is a deep cleft around each of the 12 vertices of the icosahedron (Figure 14.5). These clefts, often referred to as canyons, were discovered using X-ray crystallography and cryo-electron microscopy. They are approximately 2 nm deep, which, for a virion of this size, is a significant depth, being approximately seven per cent of the virion diameter. This is relatively much greater than the depth of the Grand Canyon on the surface of the Earth! The canyons, which are lined by the C termini of VP1 and VP3 molecules, contain the virus attachment sites.

Evolution of picornaviruses has generated a lot of variability in the capsid proteins, some of which is reflected in the existence of distinct serotypes (Table 14.2). This antigenic variability creates problems for the development of vaccines against these viruses, problems that have been overcome for poliovirus and foot and mouth disease virus, but not for the rhinoviruses.

> 'Only ... by discovering that there were three immunologically distinct forms of polio virus could we move beyond the inadequate iron lungs and truly conquer polio.'
>
> Watson J.D. (2000) *A Passion for DNA: Genes, Genomes and Society*, Oxford University Press

14.3.2 Genome

The picornavirus genome is composed of a 7–8 kb ssRNA. Covalently linked to the 5′ end of the RNA is a small (2–3 kD) protein known as VPg (virus protein, genome linked) (Figure 14.6). The covalent link is via the –OH group of a tyrosine residue at position 3 of VPg. The 3′ end of the RNA is polyadenylated.

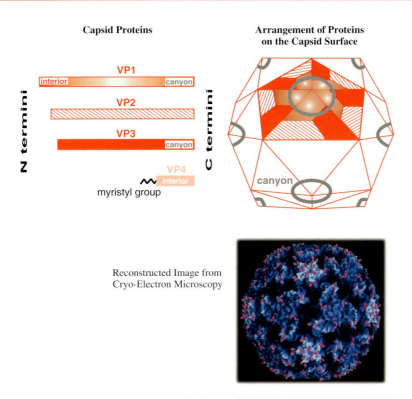

Capsid Proteins

Arrangement of Proteins on the Capsid Surface

Reconstructed Image from Cryo-Electron Microscopy

Figure 14.5 *The capsid of human rhinovirus 14.* The outer surface of the capsid is composed of regions of VP1, VP2 and VP3. Around each of the vertices is a canyon lined with the C termini of VP1 and VP3. The interior surface of the capsid is composed of VP4 and the N termini of VP1. The canyons are visible in the reconstructed image, courtesy of J.Y. Sgro, University of Wisconsin – Madison. The data for the image are from Arnold and Rossmann (1988) *Acta Crystallographica*, Section A, **44**, 270; the pink regions are neutralizing immunogenic sites (Sherry *et al.* (1986) *Journal of Virology*, **57**, 246).

Table 14.2 Serotypes of picornaviruses

Virus	Number of serotypes
Poliovirus	3
Foot and mouth disease virus	7
Human rhinoviruses	>100

The genome consists of one very large ORF flanked by untranslated regions. Within the untranslated region at the 5′ end there is much secondary structure. For example, in poliovirus RNA there are six domains (Figure 14.6), five of which form an internal ribosome entry site (IRES).

14.4 Picornavirus replication

A model of picornavirus replication has been developed, though uncertainty still surrounds many of the details. Many components of the model stem from studies with poliovirus, which replicates in a much wider range of cell types in culture than it does in the body. Information derived from studies with other picornaviruses will also be mentioned.

14.4.1 Attachment

The cell receptors for a number of picornaviruses have been characterized, including those for the rhinoviruses and poliovirus (Section 5.2.1). The poliovirus receptor is the glycoprotein CD155 (Figure 14.7). Its function is not known, but it is expressed on many celltypes.

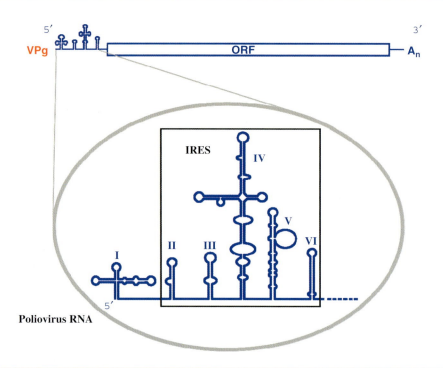

Figure 14.6 *The picornavirus genome.* VPg: virus protein, genome linked. Inset: 5′ end of poliovirus RNA. IRES: internal ribosome entry site. I-VI: secondary structure domains.

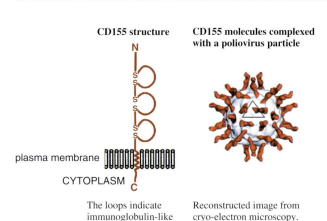

The loops indicate immunoglobulin-like domains. Each domain is stabilized by one or two disulphide (-S-S-) bonds.

Reconstructed image from cryo-electron microscopy. From He *et al.* (2003) *Journal of Virology,* **77**, 4827. Reproduced by permission of the American Society for Microbiology and the authors.

Figure 14.7 *The poliovirus receptor.*

CD155 is a member of the immunoglobulin super-family of molecules with three immunoglobulin-like domains; the virus attachment site is located in the outermost domain.

CD155 is found only in humans and some other primate species. In the early days of poliovirus research none of the common laboratory animals could be infected with the virus, but in the early 1990s transgenic mice expressing CD155 were developed. These animals were found to be susceptible to infection with all three serotypes of poliovirus and they have been used in studies of replication and pathogenesis.

For those picornaviruses like poliovirus with canyons on the virion surface there are virus attachment sites located in pockets at the canyon bases. The canyons are too narrow for access by antibodies so the virus attachment sites are protected from the host's immune surveillance, while the remainder of the virion surface can mutate to avoid the host's immune response. The virus attachment sites of foot and mouth

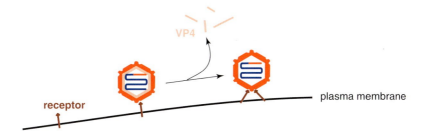

Figure 14.8 *Attachment of picornavirus virion to receptors on host cell.* The virion undergoes major conformational changes and VP4 is lost.

disease virus, however, are not in canyons but are on surface protrusions.

Binding of a virion to receptors results in major changes to the capsid structure: the N termini of VP1 move from the interior to the exterior surface of the capsid and VP4 is lost from the virion (Figure 14.8). Binding of free CD155 to poliovirus particles (Figure 14.7) also causes loss of VP4.

14.4.2 Entry

Several modes of genome entry into the host cell have been proposed for picornaviruses. One mode involves transfer of the RNA from the virion into the cytoplasm at the plasma membrane, leaving the capsid at the cell surface. Other modes involve endocytosis (Figure 14.9), followed by either release from the capsid after disruption of the endosome membrane, or release of RNA from the capsid then transport across the endosome membrane. There is evidence that foot and mouth disease virus and some rhinoviruses enter the cell by endocytosis. Once the virus genome is free in the cytoplasm the VPg is removed from the 5′ end by a cell enzyme.

14.4.3 Translation and post-translational modifications

The virus genome functions as mRNA, but as the RNA is not capped the preliminary stages of translation are different to those for capped mRNAs. The 40S ribosomal subunit binds not at the 5′ end of the RNA, but internally at the IRES (Figure 14.6). Most of the eukaryotic initiation factors (eIFs, normally involved in

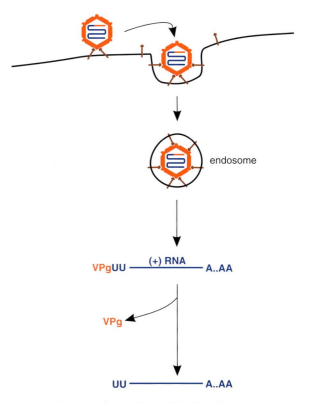

Figure 14.9 *Entry of picornavirus virion by endocytosis.* After release from the endosome VPg is removed from the 5′ end of the genome.

cap-dependent translation) are involved, but a notable exception is the cap-binding protein eIF4E.

Early investigators found that the number of virus proteins synthesized in the infected cell seemed to be greater than could be encoded by the small genome of a picornavirus. The puzzle was solved when it was found that the genome encodes a single polyprotein, which undergoes a series of cleavages to give rise to all the structural and non-structural proteins. An overview of the process is indicated in Figure 14.10, while some of the detail specific to poliovirus is shown in Figure 14.11.

Virus-coded proteases (indicated in Figure 14.11) perform the cleavages, some of which are intramolecular. The polyprotein is first cut to yield P1, P2 and P3. P1 becomes myristylated at the N terminus before being cleaved to VP0, VP3 and VP1, the proteins that will form procapsids; VP0 will later be cleaved to produce VP2 and VP4. Other cleavage products include 3B (VPg), 2C (an ATPase) and 3D (the RNA

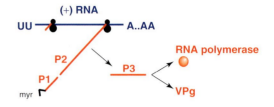

Figure 14.10 *Picornavirus translation and post-translational modifications.* A polyprotein is synthesized and rapidly cleaved into P1, P2 and P3. The N terminus of P1 is myristylated. Further proteolytic cleavages take place producing a number of proteins, including VPg and the RNA polymerase, which are among the products of P3 proteolysis.
myr: myristyl group.

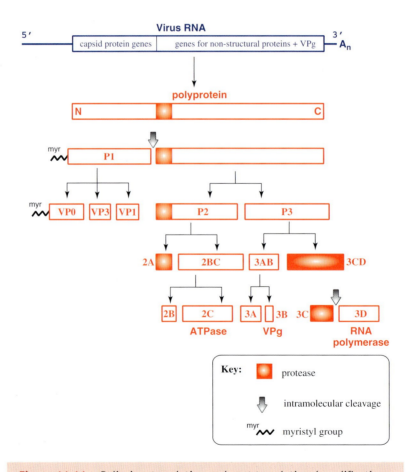

Figure 14.11 *Poliovirus translation and post-translational modifications.*

polymerase). The production of the proteins via a polyprotein precursor might suggest that equimolar amounts result, but not all cleavages take place with equal efficiency. The amounts of the proteins therefore vary; for example, greater amounts of the capsid proteins are produced than of 3D (the RNA polymerase).

Once RNA replication is under way (Section 14.4.4), significant amounts of (+) RNA are produced; some of this RNA functions as mRNA and large quantities of virus proteins are then synthesized. Because the RNA is large (7–8 kb) there may be up to 60 ribosomes translating from each mRNA.

14.4.4 Transcription/genome replication

RNA synthesis occurs in replication complexes that contain cell proteins, as well as virus proteins and RNA. Using immune electron microscopy it has been shown that the replication complexes are associated with the cytoplasmic surfaces of membranous vesicles that develop in the cytoplasm of infected cells (Figure 14.12). Large numbers of vesicles form, derived mainly from the endoplasmic reticulum. Each vesicle is 200–400 nm in diameter and is bounded

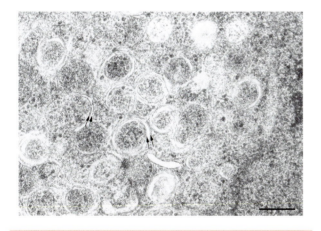

Figure 14.12 *Poliovirus-induced vesicles in the cytoplasm.* The electron micrograph shows a negatively-stained section of a HeLa cell 5 hours after infection with poliovirus. Paired arrows indicate two adjacent lipid bilayers. The bar represents 200 nm. From Schlegel *et al.* (1996) *Journal of Virology*, **70**, 6576. Reproduced by permission of the American Society for Microbiology and the author.

by two lipid bilayers. All known class IV viruses of eukaryotes replicate their RNA in similar membrane-associated complexes.

In order for the infecting (+) RNA to be replicated, multiple copies of (−) RNA must be transcribed and then used as templates for (+) RNA synthesis (Figure 14.13).

The primer for both (+) and (−) strand synthesis is the small protein VPg, which is uridylylated at the hydroxyl group of a tyrosine residue by the poliovirus RNA polymerase (Figure 14.14).

These reactions take place at a *cis*-acting *r*eplication *e*lement (*cre*) located in a stem-loop in the virus genome; an AA sequence within the loop forms the template for UU synthesis. There is evidence that UU-VPg acts as the primer for (−) RNA synthesis, but it is uncertain whether the primer for (+) RNA synthesis is VPg or a uridylylated derivative.

Much of the RNA in the replication complexes is present as replicative intermediates (RIs). An RI is an association of RNA molecules consisting of a template RNA and several growing RNAs of varying length. Thus some RIs consist of a (+) RNA associated with growing (−) RNAs, while other RIs consist of a (−) RNA associated with growing (+) RNAs (Figure 14.13). Also associated with RIs are copies of the RNA polymerase.

The poly(A) sequence at the 3′ end of the plus strand is not synthesized in the usual way. It is transcribed from a poly(U) sequence at the 5′ end of the minus strand.

It has been estimated that in a poliovirus-infected cell up to 2500 RNA molecules are synthesized per minute. The virus controls RNA synthesis in such a way that about 50 times more (+) RNA than (−) RNA is made. Some of the (+) RNA molecules are used as templates for further (−) RNA synthesis, some function as mRNA and some are destined to be the genomes of progeny virions (Figure 14.13).

14.4.5 Assembly and exit

Five copies each of VP0, VP3 and VP1 assemble into a 'pentamer' and 12 pentamers form a procapsid (Figure 14.15).

Each procapsid acquires a copy of the virus genome, with VPg still attached at the 5′ end (Figure 14.16). At around this time the 60 copies of VP0 are cleaved into VP4 and VP2.

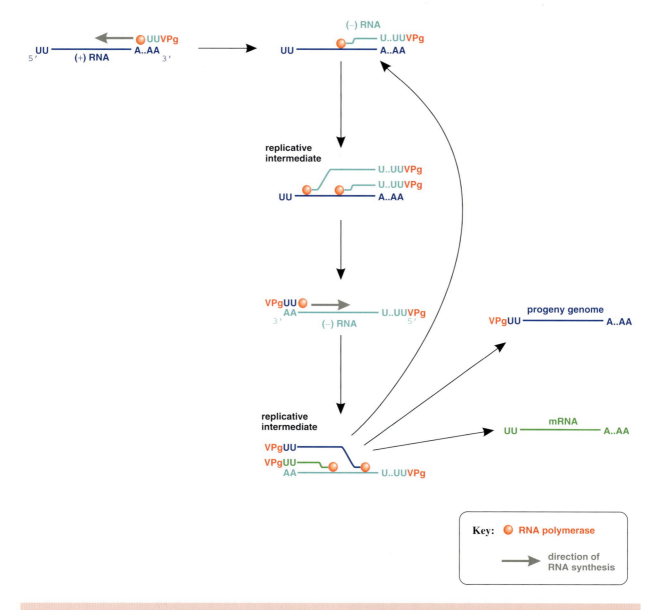

Figure 14.13 *Picornavirus transcription and genome replication.* The genome acts as a template for synthesis of (−) RNA and VPg functions as a primer to initiate synthesis. (−) RNA acts as the template for synthesis of (+) RNA, some of which functions as mRNA, some as progeny genomes and some as templates for further (−) RNA synthesis.

Infected cell cultures develop cytopathic effects; for example, Vero cells (a monkey cell line) infected with poliovirus become rounded (Figure 2.11). Lysis of the cell releases the virions; approximately 10^5 virions are produced in each poliovirus-infected cell.

14.4.6 Inhibition of host gene expression

Soon after infection, the expression of host cell genes is inhibited. In poliovirus-infected cells it has been demonstrated that all three host RNA polymerases are

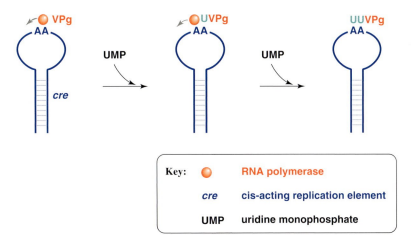

Key:
- ● RNA polymerase
- *cre* cis-acting replication element
- UMP uridine monophosphate

Figure 14.14 *Uridylylation of poliovirus VPg.* The virus RNA polymerase adds UU to a VPg, using AA in *cre* as the template.

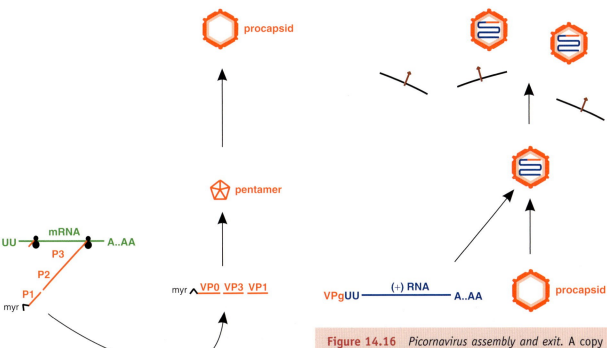

Figure 14.16 *Picornavirus assembly and exit.* A copy of the genome enters a procapsid to form a virion. Progeny virions leave the cell when it lyses.

Figure 14.15 *Procapsid assembly.* The cleavage products of protein P1 assemble in groups of five to form pentamers. A procapsid is constructed from 12 pentamers. myr: myristyl group.

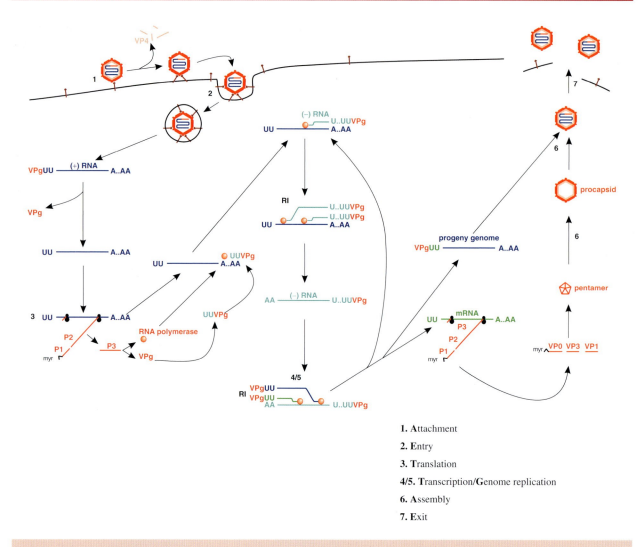

Figure 14.17 *The picornavirus replication cycle.* Note that translation gets under way before transcription, and that transcription and genome replication involve a single process (synthesis of (+) RNA). myr: myristyl group. RI: replicative intermediate.

1. **A**ttachment

2. **E**ntry

3. **T**ranslation

4/5. **T**ranscription/**G**enome replication

6. **A**ssembly

7. **E**xit

rapidly inhibited, thus terminating transcription. Translation from pre-existing cell mRNAs is inhibited as a result of the cleavage of a translation initiation factor by the virus protease 2A. Virus protein synthesis is unaffected as it is initiated by a cap-independent mechanism.

14.4.7 Overview of picornavirus replication

A model for picornavirus replication is outlined in Figure 14.17. The cell nucleus is not shown, as all stages of virus replication take place in the cytoplasm. It has been demonstrated that poliovirus can replicate in enucleated cells, indicating that the virus has no requirement for the nucleus. It is almost certain, however, that some virus proteins enter the nucleus and are involved in the inhibition of cell transcription (Section 14.4.6).

The following points should be noted for class IV viruses in general.

- Translation occurs before transcription.

- RNA replication takes place on membranes.

- The processes of transcription and genome replication are one and the same, resulting in the production of (+) RNA.

- (+) RNA has three functions:
  - templates for (−) RNA synthesis
  - mRNA
  - genomes of progeny virions.

  There must be mechanisms to ensure that a (+) RNA molecule engaged in one of these activities is precluded from commencing either of the other two.

14.5 Picornavirus recombination

If a cell becomes infected with two strains of a picornavirus, then recombinant viruses may be present amongst the progeny virus from that cell (Figure 14.18). A recombinant virus is one that has part of its genome derived from one virus and part derived from another virus. A possible mechanism whereby recombination takes place for a ssRNA virus is described in Section 20.3.3.b.

The oral polio vaccine contains attenuated strains of all three poliovirus serotypes. Administration of this vaccine can lead to co-infections of gut cells with two serotypes, and recombinants between serotypes may be formed. If a wild type poliovirus is also present in the gut then recombination may occur between the wild type virus and one of the vaccine strains. There have been a number of documented cases where recombinants have caused polio.

14.6 Picornavirus experimental systems

This section introduces some techniques that have been developed during work with picornaviruses. Reverse genetics techniques that permit manipulation of RNA genomes are discussed first, followed by techniques that enable the synthesis of infectious virus particles in a cell-free system.

14.6.1 Reverse genetics

Techniques that have been developed for manipulating DNA genomes can be applied to RNA genomes if the RNA is reverse transcribed to DNA. A defined

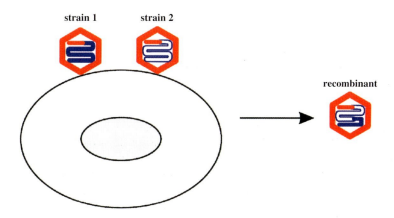

Figure 14.18 *Formation of a picornavirus recombinant.* A cell is infected with two strains of a virus. Amongst the progeny virions there may be some that have part of the genome of strain 1 and part that of strain 2.

mutation can be introduced into a gene in the DNA to investigate the function of the gene. After mutation the DNA is transcribed back to RNA, which is then transfected into cells in culture. Virus replication in the transfected cells leads to the production of mutant virus (Figure 14.19).

In one application of this technique with poliovirus, it was demonstrated that myristylation of VP0 is essential for virion assembly to proceed. A genome was produced in which the glycine residue at the N terminus of VP0 was mutated to alanine; VP0 of the mutant virus could not be myristylated and virion assembly was inhibited.

14.6.2 Cell-free synthesis of infectious virus

In 1991 a group of workers recovered infectious poliovirus from a HeLa cell extract to which they had added purified poliovirus RNA. This was the first

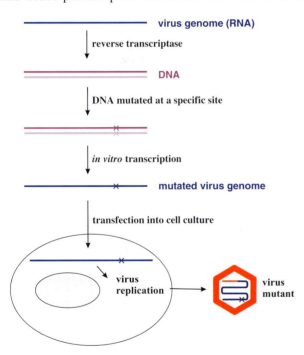

Figure 14.19 *Production of a picornavirus mutant by reverse genetics.* Virus RNA is reverse transcribed to DNA, which is genetically manipulated and cloned in bacterial cells. The DNA is transcribed to RNA, which is transfected into cells. Virus replication in the cells leads to the production of mutant virus.

report of the production of infectious virus outside cells. Cell-free systems based on cytoplasmic extracts have subsequently been used to investigate features of the poliovirus replication cycle that are impossible to dissect using cell-based systems.

In 2002 it was reported that infectious poliovirus had been produced in a process that started with the synthesis of a copy DNA (cDNA) of the virus genome (Figure 14.20).

The process started with the synthesis of discrete sequences of DNA, which together spanned the sequence of a poliovirus genome, then the sequences were ligated to form a cDNA of the virus genome. RNAs (virus genomes) were transcribed from the cDNA and inoculated into a cell extract; infectious poliovirus was subsequently found in the cell extract. This was the first time that infectious virus had been produced from molecular building blocks in the absence of cells.

14.7   Other plus-strand RNA viruses

The genomes of plus-strand RNA viruses are translated in the infected cell prior to their transcription. Because the ribosomes in eukaryotic cells usually terminate after translation of the first ORF, the viruses need strategies to ensure that all the proteins encoded in the infecting genome are translated. One strategy is that described above for the picornaviruses.

- *Polyprotein.* All the genetic information is encoded in one ORF; this is translated to produce a polyprotein which is cleaved to produce the individual virus proteins.

Two other strategies are used by plus-strand RNA viruses.

- *Subgenomic mRNAs.* The genome has two or more ORFs. The RNA-dependent RNA polymerase is encoded by the ORF at the 5′ end of the genome so that it can be translated from the infecting genome. The other ORF(s) are transcribed into subgenomic mRNA(s) that have the same 3′ end as the genome.

- *Segmented genome.* There is one ORF in each RNA segment.

Some viruses use a combination of two of these strategies. Some examples of class IV viruses and their coding strategies are given in Table 14.3.

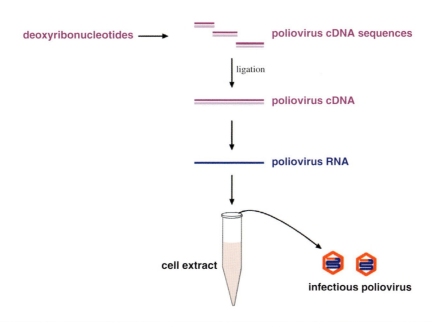

Figure 14.20 *From nucleotides to infectious poliovirus*. Short poliovirus cDNA sequences are synthesized and ligated to form a DNA copy of the genome, from which poliovirus RNA is transcribed. The RNA is introduced into a cell extract, where it directs virus replication, culminating in the assembly of infectious virus.

Table 14.3 Examples of plus-strand viruses and their coding strategies

	Family	Virus example(s)	Coding strategy
Animal viruses	*Flaviviridae*	West Nile virus (Section 21.3.1) Hepatitis C virus (Section 22.7)	Polyprotein
	Coronaviridae	Severe acute respiratory syndrome (SARS) virus (Section 21.5.1)	Subgenomic mRNAs
	Togaviridae	Rubella virus	Polyprotein plus subgenomic mRNAs
Plant viruses	*Potyviridae*	Potato virus Y	Polyprotein
	Flexiviridae	Potato virus X	Subgenomic mRNAs
	Comoviridae	Cowpea mosaic virus	Polyprotein plus segmented genome

The only ssRNA phages that are known are classified in the family *Leviviridae* (Section 19.2). Viruses in this family, such as phage MS2, have genomes with characteristics of bacterial mRNA; the genomes are polycistronic, with all of the ORFs translated by internal initiation (Section 6.6).

Learning outcomes

By the end of this chapter you should be able to

- give examples of picornaviruses and explain their importance;

- describe the picornavirus virion;

- describe the picornavirus replication cycle;

- discuss picornavirus recombination;

- describe experimental systems used for picornavirus studies.

Sources of further information

Books

Needham C. A. and Canning R. (2003) Polio: the rise and fall of a disease, in *Global Disease Eradication: the Race for the Last Child*, American Society for Microbiology, pp. 77–115

Tyrrell D. and Fielder M. (2002) *Cold Wars: the Fight Against the Common Cold*, Oxford University Press

Historical papers

Dulbecco R. and Vogt M. (1954) Plaque formation and isolation of pure lines with poliomyelitis viruses *Journal of Experimental Medicine*, **99**, 167–182

Enders J. F., Weller T. H. and Robbins F. C. (1949) Cultivation of the Lansing strain of poliomyelitis virus in cultures of various human embryonic tissues *Science*, **109**, 85–87

Recent papers

Ahlquist P. *et al.* (2003) Host factors in positive-strand RNA virus genome replication *Journal of Virology*, **77**, 8181–8186

Blomqvist S. *et al.* (2003) Characterization of a recombinant type 3/type 2 poliovirus isolated from a healthy vaccinee and containing a chimeric capsid protein VP1 *Journal of General Virology*, **84**, 573–580

Buenz E. J. and Howe C. L. (2006) Picornaviruses and cell death *Trends In Microbiology*, **14**, 28–36

Haydon D. T., Kao R. R. and Kitching R. P. (2004) The UK foot-and-mouth disease outbreak – the aftermath *Nature Reviews Microbiology*, **2**, 675–681

Racaniello V. R. (2006) One hundred years of poliovirus pathogenesis *Virology*, **344**, 9–16

Whitton J. L., Cornell C. T. and Feuer R. (2005) Host and virus determinants of picornavirus pathogenesis and tropism *Nature Reviews Microbiology*, **3**, 765–776

Wimmer E. (2006) The test-tube synthesis of a chemical called poliovirus: the simple synthesis of a virus has far-reaching societal implications *EMBO Reports*, **7** (SI), S3–S9

Xing L. *et al.* (2003) Structural analysis of human rhinovirus complexed with ICAM-1 reveals the dynamics of receptor-mediated virus uncoating *Journal of Virology*, **77**, 6101–6107

15

Rhabdoviruses (and other minus-strand RNA viruses)

At a glance

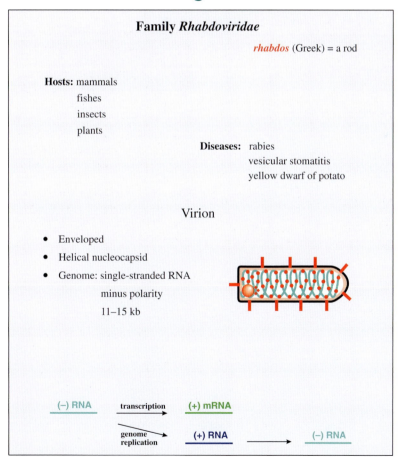

Family *Rhabdoviridae*

rhabdos (Greek) = a rod

Hosts: mammals
fishes
insects
plants

Diseases: rabies
vesicular stomatitis
yellow dwarf of potato

Virion

- Enveloped
- Helical nucleocapsid
- Genome: single-stranded RNA
  minus polarity
  11–15 kb

(−) RNA → transcription → (+) mRNA

genome replication → (+) RNA → (−) RNA

Virology: Principles and Applications   John B. Carter and Venetia A. Saunders
© 2007 John Wiley & Sons, Ltd   ISBNs: 978-0-470-02386-0 (HB); 978-0-470-02387-7 (PB)

15.1 Introduction to rhabdoviruses

The rhabdoviruses have minus-strand RNA genomes in the size range 11–15 kb. The name of these viruses is derived from the Greek word *rhabdos*, which means a rod. The virions of some rhabdoviruses, especially those infecting plants, are in the shape of rods with rounded ends, while others, especially those infecting animals, are bullet shaped (Figure 15.1).

Rhabdoviruses are found in a wide range of hosts, including mammals, fish, plants and insects, and many rhabdoviruses are important pathogens of animals and plants. The rhabdoviruses constitute the family *Rhabdoviridae*, which contains a number of genera. Some of the genera and some of the viruses in the family are listed in Table 15.1.

Many rhabdoviruses have very wide host ranges and replicate in the cells of diverse types of host, especially the so-called 'plant' rhabdoviruses, which replicate in their insect vectors as well as in their plant hosts (Chapter 4).

Festuca leaf streak virus

The virion is rounded at both ends.

The bar represents 200 nm.

Courtesy of Thorben Lundsgaard and ICTVdB.

Vesicular stomatitis virus

The virion is bullet shaped.

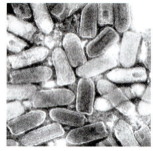

Courtesy of Professor Frederick A. Murphy, The University of Texas Medical Branch.

Figure 15.1 *Negatively stained virions of two rhabdoviruses.*

Table 15.1 Examples of rhabdoviruses

Genus	Name derivation	Hosts	Example(s)
Vesiculovirus	Vesicle = blister	Mammals, fish	Vesicular stomatitis virus
Lyssavirus	*Lyssa* (Greek) = rage, fury, canine madness	Mammals	Rabies virus
Novirhabdovirus	A *non-vi*rion protein is encoded.	Fish	Infectious haematopoietic necrosis virus
Nucleorhabdovirus	Replication cycle includes a nuclear phase.	Plants, insects	Potato yellow dwarf virus

Before looking at the structure and replication of rhabdoviruses, we consider two important rhabdoviruses, rabies virus and vesicular stomatitis virus (VSV).

15.2 Some important rhabdoviruses

15.2.1 Rabies virus

Rabies virus, like many rhabdoviruses, has an exceptionally wide host range. In the wild it has been found infecting many mammalian species, while in the laboratory it has been found that birds can be infected, as well as cell cultures from mammals, birds, reptiles and insects.

Infection with rabies virus normally occurs as a result of virus in saliva gaining access to neurones through damaged skin. The infection spreads to other neurones in the central nervous system, then to cells in the salivary glands, where infectious virus is shed into the saliva (Figure 15.2).

Each year rabies kills large numbers of humans, dogs, cattle and other animal species; precise numbers are not known, but for humans it is estimated that rabies causes about 60 000 deaths annually. Most rabies infections of humans are acquired via bites from rabid dogs, though a few people have become infected after receiving an organ transplant from a rabies-infected individual.

Rabies is endemic in wild animals in many parts of the world. In many regions a single animal species serves as the major reservoir (Figure 15.3); in Western Europe the major reservoir is the red fox.

Vaccines have been developed to provide protection to humans (e.g. veterinary surgeons), domestic animals (especially dogs) and wild animals (e.g. foxes) at risk from rabies virus infection. Rabies vaccines have been incorporated into food baits (Figure 15.4) attractive to wild mammals, and dropped from aircraft over fox-inhabited regions in Europe and over coyote- and raccoon-inhabited regions in the US. The first vaccine to be used was an attenuated vaccine, but more recent vaccines have contained a recombinant vaccinia virus that expresses the rabies virus G protein. Vaccination of wild mammals has been very successful in bringing rabies under control in a number of countries.

Rabies is normally absent from the UK. In the past, this status was maintained through the requirement for a quarantine period for certain animal species, including dogs, on entry to the country. That policy has been largely replaced with a 'pet passport scheme', which involves giving rabies vaccine to animals prior to entry, and implanting an identifying microchip in each vaccinated animal.

Many viruses related to rabies virus have been found in bats around the world, and have been classified in the genus *Lyssavirus* along with the original rabies strains. There are occasional cases of human rabies resulting from bites from infected bats. One such victim was David McRae, a licensed bat handler in Scotland, who died in 2002 after being bitten by an insectivorous bat.

15.2.2 Vesicular stomatitis virus

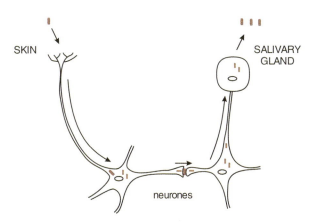

Figure 15.2 *Rabies virus infection of the animal body.* After entering the body through damaged skin a virion infects a neurone via the nerve endings and is transported to the cell body, where virus replication takes place. The infection spreads to other neurones and to salivary gland cells, which shed virions into the saliva.

Vesicle = blister
Stomatitis = inflammation of mucous membrane in the mouth

VSV causes disease in a variety of animals, including cattle, horses, sheep and pigs, affected

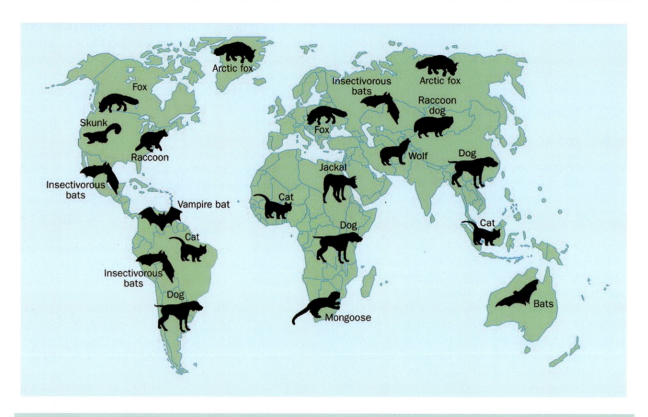

Figure 15.3 *Rabies virus reservoirs.* From Rupprecht *et al.* (2002) *The Lancet Infectious Diseases*, **2**, 327. Reproduced by permission of Elsevier Limited and the authors.

Figure 15.4 *Wild mammal bait containing rabies vaccine.* Courtesy of Michael Rolland, Pinellas County, Florida, US.

animals developing lesions on the feet and in the mouth similar to those in foot and mouth disease (Section 14.2.5). The disease can result in significant economic damage due to decreased milk and meat production, and the imposition of quarantines and trade barriers. Vesicular stomatitis is endemic in the tropics and there are cyclic epidemics in some temperate areas, but it has never been found in the UK.

VSV has a very wide host range. As well as infecting domestic livestock there is evidence of infection in wild animals including bats, deer and monkeys. This evidence is the presence in these animals of neutralizing antibodies to the virus. VSV has been isolated from a number of insect species, including mosquitoes, sand flies and black flies. Its natural cycle is unknown, but it is possible that it is transmitted between mammals by one or more of these types of insect.

In the laboratory VSV can replicate in cell cultures derived from mammals, birds, fish, insects and nematode worms. Much of our understanding of rhabdovirus structure and replication comes from studies with VSV, which is much safer than rabies virus to work with. Three species of VSV are recognized.

15.3 The rhabdovirus virion and genome organization

The rhabdovirus virion is an enveloped, rod- or bullet-shaped structure containing five protein species (Figures 15.1 and 15.5).

The nucleoprotein (N) coats the RNA at the rate of one monomer of protein to nine nucleotides, forming a nucleocapsid with helical symmetry. Associated with the nucleocapsid are copies of P (phosphoprotein) and L (large) protein. The L protein is well named, its gene taking up about half of the genome (Figure 15.5). Its large size is justified by the fact that it is a multifunctional protein, as will be described later. The M (matrix) protein forms a layer between the nucleocapsid and the envelope, and trimers of G (glycoprotein) form spikes that protrude from the envelope.

The genomes of all rhabdoviruses encode these five proteins. Many rhabdoviruses encode one or more proteins in addition to these.

15.4 Rhabdovirus replication

15.4.1 Attachment and entry

A rhabdovirus virion attaches to receptors at the cell surface and is then taken into the cell by clathrin-mediated endocytosis (Figure 15.6). The G protein spikes are involved both in the attachment to cell receptors and in the membrane fusion. The nucleocapsid is released into the cytoplasm after the membranes of the virion and the endosome have fused.

15.4.2 Transcription

Once the RNA and its associated proteins (N, P and L) are free in the cytoplasm transcription of the virus genome can begin (Figure 15.7). A plus-strand leader RNA, the function of which is uncertain, and five mRNAs are synthesized.

Transcription is carried out by an RNA-dependent RNA polymerase activity that resides, along with four

Virion components

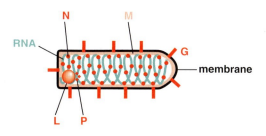

Genome organization and gene products

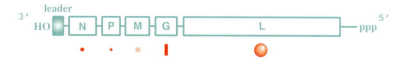

Figure 15.5 *Rhabdovirus virion and genome organization.* The genome has a leader sequence and the genes for the five structural proteins. The genes are separated by short intergenic sequences.

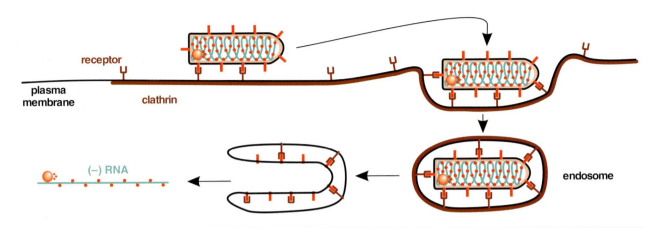

Figure 15.6 *Attachment and entry of a rhabdovirus virion.* After endocytosis the nucleocapsid is released into the cytoplasm by fusion between the membranes of the virion and the endosome.

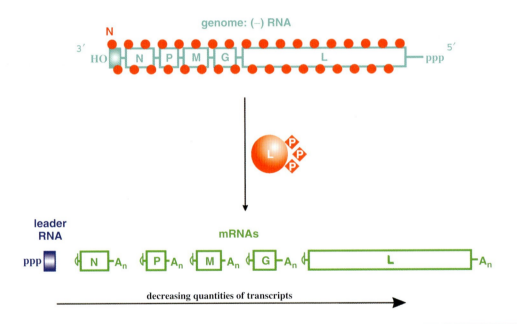

Figure 15.7 *Rhabdovirus transcription.* The minus-strand genome is transcribed into six plus-strand RNAs: a leader RNA and five mRNAs. The transcriptase is a complex of L protein with three copies of P protein.

other enzyme activities, in the L protein (Table 15.2). The polymerase is active only when L is complexed with P protein in the ratio 1L:3P. The requirement for P was demonstrated in experiments with VSV in which P and L were purified. The individual purified proteins were found to be lacking in polymerase activity, which was restored when the two proteins were mixed.

Also associated with the L protein are enzyme activities that cap and polyadenylate the mRNAs (Table 15.2). The virus supplies these activities, as the

Table 15.2 Enzyme activities associated with the rhabdovirus L protein

Enzyme	Role
RNA-dependent RNA polymerase	RNA replication
Methyl transferase	Capping mRNAs
Guanylyl transferase	Capping mRNAs
Poly(A) polymerase	Polyadenylation of mRNAs
Kinase	Phosphorylation of P

cell enzymes are present only in the nucleus. Capping and polyadenylation of the virus mRNAs proceed by mechanisms different to those carried out by the cell, though the end results are the same: each mRNA is capped at the 5′ end and polyadenylated at the 3′ end.

As the L–P complex moves along the template RNA from the 3′ end to the 5′ end a newly synthesized RNA molecule is released at each intergenic sequence (Figure 15.7). The first RNA synthesized is the leader RNA and the remainder are mRNAs. Before release, each mRNA is polyadenylated by the poly(A) polymerase activity of L. It is thought that the L–P complex

'stutters' at the 5′ end of each gene, where the sequence UUUUUUU is transcribed as about 150 adenylates. The enzyme resumes transcription when it recognizes the start of the next gene.

The virus does not need equal amounts of all the gene products; for example, it needs many copies of N protein to coat new RNA, but relatively few copies of L protein. It controls the expression of its genes by controlling the relative quantities of transcripts synthesized. As the enzyme complexes move along the (−) RNA approximately 30% of them detach at the end of each gene, so that fewer and fewer enzyme complexes remain associated with the template as they progress towards the 5′ end. Thus, many copies of N are transcribed, but relatively few copies of L (Figure 15.7).

15.4.3 Translation

The virus proteins are translated on free ribosomes, except for the G protein, which is translated in the rough endoplasmic reticulum (Figure 15.8).

As their names imply, the phosphoprotein (P) and the glycoprotein (G) undergo post-translational modification. One-sixth of the residues in VSV P protein are serine and threonine, and many of these are phosphorylated. The phosphorylation takes place in two steps, the first performed by a cell kinase and the second by the

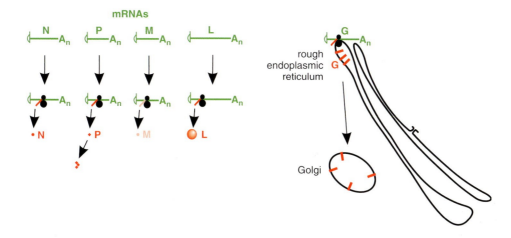

Figure 15.8 *Rhabdovirus translation and post-translational modifications of proteins.* Trimers of P protein are formed after phosphorylation. The G protein is glycosylated in the rough endoplasmic reticulum and the Golgi complex.

kinase activity of the L protein. After phosphorylation, trimers of P are formed. Glycosylation of G protein commences in the rough endoplasmic reticulum, where core monosaccharides are added, and is completed in the Golgi complex.

15.4.4 Genome replication and secondary transcription

The minus-strand virus genome is replicated via the synthesis of complementary (+) RNA molecules, which then act as templates for the synthesis of new copies of (−) RNA (Figure 15.9). Replicative intermediates can be detected in infected cells, as with the plus-strand RNA viruses (Chapter 14). The initiation of RNA synthesis does not require a primer.

We noted earlier that the leader RNA and the individual mRNAs are produced as a result of the RNA polymerase recognizing a termination signal at each intergenic sequence of the template and at the end of the L gene (Figure 15.7). During genome replication, however, the enzyme must remain associated with the

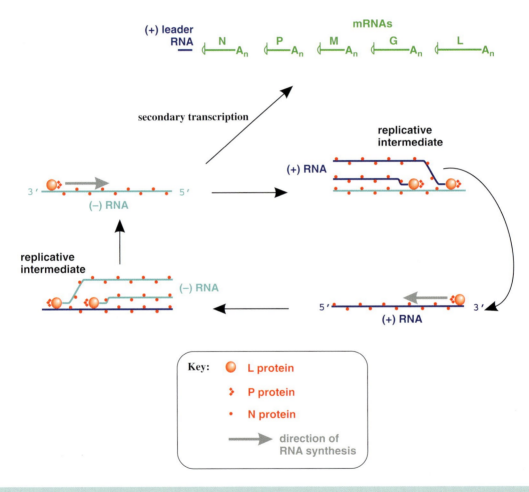

Figure 15.9 *Rhabdovirus genome replication and secondary transcription.* The (−) RNA genome is the template for genome-length (+) RNA synthesis, which in turn is the template for further (−) RNA synthesis. (−) RNAs serve as templates for further (+) RNA synthesis and for secondary transcription. (−) RNAs and genome-length (+) RNAs become coated with N protein shortly after synthesis.

template to produce genome-length (+) RNA. Another difference between the two processes is that during genome replication the newly synthesized (+) RNA quickly becomes coated with N protein, whereas the mRNAs are not coated. The genome and the genome-length (+) RNA are never present in the cell as naked molecules, but are always associated with N protein, which protects them from ribonucleases. This is true for all minus-strand RNA viruses.

The differences between the processes that result in synthesis of mRNAs and genome-length (+) RNA from the same template are not understood. One suggestion is that there may be differences in the components of the enzyme complexes involved in the two processes, one complex acting as a 'transcriptase' and the other acting as a 'replicase'.

A rhabdovirus-infected cell synthesizes about 4 to 10 times more copies of (−) RNA than genome-length (+) RNA. Some copies of the (−) RNA are used as templates for further transcription (secondary transcription; Figure 15.9) so that the amounts of virus gene products in the cell can be boosted, while some become the genomes of progeny virions.

15.4.5 Assembly of virions and exit from the cell

It was noted above that both minus strands and plus strands of genome-length RNA are coated with N protein. Only coated minus strands, however, are selected to form virions, because of the presence of a packaging signal at the 5' end of the minus strand.

The M protein plays several important roles in the assembly process. It condenses the nucleocapsid into a tightly coiled helix and it links the nucleo-capsid with a region of the plasma membrane into which copies of the G protein have been inserted (Figures 8.4(a) and 15.10). Virions bud from these regions of the plasma membrane, acquiring their envelopes in the process. The M protein has a late (L) domain that binds cell proteins involved in the budding process (Section 8.3.1).

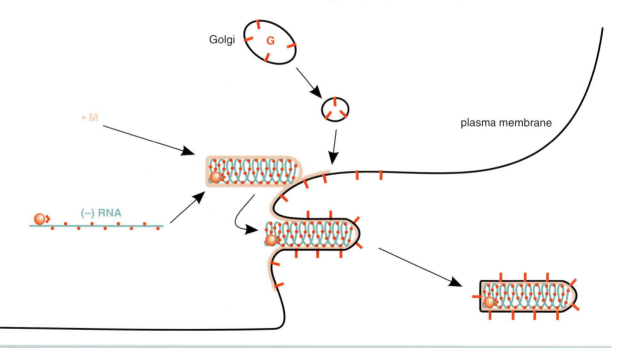

Figure 15.10 *Rhabdovirus assembly and exit.* M protein coats nucleocapsids, which then bud from regions of the plasma membrane that have been modified by the insertion of G protein.

15.4.6 Inhibition of host gene expression

Rhabdovirus infection of a cell results in strong inhibition of host gene expression. The M protein, whose important roles in virion assembly have just been described, appears to play major roles in this inhibition. There is evidence that the M protein inhibits transcription by all three host RNA polymerases and that it blocks intracellular transport of cell RNAs and proteins. One effect of these activities in animal cells is that the synthesis of interferon (Section 9.2.1.a) is inhibited.

15.4.7 Overview of rhabdovirus replication

The rhabdovirus replication cycle is summarized in Figure 15.11.

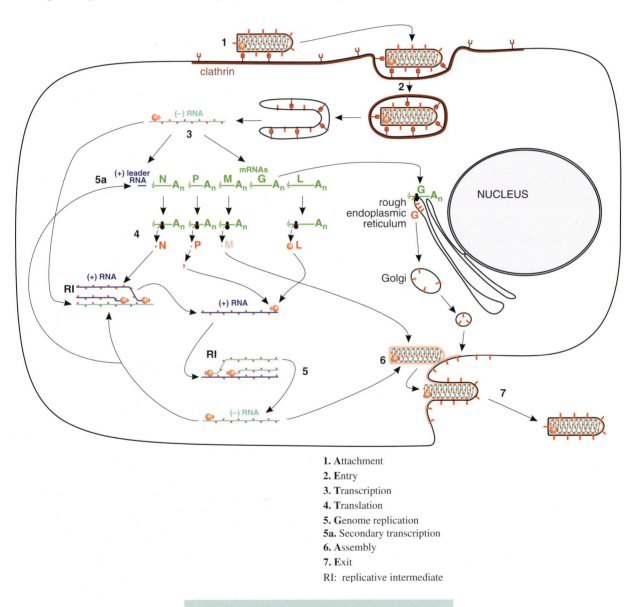

1. Attachment
2. Entry
3. Transcription
4. Translation
5. Genome replication
5a. Secondary transcription
6. Assembly
7. Exit
RI: replicative intermediate

Figure 15.11 *The rhabdovirus replication cycle.*

There appears to be no significant role for the nucleus in the replication of most rhabdoviruses. Enucleated cells support VSV replication, with only a small drop in yield compared with normal cells. This is not true for all rhabdoviruses, however, as members of the genus *Nucleorhabdovirus* replicate in the nucleus of their plant and insect hosts, with virions budding through the inner nuclear membrane.

15.5 Other minus-strand RNA viruses

Most viruses with minus-strand RNA genomes have animal hosts, while a few have plant hosts. Some examples of minus-strand RNA viruses are given in Table 15.3. Amongst them are three of the world's major human pathogens: influenza, measles and respiratory syncytial viruses. These three viruses cause millions of cases of serious disease and approximately a million deaths each year.

Measles virus is one of the main causes of childhood death in developing countries and is still responsible for some deaths in industrialized countries (Section 21.7.1). Infection results in immunosuppression, which renders the host more susceptible to secondary infections with a range of bacterial and viral pathogens, and these cause most measles-associated

Table 15.3 Examples of minus-strand RNA viruses

Family	Virus examples
Orthomyxoviridae	Influenza virus[1]
Paramyxoviridae	Measles virus
	Respiratory syncytial virus
Filoviridae	Ebolavirus (Section 21.4.1)
Bunyaviridae	Hantaan virus (Section 21.2.1)

[1] A number of aspects of influenza virus are dealt with elsewhere:

- virion structure (Section 3.5.1)
- evolutionary mechanisms (Section 20.3.3.c) and the evolution of new virus strains (Section 21.5.2)
- vaccines (Chapter 24)
- anti-viral drugs (Section 25.3.5).

deaths. Measles virus can also cause lethal infections of the central nervous system.

Viruses in the families *Paramyxoviridae* and *Filoviridae*, like those in the *Rhabdoviridae*, have non-segmented genomes that are transcribed to mRNAs by their RNA-dependent RNA polymerases terminating and reinitiating at intergenic sequences. These families form part of the order *Mononegavirales*, which are viruses with *mono*partite genomes composed of *nega*tive sense RNA. Viruses in the families *Orthomyxoviridae* and *Bunyaviridae*, on the other hand, have segmented genomes.

15.6 Viruses with ambisense genomes

The RNA segments of some of the segmented genome viruses are ambisense, where one or more of the RNA segments each encodes two genes, one in the plus sense and one in the minus sense. This is the case for the two genome segments of viruses in the family *Arenaviridae*. It is also the case for some of the three genome segments of some members of the family *Bunyaviridae*, such as tomato spotted wilt virus; two of the three genome segments of this virus are ambisense.

The minus-sense gene of an ambisense RNA is expressed by transcription of a mRNA. The plus-sense gene is expressed by synthesis of an RNA complementary to the genome, followed by transcription of the mRNA for that gene (Figure 15.12).

15.7 Reverse genetics

Reverse genetics have been used to investigate rhabdoviruses and other minus-strand RNA viruses, though the procedures are not as straightforward as with the plus-strand RNA viruses (Section 14.6.1). Minus-strand RNA is not infectious, so techniques had to be devised that will not only generate virus genomes from cDNA, but also supply the RNA polymerase and the nucleoprotein that coats the newly synthesized RNA (Section 15.4.4). After much painstaking work, there are now procedures that enable the recovery of infectious minus-strand RNA virus from a cDNA.

As with the plus-strand RNA viruses, a defined mutation can be introduced into a gene in the cDNA

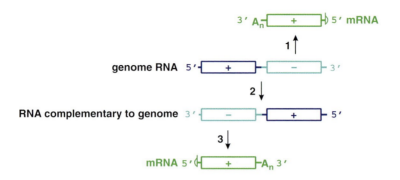

Figure 15.12 *Expression of genes encoded in an ambisense RNA.* The genome RNA encodes one gene in the plus sense and one in the minus sense.

1. The gene encoded in the minus sense is transcribed into mRNA.

2. The genome is transcribed into a complementary RNA.

3. The other gene is transcribed into mRNA.

to investigate its function in the replication cycle and its role, if any, in virus virulence. Reverse genetics are also being explored as a tool to engineer virus strains with reduced virulence for use in vaccines.

Learning outcomes

By the end of this chapter you should be able to

- discuss the importance of rabies virus, vesicular stomatitis virus and other minus-strand RNA viruses;

- describe the rhabdovirus virion;

- outline the main characteristics of the rhabdovirus genome;

- discuss the replication cycle of rhabdoviruses;

- explain the term 'ambisense genome';

- discuss the development of reverse genetics procedures for minus-strand RNA viruses.

Sources of further information

Books

Fu Z. F., editor (2005) *The World of Rhabdoviruses*, Springer
Yawaoka Y., editor (2004) *Biology of Negative Strand RNA Viruses: The Power of Reverse Genetics*, Springer

Journals

Barr J. N. *et al.* (2002) Review: transcriptional control of the RNA-dependent RNA polymerase of vesicular stomatitis virus *Biochimica et Biophysica Acta – Gene Structure and Expression*, **1577**, 337–353
Finke S. and Conzelmann K.-K. (2005) Replication strategies of rabies virus *Virus Research*, **111**, 120–131
Finke S. *et al.* (2003) Rabies virus matrix protein regulates the balance of virus transcription and replication *Journal of General Virology*, **84**, 1613–1621
Nguyen M. and Haenni A.-L. (2003) Expression strategies of ambisense viruses *Virus Research*, **93**, 141–150
Walpita P. and Flick R. (2005) Reverse genetics of negative-stranded RNA viruses: a global perspective *FEMS Microbiology Letters*, **244**, 9–18
Warrell M. J. and Warrell D. A. (2004) Rabies and other lyssavirus diseases *The Lancet*, **363**, 959–969

16

Retroviruses

At a glance

Family *Retroviridae*

retro (**Latin**) = backwards

Hosts: mammals
birds
other vertebrate animals

Diseases: immunodeficiency diseases
leukaemias
solid tumours

Virion

- Enveloped
- 80–110 nm diameter
- Genome: single-stranded RNA
plus polarity
9–10 kb
- Contains reverse transcriptase

(+) RNA → (−) DNA → ± dsDNA → transcription → (+) mRNA
reverse transcription → genome replication → (+) RNA

Virology: Principles and Applications   John B. Carter and Venetia A. Saunders
© 2007 John Wiley & Sons, Ltd   ISBNs: 978-0-470-02386-0 (HB); 978-0-470-02387-7 (PB)

16.1   Introduction to retroviruses

The retroviruses are RNA viruses that copy their genomes into DNA during their replication. Until the discovery of these viruses it had been dogma that the transfer of genetic information always occurs in the direction of DNA to RNA, so finding that some viruses carry out 'transcription backwards' (reverse transcription) caused something of a revolution. We now know that reverse transcription is carried out, not only by these RNA viruses, but also by some DNA viruses (see Chapter 18) and by uninfected cells.

The aim of the current chapter is to provide a general introduction to the retroviruses, which have been found in all classes of vertebrate animal, including fish, amphibians, birds and mammals. The human immunodeficiency viruses (HIV-1 and HIV-2) are retroviruses and the next chapter is devoted entirely to these viruses. Many retroviruses can cause cancer in their hosts, and some aspects of this are discussed in Chapter 22.

16.2   Retrovirus virion

The virion contains two copies of the RNA genome, hence the virion can be described as diploid. The two molecules are present as a dimer, formed by base pairing between complementary sequences (Figure 16.1(a)). The regions of interaction between the two RNA molecules have been described as a 'kissing-loop complex'.

As well as the virus RNA, the virion also contains molecules of host cell RNA that were packaged during assembly. This host RNA includes a molecule of transfer RNA (tRNA) bound to each copy of the virus RNA through base pairing. The sequence in the virus RNA that binds a tRNA is known as the primer binding site (PBS) (Figure 16.1(b)). Each retrovirus binds a specific tRNA (Table 16.1).

A number of protein species are associated with the RNA. The most abundant protein is the nucleocapsid (NC) protein, which coats the RNA, while other proteins, present in much smaller amounts, have enzyme activities (Table 16.2).

Reverse transcriptases are used in molecular biology

RTs have the potential to copy any RNA into DNA, even in the absence of specific tRNA primers. This has made them indispensable tools in molecular biology, where they have a number of applications, including the production of cDNA libraries and the reverse transcription–polymerase chain reaction (RT-PCR).

Two commonly used RTs are those from avian myeloblastosis virus and Moloney murine leukaemia virus.

Encasing the RNA and its associated proteins is the capsid, which appears to be constructed from a lattice of capsid (CA) protein. The shape of the capsid is spherical, cylindrical or conical depending on the virus. A layer of matrix (MA) protein

---

Table 16.1   Examples of tRNAs used by retroviruses as primers

tRNA	Retrovirus
tRNA^{pro}	Human T-lymphotropic viruses 1 and 2
	Murine leukaemia virus
tRNA^{lys-3}	HIV-1 and 2
	Mouse mammary tumour virus

Table 16.2   Enzyme activities present in the retrovirus virion

RNA-dependent DNA polymerase (reverse transcriptase; RT)
DNA-dependent DNA polymerase
Ribonuclease H (RNase H)
Integrase
Protease

(a) Virion components

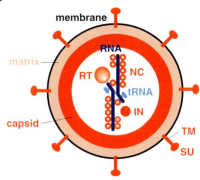

(b) Genome organization and gene products

Figure 16.1 *Retrovirus virion and genome organization*

Genes	gag	group-specific antigen
	pol	polymerase
	env	envelope

Proteins	CA	capsid
	IN	integrase
	MA	matrix
	NC	nucleocapsid
	PR	protease
	RH	ribonuclease H
	RT	reverse transcriptase
	SU	surface glycoprotein
	TM	transmembrane glycoprotein

Non-coding sequences	PBS	primer-binding site
	R	repeat sequence
	U3	unique sequence at 3′ end of genome
	U5	unique sequence at 5′ end of genome

lies between the capsid and the envelope. Associated with the envelope are two proteins: a transmembrane (TM) protein bound to a heavily glycosylated surface (SU) protein. In most retroviruses the bonds between the TM and SU proteins are non-covalent.

The genes encoding the virus proteins are organized in three major regions of the genome (Figure 16.1(b)):

• *gag* (*g*roup-specific *a*nti*g*en) – internal structural proteins

- *pol* (*pol*ymerase) – enzymes

- *env* (*env*elope) – envelope proteins.

16.3 Retrovirus replication

16.3.1 Attachment and entry

The virion binds to cell receptors via the virus attachment site, which is located on the SU protein. This interaction causes a conformational change in the TM protein that allows a hydrophobic fusion sequence to fuse the virion membrane and a cell membrane (Section 5.2.4.b). Most retroviruses fuse their membrane with the plasma membrane of the cell (Figure 16.2), though some are endocytosed and fuse their membrane with an endosome membrane. The structure that is released into the cytoplasm loses some proteins and a reverse transcription complex is formed.

16.3.2 Reverse transcription

Reverse transcription takes place within the reverse transcription complex and some of the detail is indicated in Figure 16.3. Synthesis of both the (−) DNA and the (+) DNA begins at the 3′–OH of a primer RNA. The primer for synthesis of the (−) DNA is the tRNA bound to the genome, while the primer for synthesis of the (+) DNA is a polypurine tract (PPT) in the virus genome. The latter becomes accessible as a result of hydrolysis of the genome RNA from the 3′ end by the RNase H, which is an enzyme that specifically digests RNA in RNA–DNA duplexes.

During synthesis of the two DNA strands, each detaches from its template and re-attaches at the other end of the template through base pairing. The DNA that results from reverse transcription (the provirus) is longer than the RNA genome. Each of the termini has the sequence U3–R–U5, known as a long terminal repeat (LTR), one terminus having acquired a U3 sequence and the other a U5 sequence.

16.3.3 Integration of the provirus

The provirus, still associated with some virion protein, is transported to the nucleus as a pre-integration complex (Figure 16.4). For most retroviruses this can occur only if the cell goes into mitosis, and it is likely that mitosis-induced breakdown of the nuclear membranes is necessary for the pre-integration complex to enter the nucleus. This means that there can be a productive infection only in dividing cells. HIV and related viruses, however, can productively infect resting cells, as the pre-integration complexes of these viruses are able to enter intact nuclei (Section 17.4.2).

One of the virus proteins still associated with the provirus is the integrase; this enzyme cuts the DNA of a cell chromosome and seals the provirus into the gap. The integrated provirus genes may be expressed immediately, or there may be little or no expression of viral genes, in which case a latent infection has been initiated. If a latently infected cell divides, the provirus is copied along with the cell genome and each of the daughter cells has a copy of the provirus.

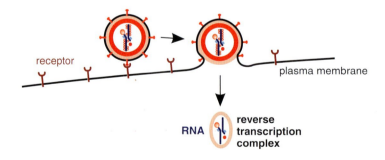

Figure 16.2 *Retrovirus attachment and entry.* Fusion of the virion membrane with the plasma membrane of the cell releases the virion contents, which undergo modification to form a reverse transcription complex.

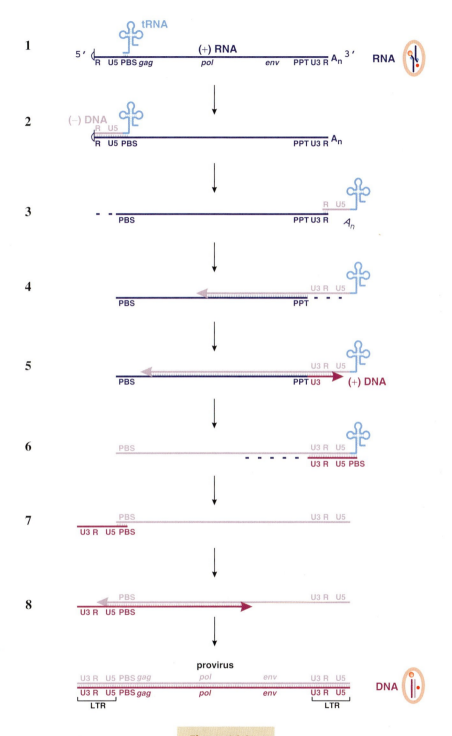

Figure 16.3

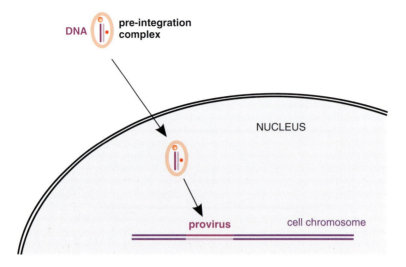

Figure 16.4 *Transport of pre-integration complex to the nucleus and integration of the provirus into a cell chromosome.*

16.3.4 Transcription and genome replication

The two LTRs of the provirus have identical sequences, but are functionally different; transcription is initiated in one and terminated in the other. Transcription factors bind to a promoter in the upstream LTR, then the cell RNA polymerase II starts transcription at the U3–R junction. Transcription continues into the downstream LTR. There is a polyadenylation signal in the R region and transcription terminates at the R–U5 junction (Figure 16.5). Each transcript is capped and polyadenylated. Some transcripts will function as

Figure 16.3 *Retroviral reverse transcription.* LTR: long terminal repeat. PBS: primer binding site. PPT: polypurine tract (a sequence made up entirely, or almost entirely, of purine residues). R: repeat sequence. U3: unique sequence at 3′ end of genome. U5: unique sequence at 5′ end of genome.

1. A copy of the virus genome with a tRNA bound at the PBS.

2. The reverse transcriptase begins (−) DNA synthesis at the 3′ end of the tRNA.

3. The RNase H digests the RNA from the RNA-DNA duplex. The (−) DNA attaches at the 3′ end of either the same strand or the second copy of the genome.

4. Elongation of the (−) DNA continues, while the RNase H degrades the template RNA from the 3′ end as far as the PPT.

5. Synthesis of (+) DNA begins.

6. The remaining RNA is degraded.

7. The (+) DNA detaches from the 5′ end of the (−) DNA template and attaches at the 3′ end.

8. Synthesis of both DNA strands is completed.

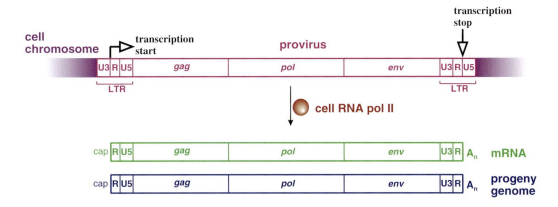

Figure 16.5 *Retrovirus transcription.* Genome-length RNAs are synthesized. Some will function as mRNAs (shown in green) while some will become the genomes of progeny virions (shown in blue). These RNAs are identical; they are shown in different colours to emphasize the two functions.

mRNA and a proportion of these become spliced; others will become the genomes of progeny virions.

16.3.5 Translation and post-translational modifications

The *env* gene is translated from spliced mRNAs in the rough endoplasmic reticulum, where glycosylation commences (Figure 16.6). The Env protein molecules are transported to the Golgi complex, where they are cleaved by a host protease into SU and TM molecules. The two cleavage products remain in close association, and after further glycosylation they are transported to the plasma membrane.

The proteins encoded by the *gag* and *pol* genes are translated from genome-length mRNAs into Gag and Gag–Pol polyproteins (Figure 16.6). Retroviruses require much greater quantities of the Gag proteins than of the Pol proteins, and have evolved mechanisms to synthesize the required amount of each. These mechanisms involve approximately 95 per cent of ribosomes terminating translation after the synthesis of Gag, while the other ribosomes continue translation to synthesize Gag–Pol.

One mechanism, used by murine leukaemia virus, involves reading through a stop codon at the end of *gag* (Figure 16.7). A 'suppressor tRNA' incorporates an amino acid at the stop codon and translation continues into *pol*, which is in the same reading frame as *gag*.

The majority of retroviruses, however, ensure the correct proportions of the Gag and Pol proteins by a ribosomal frameshifting mechanism (Figure 16.8). Here *gag* and *pol* are in different reading frames and there is a −1 shift in reading frame before the *gag–pol* junction in about five per cent of translations. This mechanism is used by HIV-1.

The Gag and Gag–Pol proteins of most retroviruses are myristylated at their N termini.

16.3.6 Assembly and release of virions

Some retroviruses form immature particles in the cytoplasm that are then transported to the plasma membrane, but most retroviruses assemble their components on the inner surface of the plasma membrane (Figure 16.9).

The N termini of the Gag and Gag–Pol proteins become anchored to the plasma membrane by the myristyl groups, and the association is stabilized through electrostatic interactions between positive charges in the MA domains and negatively charged phosphate groups in the membrane. The MA domains also bind to the cytoplasmic tails of TM proteins in the membrane.

The NC domains of Gag and Gag–Pol bind the polyproteins to the virus RNA and mediate the formation of the genome dimer. The proteins bind first to a packaging signal near the 5′ end of each

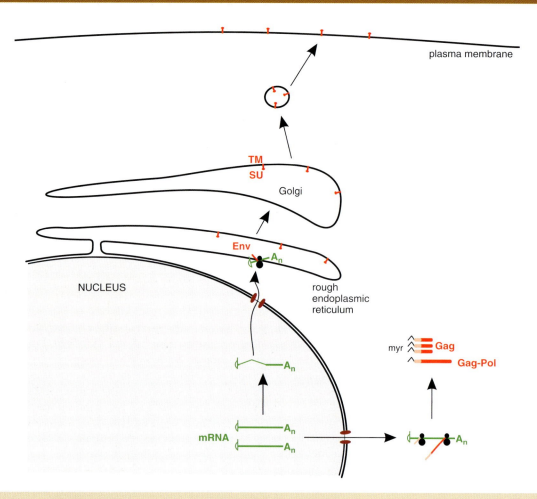

Figure 16.6 *Retrovirus translation and post-translational modifications.* Gag and Gag–Pol are translated on free ribosomes, and then myristylated (most retroviruses). Env is translated in the rough endoplasmic reticulum and transported to the plasma membrane via the Golgi complex; en route it is glycosylated and cleaved to SU and TM.

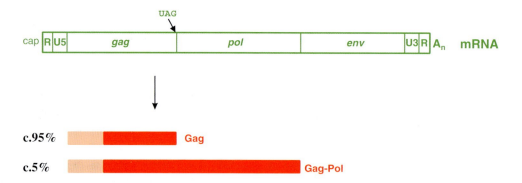

Figure 16.7 *Translation of Gag–Pol by reading through a stop codon (UAG).* Approximately five per cent of ribosomes incorporate an amino acid at the *gag* stop codon and translate *pol* too.

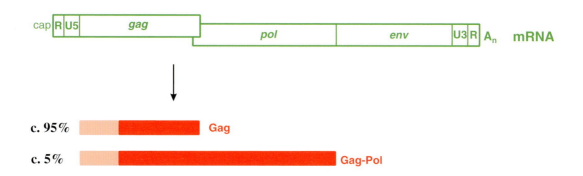

Figure 16.8 *Translation of Gag–Pol by ribosomal frameshifting.* Approximately five per cent of ribosomes shift into a different reading frame before the *gag* stop codon and translate *pol* too.

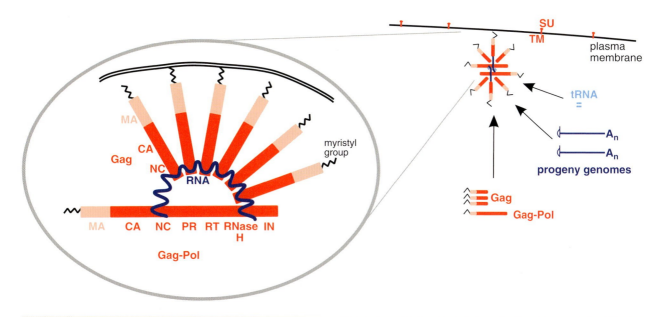

Figure 16.9 *Retrovirus assembly – early stages.* Two copies of the genome associate with cell tRNAs and with Gag and Gag–Pol proteins. The domains of Gag and Gag–Pol are indicated in the inset. The order of the Gag domains MA–CA–NC is the same as the exterior-to-interior order of the proteins in the virion (Figure 16.1(a)).

RNA molecule, and a tRNA binds to the PBS (Figure 16.1(b)). The RNA then becomes coated with many copies of Gag and a few copies of Gag–Pol.

The immature virion acquires its envelope by budding from the cell surface (Figure 16.10). At this stage multiple copies of Gag and Gag–Pol are arranged radially with their N termini facing outward and their C termini inward. The late (L) domains of Gag bind host cell factors that are involved in the budding process (Section 8.3.1). During and/or after the budding process the Gag and Gag–Pol polyproteins are cleaved by the virus protease. The cleavage products of Gag form the matrix, the capsid and the protein component of the nucleocapsid, while the Pol cleavage products are the virion enzymes.

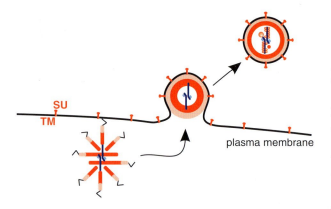

SU
TM

plasma membrane

Figure 16.10 *Retrovirus assembly – late stages.* The envelope is acquired by budding from the plasma membrane. During and after budding Gag and Gag–Pol are cleaved to form the virion proteins.

16.3.7 Overview of retrovirus replication

The retrovirus replication cycle is summarized in Figure16.11.

16.4 Examples of retroviruses

Retroviruses can be classed as either simple or complex, depending on the complexity of their genomes. The simple retroviruses are those that have only the three standard retrovirus genes (*gag*, *pol*, *env*), or in some cases one additional gene, called an oncogene because its expression might result in its host cell developing into a tumour cell. An example of an oncogene is *src* in the genome of Rous sarcoma virus, which infects chickens (Figure 16.12).

The complex retroviruses have additional genes, the products of which have a variety of functions in the replication cycle. The human immunodeficiency viruses are complex retroviruses; because of their importance the next chapter is devoted entirely to them.

The genera of the family *Retroviridae* and some representative viruses are listed in Table 16.3. As the names of the viruses imply, many of them of them are causative agents of disease in mammals (feline leukaemia virus), birds (Rous sarcoma virus) and fish (walleye dermal sarcoma virus).

16.5 Retroviruses as gene vectors

Some genetically modified retroviruses are used as gene vectors. They can introduce genes into a variety of cell types, where the genes are expressed at high levels after integration into the cell genomes. Technologies have been developed for expression of genes in cell cultures and for clinical treatments of genetic disorders and cancers. One of the most commonly used viruses in these applications is murine leukaemia virus. Lentiviral vectors have also been developed; these have the advantage that they can deliver genes into non-dividing cells and tissues.

Patients with the genetic disorder X-linked severe combined immunodeficiency (X-SCID) have been successfully treated with retroviral vectors. The procedure involves taking stem cells from the patient and infecting them with a recombinant retrovirus, the genome of which contains a good copy of the gene that is defective. If the vector provirus is integrated into the cell genome, and if the good copy of the gene is expressed, then the patient is able to develop a normal immune system. The successes, however, have been tempered by the development of cancer in a few treated patients.

16.6 Endogenous retroviruses

It has been known for some time that the genomes of vertebrate animals contain retroviral sequences. The genomes of most of these endogenous retroviruses (ERVs) are defective. Sequencing the human genome has revealed the presence of almost 100 000 human ERV (HERV) sequences, and ERVs have been found in the genomes of other species as they have been sequenced.

Some ERVs are closely related to normal retroviruses (exogenous retroviruses); for example, there are ERVs in mice that have very similar sequences to the genome of mouse mammary tumour virus. It is highly likely that ERVs originated as a result of exogenous retroviruses infecting germ line cells (sperm and/or egg). If one of these cells with an integrated provirus survived to be involved in the reproductive process, then each cell in the body of the offspring would contain a copy of the provirus. Over time ERVs have copied themselves to other locations in the genome, giving rise to families of related ERV elements.

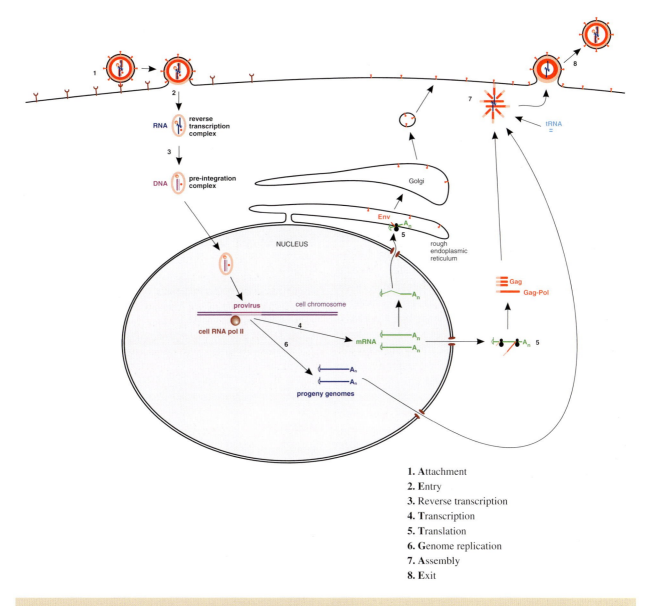

1. Attachment
2. Entry
3. Reverse transcription
4. Transcription
5. Translation
6. Genome replication
7. Assembly
8. Exit

Figure 16.11 *The retrovirus replication cycle.* Note that there is an additional step in retrovirus replication: reverse transcription, which takes place between entry and transcription.

Figure 16.12 *Rous sarcoma virus genome.* There is an oncogene (*src*) in addition to the three standard retrovirus genes.

Table 16.3 Examples of retroviruses

Simple retroviruses		Complex retroviruses	
Genus	Virus examples	Genus	Virus examples
Alpharetrovirus	Rous sarcoma virus	*Deltaretrovirus*	Human T-lymphotropic viruses 1 & 2
Betaretrovirus	Mouse mammary tumour virus	*Epsilonretrovirus*	Walleye dermal sarcoma virus
		Lentivirus	Human immunodeficiency virus 1
Gammaretrovirus	Murine leukaemia virus	*Spumavirus*	Chimpanzee foamy virus
	Feline leukaemia virus		

As stated above, most ERVs are defective, so they do not normally replicate. There are circumstances, however, when some ERVs can replicate. Missing functions may be supplied by another ERV or an exogenous retrovirus. Some ERVs do not replicate in cells of the species in which they occur, but are able to replicate in the cells of other species; e.g., some mouse ERVs and some pig ERVs can replicate in human cells. There are also some ERVs that are not defective; they have an intact genome (*gag*, *pol* and *env* genes) and can initiate a productive infection.

Because of these findings concern has been expressed that there may be a risk of transmitting retroviruses from pigs into humans if pigs are used as sources of cells, tissues and organs because of shortages of their human counterparts for transplant purposes.

Learning outcomes

By the end of this chapter you should be able to

- describe the retrovirus virion;

- describe the main features of the retrovirus genome;

- explain the main features of the retrovirus replication cycle;

- give examples of retroviruses and explain their importance;

- discuss endogenous retroviruses.

Sources of further information

Books

Flint S. J. *et al.* (2004) Chapter 7 in *Principles of Virology: Molecular Biology, Pathogenesis and Control of Animal Viruses*, 2nd edn, ASM Press

Knipe D. M. and Howley P. M. (2001) Chapter 27 in *Fundamental Virology*, 4th edn, Lippincott, Williams and Wilkins

Lewin B. (2004) Chapter 17 in *Genes VIII*, Pearson Prentice-Hall

Journals

Baillie G. J. *et al.* (2004) Multiple groups of endogenous betaretroviruses in mice, rats, and other mammals *Journal of Virology*, **78**, 5784–5798

Baum C. *et al.* (2006) Retrovirus vectors: toward the plentivirus? *Molecular Therapy*, **13**, 1050–1063

Bushman F. *et al.* (2005) Genome-wide analysis of retroviral DNA integration *Nature Reviews Microbiology*, **3**, 848–858

D'Souza V. and Summers M. F. (2005) How retroviruses select their genomes *Nature Reviews Microbiology*, **3**, 643–655

Muriaux D. *et al.* (2001) RNA is a structural element in retrovirus particles *Proceedings of the National Academy of Sciences USA*, **98**, 5246–5251

17

Human immunodeficiency viruses

At a glance

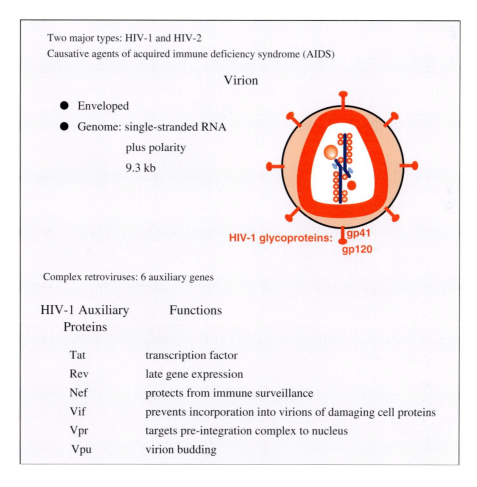

Two major types: HIV-1 and HIV-2
Causative agents of acquired immune deficiency syndrome (AIDS)

Virion

- Enveloped
- Genome: single-stranded RNA
  plus polarity
  9.3 kb

HIV-1 glycoproteins: gp41 gp120

Complex retroviruses: 6 auxiliary genes

HIV-1 Auxiliary Proteins	Functions
Tat	transcription factor
Rev	late gene expression
Nef	protects from immune surveillance
Vif	prevents incorporation into virions of damaging cell proteins
Vpr	targets pre-integration complex to nucleus
Vpu	virion budding

Virology: Principles and Applications   John B. Carter and Venetia A. Saunders
© 2007 John Wiley & Sons, Ltd   ISBNs: 978-0-470-02386-0 (HB); 978-0-470-02387-7 (PB)

17.1   Introduction to HIV

There are two types of human immunodeficiency virus (HIV-1 and HIV-2), which each evolved from a different simian immunodeficiency virus (SIV). Both viruses emerged in the late 20th century. In contrast to the SIVs, which appear not to harm their natural primate hosts, HIV infection damages the immune system, leaving the body susceptible to infection with a wide range of bacteria, viruses, fungi and protozoa. This condition is called acquired immune deficiency syndrome (AIDS). It should be noted that the word 'virus' in the phrase 'HIV virus' is superfluous and that scientists use the abbreviation 'AIDS', not 'Aids'!

HIV-1 is much more prevalent than HIV-2; it is HIV-1 that is largely responsible for the AIDS pandemic, while HIV-2 is mainly restricted to West Africa. Now, in each year of the early 21st century there are approximately 5 million new HIV infections, and approximately 3 million deaths from AIDS, which has become the fourth biggest cause of mortality in the world. The magnitude of this problem has resulted in the allocation of huge resources to the study of these viruses, major objectives being the development of anti-viral drugs and a vaccine. So far there has been qualified success in achieving the first of these objectives. The emphasis of this chapter is on HIV-1, as it has been studied more intensively than HIV-2.

17.2   HIV virion

The virion has the general characteristics of retroviruses described in the previous chapter but, in contrast to most retroviruses, the capsid is cone shaped (Figure 17.1), with a diameter of 40–60 nm at the wide end and about 20 nm at the narrow end. Generally, there is one capsid per virion, though virions with two or more capsids have been reported.

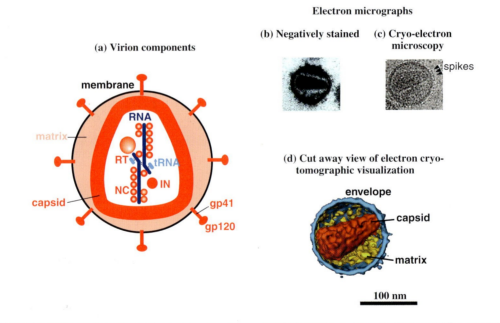

Figure 17.1 *HIV virion.* (a) Virion components. IN: integrase. NC: nucleocapsid protein. RT: reverse transcriptase. The TM and SU glycoproteins indicated are those of HIV-1 (gp41 and gp120). (b) Courtesy of the University of Otago, New Zealand. (c) From Briggs *et al.* (2003) *The EMBO Journal*, **22**, 1707. Reproduced by permission of Nature Publishing Group and the author. (d) From Grünewald and Cyrklaff (2006) *Current Opinion in Microbiology*, **9**, 437. Reproduced by permission of Elsevier Limited and the authors.

The diameter of the HIV virion measured in negatively stained preparations is in the range 80–110 nm, while results from cryo-electron microscopy are at the upper end of this range or greater.

The NC protein of HIV-1 is a typical retrovirus NC protein, being highly basic (29 per cent of the amino acid residues are basic) and having zinc fingers (Figure 17.2). The TM and SU proteins have approximate molecular weights of 41 and 120 kD, respectively, and are named gp41 and gp120 (gp = glycoprotein); gp120 is heavily glycosylated (Figure 17.2). The C

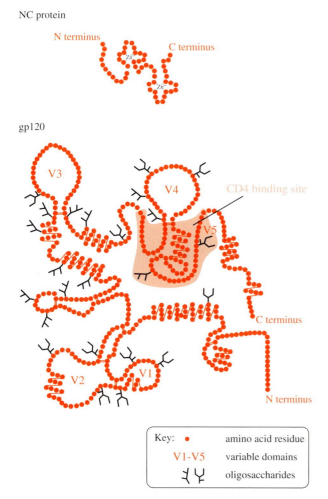

Figure 17.2 *Structures of HIV-1 NC protein and gp120. The NC protein has two zinc fingers. gp120 has five domains that are highly variable (V1-V5).*

terminus of gp41 is inside the virion, where it is bound to the MA protein. Spikes can be seen at the surface of virions in electron micrographs (Figure 17.1(b), (c)). Each spike is a gp41–gp120 trimer and there is an average of 14 spikes on each virion. The equivalent glycoproteins in HIV-2 (gp38 and gp130) are unrelated to those of HIV-1, whereas most of the internal proteins of the two viruses are related.

As well as the standard retrovirus proteins, the HIV-1 virion also contains the following virus proteins: Nef, Vpr, Vif, p1, p2, p6 and p6*. The presence of host proteins has also been reported, including major histocompatibility complex class II proteins associated with the envelope, and cyclophilin A associated with the capsid.

17.3 HIV genome

HIV-1 and HIV-2 have genomes about 9.3 kb in length. The genomes encode auxiliary genes in addition to *gag*, *pol* and *env*, and so the viruses are classed as complex retroviruses. The auxiliary genes have many roles in controlling virus gene expression, transporting virus components within the cell and modifying the host's immune response. Some of the auxiliary gene products have multiple roles.

The organization of the HIV-1 genome is shown in Figure 17.3. All three reading frames are used and there is extensive overlapping; e.g., part of *vpu* in frame 2 overlaps *env* in frame 3. The sequences for *tat* and *rev* are split, the functional sequences being formed when the transcripts are spliced.

HIV-2 has similar genes to HIV-1, except that it has no *vpu* gene, but it has a *vpx* gene which is related to *vpr*.

17.4 HIV-1 replication

A general description of the retrovirus replication cycle was given in the previous chapter. Here we shall concentrate on details specific to HIV-1.

17.4.1 Attachment and entry

The cell receptor for HIV-1 is CD4 (Figure 17.4), which is found on several cell types, including helper T cells and some macrophages; CD4 T cells are the main target cells. Attachment of the virion occurs when a site on gp120 (Figure 17.2) recognizes a site on the outer domain of CD4.

As well as binding to receptor molecules an HIV-1 virion must also attach to a co-receptor on the cell surface. The molecules that act as co-receptors have seven transmembrane domains and are chemokine receptors. During immune responses they bind chemokines and these interactions control leukocyte trafficking and T cell differentiation. Most chemokines fall into one of two major classes, determined by the arrangement of cysteine residues near the N terminus: C–C and C–X–C, where C =

cysteine and X = any amino acid. The chemokine receptors are designated CCR and CXCR, respectively. A number of these molecules on T cells act as co-receptors for HIV-1, particularly CCR5 and CXCR4 (Table 17.1).

Most HIV-1 strains use CCR5 and are known as R5 strains. It is interesting to note that in some individuals who have had multiple exposures to the virus, but have not become infected, there is a 32-nucleotide deletion in the *CCR5* gene. Individuals who are homozygous for this mutation express no CCR5 on their cells and are highly resistant to infection with HIV-1, while those who are heterozygous have increased resistance. The mutation is found mainly in Europeans.

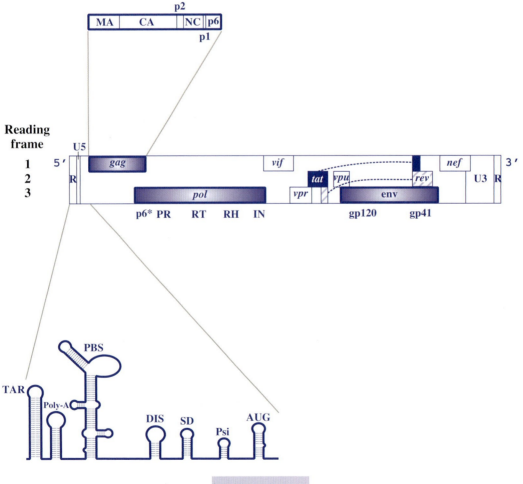

Figure 17.3

HIV-1 strains that use CXCR4 as a co-receptor are known as X4 strains, and there are some strains (R5X4 strains) that can use either co-receptor. R5 strains do not infect naïve T cells, but all three strains infect memory T cells.

The interaction of gp120 with the receptor and co-receptor results in a dramatic re-arrangement of gp41, which proceeds to fuse the membranes of the virion and the cell. The contents of the virion envelope are released into the cytoplasm and develop into the reverse transcription complex, which contains the MA, Vpr, RT and IN proteins, as well as the virus genome.

17.4.2 Reverse transcription and transport to the nucleus

The reverse transcription complex associates rapidly with microtubules (Figure 17.5). Reverse transcription is primed by tRNA^{lys-3}, and proceeds as outlined in Section 16.3.2.

Whereas the proviruses of most retroviruses are entirely dsDNA, those of HIV and other lentiviruses have a short triple-stranded sequence known as a central DNA flap. This comes about because, as well as the polypurine tract (PPT) towards the 3' end of the virus genome, there is also a central PPT that acts as a second initiation site of (+) DNA synthesis. Synthesis of the (+) DNA initiated at the 3' PPT stops soon after reaching the (+) DNA initiated at the central PPT, resulting in a short overlapping ssDNA. This DNA flap plays a vital role in the early stages of infection.

After reverse transcription has been completed, the pre-integration complex, which contains host proteins as well as virus proteins, is moved along microtubules towards the nucleus. As discussed in Chapter 16,

Figure 17.3 *HIV-1 genome organization*

Main genes	*gag*	group-specific antigen (encodes matrix, capsid, p2, nucleocapsid, p1 and p6)
	pol	polymerase (encodes p6*, protease, reverse transcriptase, RNase H, integrase)
	env	envelope
Auxiliary genes	*nef*	negative regulatory factor
	rev	regulator of expression of virion proteins
	tat	transactivator of transcription
	vif	virion infectivity factor
	vpr	viral protein R
	vpu	viral protein U
Non-coding sequences	R	repeat sequence
	U3	unique sequence at 3' end of genome
	U5	unique sequence at 5' end of genome
Domains at the 5' end of the genome	TAR	trans-acting response element
	Poly-A	polyadenylation signal
	PBS	primer-binding site
	DIS	dimerization initiation site (involved in formation of kissing loop complex)
	SD	splice donor site
	Psi (ψ)	main part of the packaging signal
	AUG	start codon of the *gag* gene

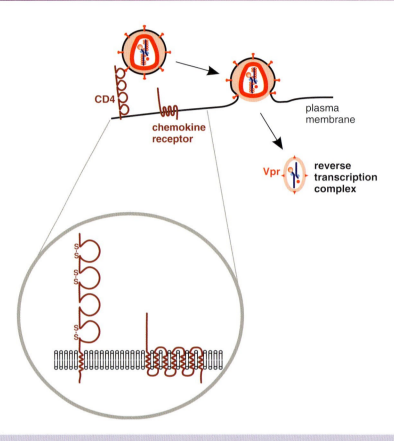

Figure 17.4 *HIV-1 attachment and entry.* The receptor (CD4) is a member of the immunoglobulin superfamily of molecules. Each of the loops represents an immunoglobulin-like domain; three of the four domains are stabilized by disulphide bonds. The co-receptor is a chemokine receptor.

Table 17.1 Co-receptors for HIV-1 strains and categories of T cell infected. Please see Chapter 9 for explanation of the different classes of T cell

HIV-1 strain	Main co-receptor(s) used		CD4 T cells infected	
	CCR5	CXCR4	Naïve cells	Memory cells
R5	+	−	−	+
R5X4	+	+	+	+
X4	−	+	+	+

most retroviruses can productively infect only if there is breakdown of the nuclear membranes. The pre-integration complex of HIV, however, can enter an intact nucleus, such as that of a resting T cell or a macrophage, and is presumably transported through a nuclear pore. Nuclear localization signals have been identified in the following pre-integration complex components: MA, Vpr and IN.

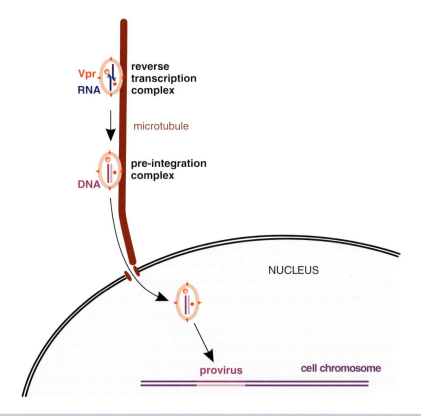

Figure 17.5 *HIV-1 reverse transcription and integration of the provirus.* The pre-integration complex is transported on a microtubule to a nuclear pore.

There is evidence that integration of the provirus in a resting memory CD4 T cell may result in a latent infection. These cells can provide a reservoir of infection that is significant for the survival of the virus in individuals receiving anti-retroviral drug therapy. In many cells, though, provirus integration is the prelude to a productive infection in which two phases of gene expression can be distinguished.

17.4.3 Early gene expression

Transcription is initiated after cell transcription factors bind to promoter and enhancer sequences in the U3 region of the upstream LTR. NF-κB plays a key role, and the LTR has binding sites for other transcription factors, including AP-1 and Sp-1. Transcription is terminated in the downstream LTR; the polyadenylation signal AATAAA is in the R region and transcripts are polyadenylated at the R–U5 junction.

Many of the genome-length transcripts are spliced and three size classes of virus transcript can be detected in infected cells using northern blotting (Figure 17.6). The largest RNAs are genome length (about 9.3 kb), while the other two size classes are each made up of a number of mRNA species that have undergone splicing; mRNAs that have been spliced once are around 4.5 kb, while mRNAs that have undergone two or more splicing events ('multiply spliced' transcripts) are around 2 kb. The virus genome has a number of splice donor sites (one is indicated in Figure 17.3) and acceptor sites; these enable splicing events that result in more than 30 mRNA species.

Early in infection most of the primary transcripts are multiply spliced and these RNAs are translated into the Nef, Tat and Rev proteins.

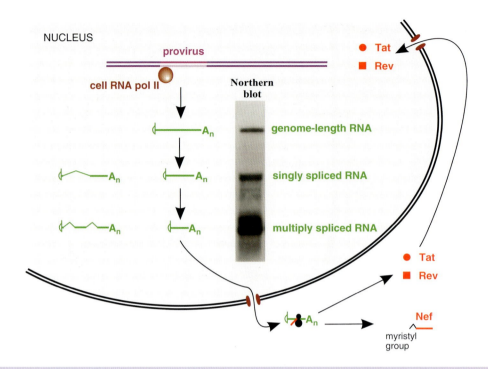

Figure 17.6 *HIV-1 early gene expression.* Genome-length RNA is transcribed then much of it is spliced, giving rise to two further size classes of RNA that can be detected in northern blots of RNA from infected cells. Early in infection most of the RNA is multiply spliced and is transported to the cytoplasm, where the Tat, Rev and Nef proteins are translated. Nef is myristylated and performs a number of roles in the cytoplasm, while Tat and Rev are transported to the nucleus. Northern blot from Malim M.H. *et al.* (1990) *Cell,* **60,** 675; reproduced by permission of Elsevier Limited and the authors.

17.4.3.a Roles of Nef, Tat and Rev

The Nef (Negative regulatory factor) protein acquired its name because it was originally thought to have an inhibitory effect on HIV replication, though later work showed that this protein stimulates replication! In infected cells Nef alters the endosome trafficking pathway, reducing expression at the cell surface of CD4 and MHC class I and II proteins. These changes can shield HIV-infected cells from immune surveillance.

The roles of the Tat and Rev proteins are summarized in Figure 17.7.

The Tat (*T*ransactivator of *t*ranscription) protein plays an important role in enhancing transcription. A nuclear localization signal (Figure 17.8) directs Tat to the nucleus, where it binds to a sequence at the 5′ end of nascent virus transcripts: this sequence is known as the transactivation response (TAR) element

(Figures 17.3 and 17.8). Cell proteins also bind to TAR, and among these proteins is a kinase, which phosphorylates components of the RNA polymerase complex. Phosphorylation increases the processivity of the enzyme along the proviral template. Tat therefore functions as a transcription factor, but an unusual one, in that it binds not to DNA, but to RNA. In the absence of Tat most transcripts are incomplete, though early in infection sufficient are completed to allow the synthesis of a small amount of Tat, which then significantly boosts the synthesis of genome-length RNA.

The other early protein, the Rev (*R*egulator of *e*xpression of *v*irion proteins) protein, has a nuclear localization signal (Figure 17.8). As Rev accumulates in the nucleus it causes a shift from early to late protein synthesis by binding to the Rev response element (RRE) in the virus RNA. The RRE is present in the

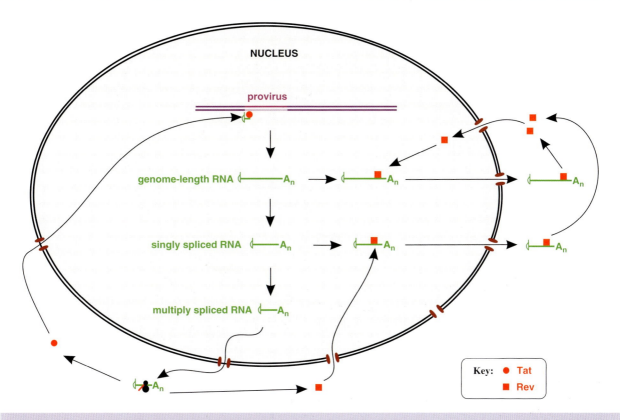

NUCLEUS

provirus

genome-length RNA

singly spliced RNA

multiply spliced RNA

Key: ● Tat
■ Rev

Figure 17.7 *Roles of Tat and Rev.* Tat binds to nascent transcripts and helps to ensure that the entire virus genome is transcribed. Rev binds to genome-length RNA and singly spliced RNA and aids their transport to the cytoplasm, where the late proteins are translated. Rev is recycled to the nucleus.

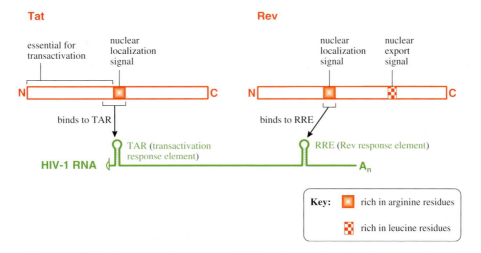

Tat

essential for transactivation

nuclear localization signal

N ☐ C

binds to TAR

Rev

nuclear localization signal

nuclear export signal

N ☐ C

binds to RRE

TAR (transactivation response element)

RRE (Rev response element)

HIV-1 RNA

A_n

Key: ▨ rich in arginine residues

▨ rich in leucine residues

Figure 17.8 *HIV-1 Tat and Rev proteins.* Features of the proteins and their binding sites in the virus RNA are indicated. The TAR and RRE regions of the RNA have complex secondary structures. The RRE is present in genome-length RNA and the singly spliced RNAs, but it is absent from the multiply spliced RNAs.

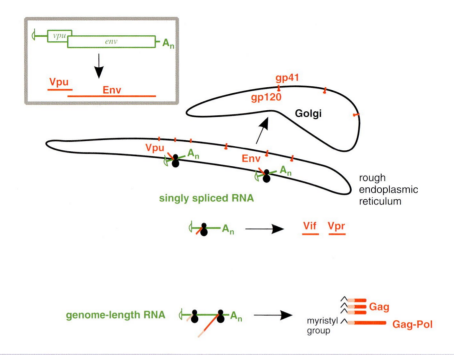

Figure 17.9 *HIV-1 late gene expression.* Vpu and Env are translated from singly spliced RNAs in the rough endoplasmic reticulum. The inset shows translation of Vpu and Env from a bicistronic mRNA. Env is synthesized when the *vpu* start codon is by-passed during leaky scanning. The remaining proteins are translated on free ribosomes: Vif and Vpr from singly spliced RNAs, and Gag and Gag–Pol from genome-length RNAs.

unspliced and singly spliced transcripts, but is absent from the multiply spliced transcripts.

The late genes are translated from genome-length and singly spliced transcripts, but these mRNAs are not transported from the nucleus until they have bound multiple copies of Rev.

17.4.4 Late gene expression

Translation of the late proteins is shown in Figure 17.9.

Gag and Gag–Pol are translated from unspliced transcripts, with Gag–Pol translated when a ribosomal frameshift takes place (Section 16.3.5). This occurs on roughly five per cent of occasions when a ribosome traverses the sequence UUUUUA at the junction of the NC and p1 domains of *gag* (Figure 17.3). This sequence (known as a slippery sequence), together with a downstream secondary structure, causes the ribosome to slip from reading frame 1 to reading frame 3 (Figure 17.10). The slippery sequence is

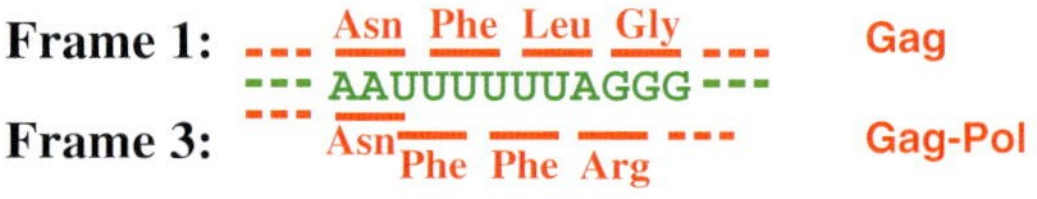

Frame 1:	--- Asn Phe Leu Gly ---	Gag
	--- AAUUUUUUAGGG ---	
Frame 3:	Asn —— Phe Phe Arg ---	Gag-Pol

Figure 17.10 *Expression of HIV-1 Gag–Pol by ribosomal frameshifting.* A ribosome reading in frame 1 shifts at the slippery sequence UUUUUUA to reading in frame 3.

reminiscent of the 7-U sequence, over which the RNA polymerase 'stutters', at the end of each rhabdovirus gene (Section 15.4.2). After the frameshift has taken place translation continues through the *pol* region, yielding the Gag–Pol polyprotein.

The remaining virus proteins (Vif, Vpr, Vpu and Env) are translated from singly spliced transcripts. Vpu and Env are translated in the rough endoplasmic reticulum from a bicistronic transcript (Figure 17.9). Env becomes heavily glycosylated, acquiring an apparent molecular weight of 160 kD. Trimers of Env are formed before the molecules are cleaved to form the envelope proteins gp120 and gp41; the cleavage is carried out by furin, a host protease located in the Golgi complex. Vpu is a membrane-associated protein and is required for efficient budding of virions from the plasma membrane.

Some of the virus proteins are phosphorylated, including Vpu and the MA region of some Gag molecules.

17.4.5 Assembly and exit of virions

General aspects of retrovirus virion assembly were described in Section 16.3.6. Here we discuss some aspects specific to the assembly of HIV-1 virions; these are summarized in Figure 17.11.

Formation of the RNA dimer that will constitute the genome of a new virion commences by base pairing between complementary sequences in the loop of the dimerization initiation site near the 5′ end of each RNA (Figure 17.3). It is thought that this leads to the formation of a 'kissing-loop complex' that stabilizes the dimer.

Molecules of Gag and Gag–Pol form an orderly arrangement, and their domains bind to the virus genome and to other proteins that will become incorporated into the virion. The basic NC domains with their zinc fingers (Figure 17.2) bind to the virus genome, initially to a domain known as ψ (Figure 17.3), which is

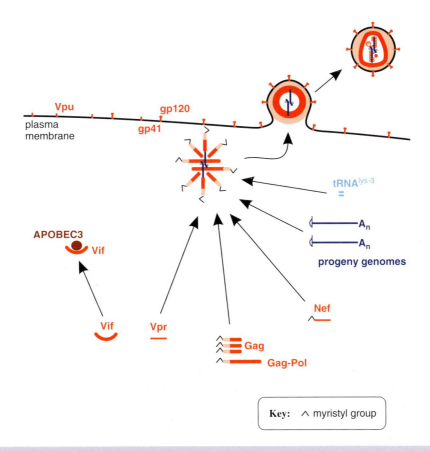

Figure 17.11 *Assembly of HIV-1 virions.* All of the virus proteins in the diagram are incorporated into new virions, along with two copies of the virus genome and cell tRNA^{lys-3}. The envelope is acquired by budding from the plasma membrane. Vif binds to APOBEC3 cell proteins and prevents their incorporation into virions.

the main part of the packaging signal. The CA domains bind the host protein cyclophilin A, while p6 domains bind the virus protein Vpr.

The p6 domain functions as the late (L) domain, responsible for the release of the budding virion from the host cell (Section 8.3.1). This domain contains the sequence proline–threonine–alanine–proline (PTAP),

which can bind cell proteins thought to be important in the pinching off process.

Some molecules of Vif protein are incorporated into virions, but Vif also plays an important role in ensuring the exclusion from progeny virions of cell enzymes (APOBEC3F and APOBEC3G) that could interfere with replication in the next host cell (Section 9.2.1.c).

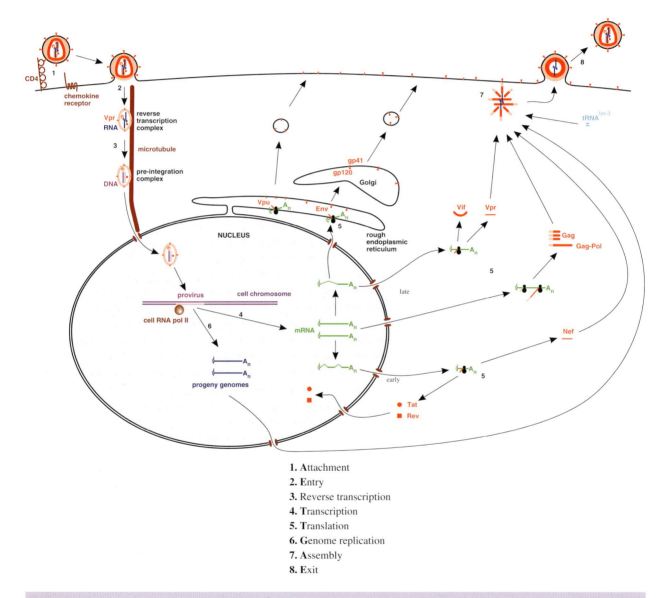

1. **A**ttachment
2. **E**ntry
3. **R**everse transcription
4. **T**ranscription
5. **T**ranslation
6. **G**enome replication
7. **A**ssembly
8. **E**xit

Figure 17.12 *The HIV-1 replication cycle.* Note the additional step in retrovirus replication: reverse transcription, which takes place between entry and transcription.

If *vif* is mutated these enzymes can be incorporated into virions and when reverse transcription is under way they can induce lethal mutations by deaminating deoxycytidine to deoxyuridine. Vif prevents the incorporation of these APOBEC3 proteins into virions by binding to them and inducing their degradation.

Gag–Pol dimers are formed; these undergo self-cleavage to form the virus enzymes, including the protease, which is a dimer. The protease then cleaves the Gag polyproteins into the constituents of the mature virion.

17.4.6 Overview of HIV-1 replication

The HIV-1 replication cycle is summarized in Figure 17.12.

17.5 HIV-1 variability

During the replication of retrovirus genomes there is a high error rate as a result of the lack of a proofreading mechanism (see Section 20.2.2.1). HIV-1 appears to be a particularly variable virus, having evolved into a number of groups and subtypes. The variability of HIV-1 is manifest in characters such as the antigens, the host cell range and resistance to drugs.

17.5.1 Antigens

HIV-1 antigens show a high degree of variability. The surface protein gp120 is one of the most variable, in spite of the constraint imposed by the overlap between the *env* and *vpu* genes (Figure 17.3). There are five domains of gp120 that are especially variable (Figure 17.2). The exceptionally high variability of gp120 presumably results from evolutionary pressure exerted by the immune response of the host. Interestingly, the Nef protein, which is not on the virion surface, is also very highly variable.

17.5.2 Host cell range

The existence of HIV-1 strains with different co-receptor preferences was mentioned in Section 17.4.1. Transmission of HIV-1 to a new host is almost always associated with R5 strains, and these predominate during the acute and asymptomatic phases of infection. In about 50 per cent of infected individuals there

is evolution towards X4 and R5X4 strains as AIDS develops.

17.5.3 Resistance to drugs

The presence of anti-retroviral drugs in the body of an infected host exerts an evolutionary pressure on the virus and drug-resistant variants can quickly emerge. HIV-1 drug resistance is discussed further in Chapter 25.

17.6 Progression of HIV infection

Shortly after a person becomes infected with HIV there is a huge rise in viraemia (concentration of virus in the blood; Figure 17.13), and in some people there is an illness resembling glandular fever or influenza. The host's immune responses then control virus replication to some extent and there is a period of asymptomatic infection. In the absence of intervention with drugs this period typically lasts for 8–10 years, but it may be significantly shorter or longer than this, depending on characteristics of both the host and the virus.

Extensive virus replication continues throughout the asymptomatic period, with estimates of more than 10^{10} HIV-1 virions produced each day. The infection persists in spite of immune responses against the virus and there are several reasons for this. One reason is that the cell types killed by HIV infection (CD4 T cells and macrophages) are those involved in the immune responses; there is also evidence that non-infected CD4 T cells are killed by apoptosis. CD4 T cells play pivotal roles as helpers for several cell types including B cells, cytotoxic T cell precursors, natural killer cells and macrophages, hence immune responses are impaired.

Another reason why an HIV infection is not cleared is that the virus evolves as the infection proceeds, producing new antigenic variants that may not be recognized by the antibodies and T cells present. Furthermore, in latently infected cells the virus is shielded from the immune system.

In about 50 per cent of infected individuals X4 and R5X4 strains of HIV-1 emerge during the asymptomatic period and the co-receptor preference shifts from CCR5 to CXCR4 (Section 17.5.2). Individuals in whom X4 strains emerge at an early stage are likely to progress more rapidly to AIDS.

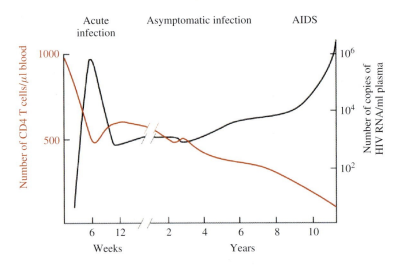

Figure 17.13 *Relative levels of viraemia and CD4 T cells at stages of HIV infection.* Shortly after infection the concentration of HIV in the blood rises to a high level, then it falls off and relatively low levels are detectable throughout the asymptomatic period. A rise in viraemia heralds the onset of AIDS.

For a while, the body is able to tolerate the onslaught on the immune system by rapidly replacing most of the cells that have been destroyed, but the concentration of CD4 T cells in the blood steadily declines until a point is reached where the level of viraemia rises sharply and AIDS develops (Figure 17.13). AIDS is characterized by infections with pathogenic micro-organisms; brain disease and/or cancer may also develop. The most common types of cancer are those that are associated with viruses, such as Kaposi's sarcoma and non-Hodgkin's lymphoma (Chapter 22).

In comparison with HIV-1, HIV-2 infections are associated with longer asymptomatic periods, slower progression of disease and lower rates of transmission.

17.7 Prevention of HIV transmission

As well as the measures that individuals can take to prevent transmission of HIV, there are also measures that society can take using a variety of tools and procedures, some of them developed by virologists. It is important that potential donors of blood, organs and semen are screened for HIV infection; this can be done by testing their blood for HIV-specific antibodies (Chapter 2). The preparation of blood products for haemophiliacs can include treatment with lipid solvents and detergents to destroy the virions of HIV (and other enveloped viruses, such as hepatitis B virus).

There is a risk of transmitting HIV, and several other viruses (e.g. hepatitis B and C viruses), if syringes and needles are used to inject more than one person. The use of 'auto-disable' syringes ensures that this cannot happen. This simple, but ingenious, invention results in the syringe plunger breaking if an attempt is made to use the syringe more than once.

The risk of HIV transmission from a mother to her child is between about 15 and 40 per cent, the higher levels of risk being associated with breast-feeding. This risk can be greatly reduced by anti-retroviral drug treatment of the woman before and after birth. In many countries, including the UK, treatment of HIV-infected women has resulted in a decline in the number of HIV-positive children. Although drug treatment reduces the risk of virus transmission to children, it does not cure infected individuals.

Anti-retroviral drugs are given as post-exposure prophylaxis to protect individuals from HIV infection following needle stick injuries and risky sexual activities.

The use of drugs against HIV is discussed in detail in Chapter 25.

A major goal of HIV research is the development of an effective vaccine. The deployment of such a vaccine could dramatically reduce transmission rates of the virus, but after the expenditure of much effort there is no sign on the horizon of an HIV vaccine suitable for mass immunization. There are a number of reasons for this, including the ability of the virus to rapidly evolve multiple antigenic variants as a result of its high mutation rate (Section 17.5).

Learning outcomes

By the end of this chapter you should be able to

- explain the importance of HIV;

- describe, with the aid of a labelled diagram, the HIV-1 virion;

- describe the HIV-1 genome;

- write an illustrated account of the replication cycle of HIV-1;

- discuss the variability of HIV-1;

- discuss the effects of HIV infection on the host;

- evaluate approaches to the prevention of HIV transmission.

Sources of further information

Books

Richman D. D., editor (2003) *Human Immunodeficiency Virus*, International Medical Press

Turnbull E. and Borrow P. (2005) The immune response to human immunodeficiency virus type 1 (HIV-1) pp. 23–89 in *Molecular Pathogenesis of Virus Infections*, editors Digard P., Nash A. A. and Randall R. E., 64th Symposium of the Society for General Microbiology, Cambridge University Press

Journals

Adamson C. S. and Jones I. M. (2004) The molecular basis of HIV capsid assembly – five years of progress *Reviews in Medical Virology,* **14**, 107–121

Barboric M. and Peterlin B. M. (2005) A new paradigm in eukaryotic biology: HIV Tat and the control of transcriptional elongation *PLoS Biology*, **3**, e76

Persaud D. *et al.* (2003) Latency in human immunodeficiency virus type 1 infection: no easy answers *Journal of Virology*, **77**, 1659–1665

Sierra S., Kupfer B. and Kaiser R. (2005) Basics of the virology of HIV-1 and its replication *Journal of Clinical Virology*, **34**, 233–244

Trkola A. (2004) HIV–host interactions: vital to the virus and key to its inhibition *Current Opinion in Microbiology*, **7**, 555–559

Zheng Y. H., Lovsina N. and Peterlin B. M. (2005) Newly identified host factors modulate HIV replication *Immunology Letters*, **97**, 225–234

Zhu P. *et al.* (2006) Distribution and three-dimensional structure of AIDS virus envelope spikes *Nature*, **441**, 847–852

18

Hepadnaviruses (and other reverse-transcribing DNA viruses)

At a glance

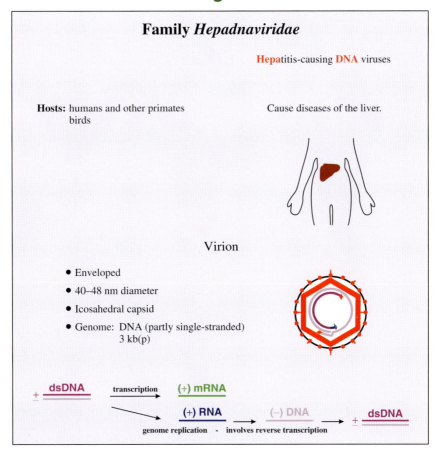

Family *Hepadnaviridae*

Hepatitis-causing **DNA** viruses

Hosts: humans and other primates
birds

Cause diseases of the liver.

Virion

- Enveloped
- 40–48 nm diameter
- Icosahedral capsid
- Genome: DNA (partly single-stranded)
  3 kb(p)

<u>dsDNA</u> —transcription→ <u>(+) mRNA</u>

→ <u>(+) RNA</u> → <u>(−) DNA</u> → <u>dsDNA</u>

genome replication - involves reverse transcription

Virology: Principles and Applications   John B. Carter and Venetia A. Saunders
© 2007 John Wiley & Sons, Ltd   ISBNs: 978-0-470-02386-0 (HB); 978-0-470-02387-7 (PB)

18.1 Introduction to hepadnaviruses

The hepadnaviruses got their name because they cause *hepa*titis and they have *DNA* genomes. They are known as hepatitis B viruses (HBVs) and are classified in the family *Hepadnaviridae*. Some members infect mammals and some infect birds; examples include woodchuck HBV and heron HBV. The best-known hepadnavirus is that which infects humans; it is commonly referred to as HBV, and is of major importance as an agent of disease and death. Duck HBV, on the other hand, is non-pathogenic in its natural host.

The hepadnaviruses are especially fascinating for two reasons. First they have very small genomes, which are used with great economy to encode the virus proteins and to control expression of the virus genes. Second, their DNA genomes are replicated via an RNA intermediate. In other words, their replication involves reverse transcription, so they are very different from DNA viruses that replicate their DNA directly to DNA. The discovery of the mode of replication of the hepadnavirus genome led to the creation of Baltimore class VII.

DNA viruses that replicate via RNA have also been found in plants; an example is cauliflower mosaic virus. These viruses, together with the hepadnaviruses, have been termed pararetroviruses. This chapter will concentrate on HBV.

18.2 Importance of HBV

No-one knows how many people are infected with HBV, but the figure is probably around 400 million.

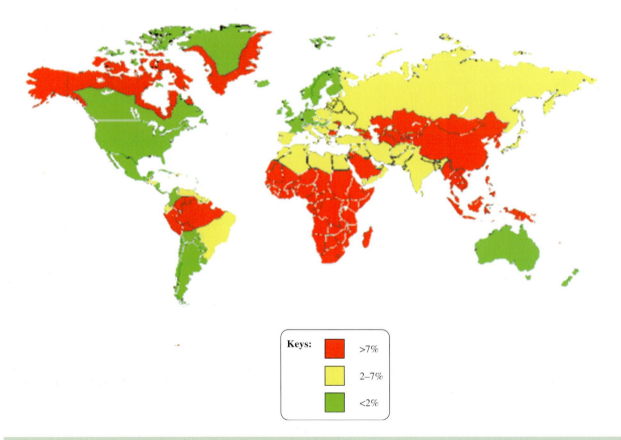

Figure 18.1 *World distribution of HBV infection.* Reproduced by permission of the US Centers for Disease Control and Prevention.

Keys:
- >7%
- 2–7%
- <2%

The majority are in Asia and many are in Africa (Figure 18.1). There are high percentages of people infected in the far north of North America and Greenland, but because of the low populations of these regions the numbers infected are relatively low.

Virus is present in the blood and semen of infected individuals and the modes of transmission generally parallel those for HIV transmission. There are over 50 million new HBV infections each year, the majority in babies who acquire the infection from their mothers. Over 8 million infections per year are thought to result from the re-use of syringes and needles for injections, mainly in the developing world.

Many HBV infections result in mild symptoms or are asymptomatic, especially in children. It is in children, however, that HBV infection is most likely to become persistent, with 90–95 per cent of those infected as newborn infants becoming long-term carriers, compared with 1–10 per cent of those becoming infected as adults.

Individuals who are persistently infected with HBV may remain healthy for much of the time, but some develop severe hepatitis, which may lead to cirrhosis and eventually to liver cancer (Chapter 22). These diseases resulting from HBV infection cause about one million deaths per year.

18.3 HBV virion

The virion is roughly spherical, with a diameter of about 42 nm. The main virion components are an envelope enclosing a capsid, inside which is the DNA and P (polymerase) protein (Figure 18.2).

18.3.1 DNA

The genome is made up of two strands of DNA, one of which is incomplete; hence the DNA is partly single-stranded and partly double-stranded. A short sequence is triple-stranded as a result of a complementary sequence at the 5′ ends, and this results in the

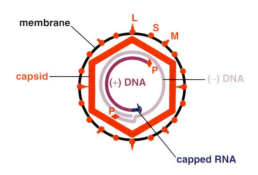

Figure 18.2 *The HBV virion.* S: small envelope protein. M: medium envelope protein. L: large envelope protein. P: polymerase (one molecule is covalently linked to the 5′ end of the (+) DNA; the virion may contain a second molecule of P, as indicated here).

DNA having a circular conformation (Figure 18.2). The genome is very small, with a length of about 3.2 kb(p).

At the 5′ end of each of the DNA strands there is a covalently linked molecule: a capped RNA on the short strand and a protein (P) on the long strand.

18.3.2 P (polymerase) protein

The virion contains at least one complete molecule of P (polymerase) protein. The N terminus of P constitutes a 'terminal protein' domain (Figure 18.3); it is not known whether the protein linked to the DNA is a complete molecule of P or only the terminal protein. The terminal protein domain is separated by a 'spacer' from a domain with reverse transcriptase activity. At the C terminus there is a domain with ribonuclease H (RNase H) activity. P also has DNA-dependent DNA polymerase activity.

18.3.3 Capsid

The capsid, which has icosahedral symmetry, has holes in it and short spikes protrude from its surface

Figure 18.3 *Domains of the HBV P (polymerase) protein.*

Figure 18.4 *HBV capsid.* Derived from cryo-electron microscopy images of capsids assembled in *E. coli* cells expressing HBV C protein. The bar represents 5 nm. From Watts *et al.* (2002) *The EMBO Journal*, **21**, 876. Reproduced by permission of Nature Publishing Group and the authors.

(Figure 18.4). It is constructed from dimers of the C (core) protein, which is largely α-helical, unlike the capsid proteins of many other viruses. The C terminus of the C protein is highly basic due to the presence of a large number of arginine residues; this region binds the virus genome.

18.3.4 Envelope

The virion envelope contains three protein species designated as small (S), medium (M) and large (L). The M and L proteins are longer versions of the S protein (Figure 18.5), which is the most abundant of the three. The surface regions of the envelope proteins constitute an antigen known as hepatitis B surface antigen (HBsAg).

Each envelope protein has one or more glycosylation sites, though not all molecules are glycosylated (Figure 18.5). The virus attachment site is near the N terminus of the L protein, but only about 50 per cent of the L molecules have the N terminus on the outside of the virion; the N termini of the remaining L molecules are on the inside bound to the capsid.

18.4   Non-infectious particles

An unusual and intriguing feature of HBV infection is the presence in the blood of not only virions, but also large quantities of non-infectious particles that have been released from infected liver cells (Figure 18.6). These particles are composed of lipid and virus envelope proteins, but they do not contain nucleocapsids. Some of the particles are spheres and some are filaments; both have diameters of 22 nm and the filaments have variable lengths up to 200 nm.

All the particles (virions and non-infectious particles) are much more abundant in the blood than in the liver, and the non-infectious particles, especially the spheres, vastly outnumber the virions. Presumably the virus has a good reason for inducing the production of such large amounts of non-infectious material, but that reason is not understood. It has been suggested that it might be a decoy for antibodies, thereby providing virions with some protection from the host's immune system.

18.5   Soluble virus protein

In addition to the virions and non-infectious particles a soluble virus protein is found in the blood of some infected individuals. This protein is known as hepatitis B e antigen (HBeAg). It is similar to the C protein, but has 10 extra amino acid residues at the N terminus and lacks 34 amino acid residues at the C terminus. The function of HBeAg, like that of the small spheres and filaments, is not known.

18.6   HBV genome

At 3.2 kb, the HBV genome is very small, though there are viruses with smaller genomes (Section 3.2.1). There are four ORFs, from which seven proteins are translated (Figure 18.7), so a large amount of coding information is packed into the small genome. The virus cleverly achieves this by using every nucleotide in the genome for protein coding and by reading more than half of the genome in two reading frames. The P ORF, which occupies about 80 per cent of the genome, overlaps the C and X ORFs and the entire S ORF

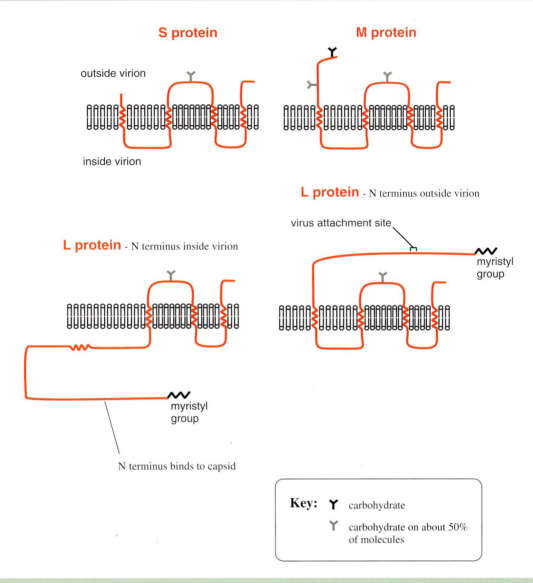

Figure 18.5 *Arrangement of HBV small (S), medium (M) and large (L) proteins in the virion envelope.* The N termini of about 50 per cent of L protein molecules are outside the membrane and about 50 per cent are inside.

is within the P ORF. A further way in which the virus maximizes its coding capacity is by expressing the L protein in two different conformations that have different functions. In one conformation L protein molecules act as virus attachment proteins, while in the other they bind the virion envelope to the capsid (Figure 18.5).

As the entire genome is involved in coding for protein, it follows that all the regulatory sequences, such as promoters, are within protein coding sequences. The genome contains direct repeats of 11 nucleotides known as DR1 and DR2.

Expression of the pre-S1–pre-S2–S region gives rise to three proteins. Translation of the S region produces the S protein, translation of pre-S2–S produces the M

Structure	Diameter (nm)	Typical concentration (μg/ml)
Virion	42	0.1
Non-infectious particles		
Sphere	22	100
Filament	22	1

Figure 18.6 *HBV structures present in the blood of an infected person.*

protein and translation of the complete ORF produces the L protein. Similarly, expression of the pre-C–C region gives rise to two proteins. Even though the virus maximizes the use of its small genome it encodes only seven proteins, so it is heavily dependent on host cell functions. Host proteins that have been demonstrated to play roles in the replication of HBV include enzymes, transcription factors and chaperones.

18.7 HBV genetic groups

Sequencing of many HBV isolates has revealed eight genetic groups (genotypes A–H) and these human viruses are related to similar viruses in other primate species (Figure 18.8).

The genetic groups are fairly restricted geographically, e.g. genotype A predominates in Northern Europe while genotypes B and C are in Asia.

18.8 HBV replication

Hepatocytes (liver cells) are the host cells for HBV in the body. In the laboratory, primary cell cultures of human hepatocytes support replication, but unfortunately none of the established cell lines derived from liver tumours can be infected by HBV virions. Some cell lines, however, can be infected using HBV DNA (a procedure known as transfection), though this does not permit study of the processes of entry and uncoating. Some hepadnaviruses, such as the woodchuck and

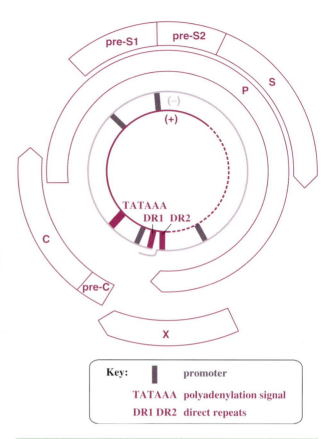

Key:	▮	promoter
	TATAAA	polyadenylation signal
	DR1 DR2	direct repeats

Figure 18.7 *HBV genome organization.* The incomplete (+) strand and the complete (−) strand of the DNA are shown surrounded by the four ORFs. The locations of the four promoters are indicated. S encodes the S protein. Pre-S2 + S encodes the M protein. Pre-S1 + pre-S2 + S encodes the L protein.

duck HBVs, replicate in cell lines. Much of our knowledge of hepadnavirus replication is therefore derived from studies with cell cultures, either transfected with human HBV DNA or infected with animal hepadnaviruses.

18.8.1 Attachment

The identity of the hepatocyte receptor to which HBV attaches is not yet known, though various candidates have been proposed, such as polymeric IgA and annexin V. The virus attachment site is on the L protein (Figure 18.9).

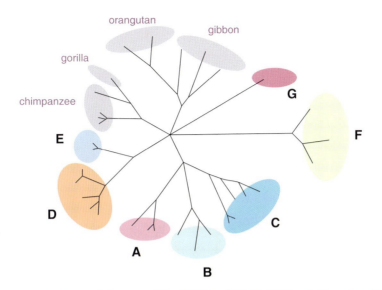

Figure 18.8 Phylogenetic tree showing relationships between HBV isolates from humans (genotypes A–G) and other primate species. From Kidd-Ljunggren, Miyakawa and Kidd (2002) *Journal of General Virology*, **83**, 1267. Reproduced by permission of the Society for General Microbiology and the authors. Since the publication of this phylogenetic tree a further HBV genotype (H) has been described.

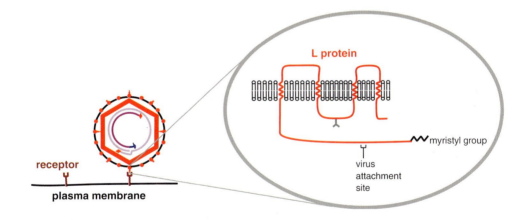

Figure 18.9 *Attachment of HBV virion to receptor on host cell.* The virus attachment site is on the L protein.

18.8.2 Entry

The virion is endocytosed then the nucleocapsid is released from the endosome by fusion of the virion and endosome membranes (Figure 18.10). The nucleocapsid is shown entering the nucleus, as evidence has been published that HBV capsids can pass through nuclear pores. It is not certain, however, whether this happens, or whether uncoating occurs within a nuclear pore, or whether the genome is released into the nucleus with the capsid remaining in the cytoplasm.

Once the virus genome is free in the nucleus it is converted into a circular DNA molecule (Figure 18.11). The covalently bound P protein is removed from the

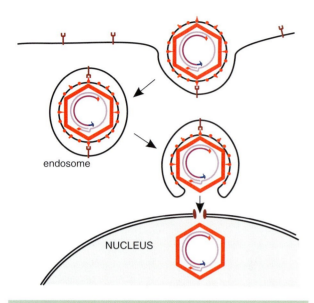

Figure 18.10 *Endocytosis of attached HBV virion followed by release of nucleocapsid and entry into the nucleus.*

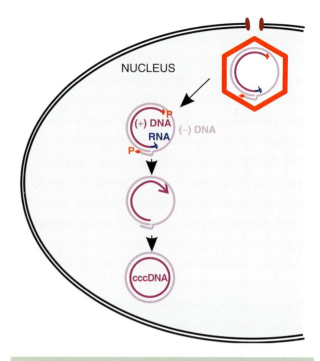

Figure 18.11 *Release of HBV genome from the capsid and conversion into cccDNA.*

5′ end of the minus strand, which is shortened to remove the third strand of the triple-stranded region. The RNA is removed from the 5′ end of the plus strand, while DNA synthesis at the 3′ end makes the entire molecule double-stranded. The ends of each strand are then ligated to form covalently closed circular DNA (cccDNA). It is uncertain whether all these modifications to the virus genome are carried out by host enzymes, or whether the virus P protein plays a role in some.

The virus DNA is not replicated in the nucleus, but more copies are brought into the nucleus later in the replication cycle.

18.8.3 Transcription

The cccDNA is the template for transcription. You will recall that the HBV genome has four promoters, sited upstream of the pre-S1, pre-S2, X and pre-C regions. Only the pre-S2 promoter has a TATA box. At least two of the promoters are highly specific to liver cells and it has been shown that some of the transcription factors involved are liver cell proteins. This is at least part of the explanation why HBV is specific to liver cells. Rates of transcription from

all four promoters are controlled by binding of cell transcription factors to two enhancer sequences present in the HBV genome. It is thought that the virus protein X is also a transcription factor, though it appears not to bind to DNA and so is probably not a typical transcription factor.

Transcription is carried out by the cell RNA polymerase II and results in the synthesis of four size classes of RNA (Figure 18.12). All the transcripts are capped at the 5′ end and polyadenylated at the 3′ end. They all use the same polyadenylation signal (TATAAA), so they have a common 3′ end.

Note that the 3.5 kb mRNAs are longer than the genome, which is 3.2 kb. During their synthesis part of the genome is transcribed twice, hence these RNAs have direct terminal repeats. In order to synthesize the 3.5 kb mRNAs the RNA polymerase must ignore the polyadenylation signal on the first pass. Studies with duck HBV indicate that this is controlled by a DNA sequence (called PET = positive effector of transcription), which somehow allows the RNA

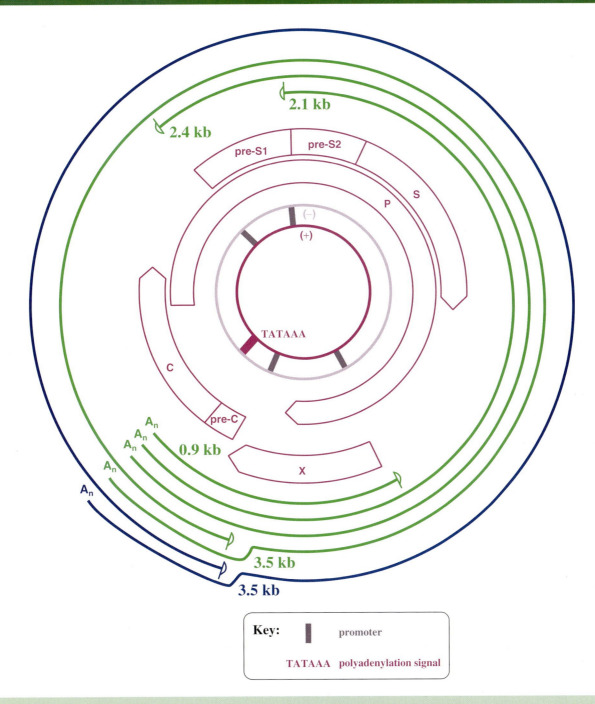

Figure 18.12 *HBV transcripts.* The four size classes of transcript are shown. Some of the larger RNAs in the 3.5 kb class will function as mRNAs and some will function as pregenomes; these functions are indicated by the blue and green colours respectively.

polymerase to continue through the polyadenylation signal on the first pass, but not on the second pass. On the first pass PET suppresses a sequence called NET (negative effector of transcription), which is needed for termination of transcription. If NET is deleted the RNA polymerase makes additional transits around the DNA, producing very long transcripts.

18.8.4 Translation

Translation of six of the HBV proteins is depicted in Figure 18.13.

Of the RNAs that are approximately 3.5 kb long there are two subsets that differ slightly in size. The shorter subset, which does not include the start codon for the pre-C sequence, acts as mRNA for the C and P proteins. The C ORF is upstream of the P ORF, and most ribosomes that bind to these mRNAs commence translation at the C start codon.

There are at least three further start codons before the P start codon. When translation of P commences it appears that the ribosome bypasses the upstream start codons by 'leaky scanning'. Thus HBV produces much more capsid protein than polymerase, just as retroviruses control the relative amounts of Gag and Pol proteins synthesized, (see Section 16.3.5). The shorter

subset of 3.5 kb RNAs also acts as pregenomes (see below).

HBeAg is translated from the longer subset of the 3.5 kb mRNAs, which include the start codon for the pre-C sequence. After translation HBeAg is secreted from the cell.

There are also subsets of the 2.1 kb mRNAs. The M protein is translated from the longest subset and the S protein from several shorter subsets.

18.8.4.a Post-translational modifications

Some of the envelope protein molecules become glycosylated and L is myristylated at the N terminus (Figure 18.5). It has been found that myristylation of L is not required for efficient virion assembly but virions produced in its absence are not infective.

Initially all the L protein N termini are on the cytoplasm side of the endoplasmic reticulum membranes, but after translation about 50 per cent of the N termini are moved through the membranes into the lumen of the endoplasmic reticulum. It is thought that the cell heat shock protein Hsc70 plays a role in this process.

18.8.5 Nucleocapsid assembly

The C protein forms dimers, which are assembled into capsids (Figure 18.14). It is known that the C protein can self-assemble, because capsids can be formed in *Escherichia coli* recombinants expressing the HBV C gene (Figure 18.4).

A molecule of P protein, along with several cell proteins, binds to a molecule of 3.5 kb RNA, which will function as pregenome RNA (Figures 18.14 and 18.15). P binds at a site within a sequence known as ε (Greek letter *epsilon*) that has a high degree of secondary structure. The ε sequence is within the terminal repeat and so is present at both ends of the pregenome RNA. For some reason though, P binds only at the 5′ end. The ε structure with its bound P protein acts as the packaging signal for incorporation of the pregenome RNA into a capsid.

18.8.6 Genome synthesis

The reverse transcriptase domain of P (Figure 18.3) carries out DNA synthesis and the terminal protein

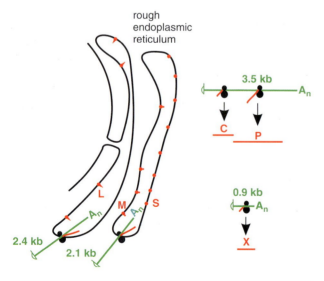

Figure 18.13   *Translation of HBV proteins.*

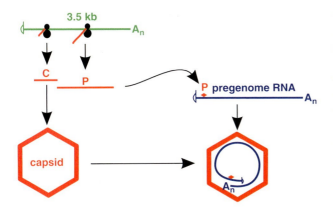

Figure 18.14 *Early stages of HBV assembly.* A capsid is assembled from C protein and acquires a copy of pregenome RNA bound to P protein and several cell proteins (not shown).

remnant of 15–18 bases including the cap. The –OH group at the 3′ end of the RNA remnant acts as the primer for the synthesis of (+) DNA.

During (+) DNA synthesis a nucleocapsid can either migrate to the nucleus to increase the pool of cccDNA or it can undergo a maturation event that enables it to bud through a membrane containing virus envelope proteins (Figure 18.17). DNA synthesis ceases on budding, as the nucleocapsid is cut off from the pool of nucleotides in the cytoplasm. This explains why the (+) DNA in the virion is incomplete.

18.8.7 Exit

The membranes through which budding occurs are part of a compartment between the endoplasmic reticulum and the Golgi complex. As budding proceeds copies of the L protein with their N termini on the cytoplasmic surfaces of the membranes bind to capsids (Figure 18.7). Virions are transported to the cell surface, where they are released from the cell. It is thought that transport occurs within vesicles, which fuse with the plasma membrane, releasing the virions by exocytosis. Non-infectious spheres and filaments are also released from the cell.

The host hepatocyte is not killed by HBV infection; the cell may survive for many months, releasing large quantities of virions and non-infectious particles. No cytopathic effect is observed in productively infected cell cultures (except for those infected with some HBV mutants). When the liver becomes damaged in an HBV-infected individual it is not as a result of cell damage caused by the virus, but it is due to the killing of HBV-infected cells by the body's immune system.

domain of P acts as the primer for the initiation of minus-strand DNA synthesis. A covalent bond is formed between the –OH group of a tyrosine residue near the N terminus of P and the first nucleotide. There are conflicting reports as to whether the terminal protein becomes cleaved from the reverse transcriptase after the initiation of DNA synthesis.

The pregenome RNA acts as template for DNA synthesis. A model for the process is outlined in Figure 18.16. Initially a 4-nucleotide (−) DNA is synthesized then it is transferred to a complementary sequence in DR1 near the 3′ end of the pregenome. DNA synthesis continues to the 5′ end of the pregenome template. The RNase H activity of the P protein degrades the pregenome RNA from the RNA-DNA duplex. All the RNA is removed except for a

Figure 18.15 *HBV pregenome with P protein bound near the 5′ end.* Not to scale. DR1, DR2: direct repeats. The pregenome has terminal repeats, hence there is a copy of DR1 and ε at each end of the RNA.

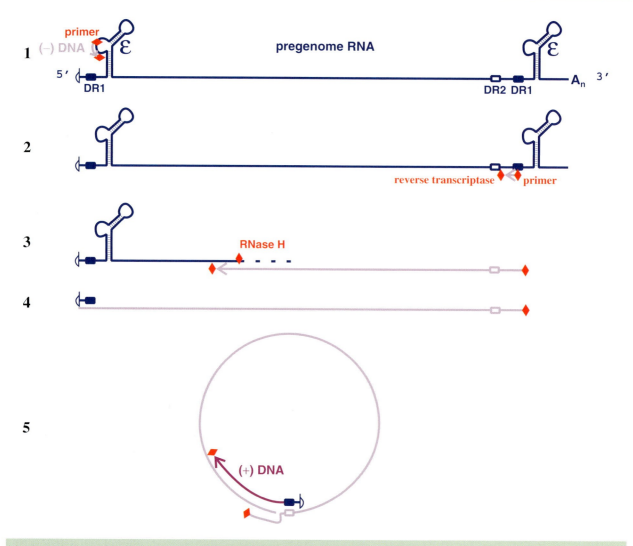

Figure 18.16 *Synthesis of HBV DNA by reverse transcription (model).*

1. Initiation of (−) DNA synthesis near the 5′ end of the pregenome RNA.

2. Transfer of DNA to the same sequence near the 3′ end of the pregenome.

3. Continuation of (−) DNA synthesis and degradation of the pregenome RNA by RNase H.

4. Completion of (−) DNA synthesis. All of the pregenome RNA has been degraded, except for a short sequence at the 5′ end.

5. Synthesis of (+) DNA on the circularized (−) DNA template.

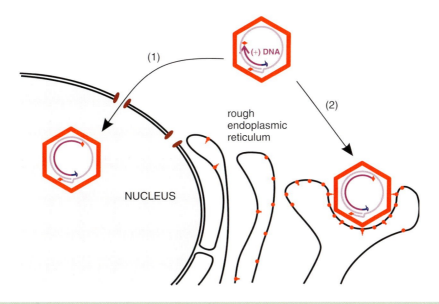

Figure 18.17 *Roles of progeny HBV nucleocapsids.* During DNA synthesis a nucleocapsid may either (1) move to the nucleus where its DNA boosts the pool of virus DNA or (2) bud through a membrane containing virus envelope proteins to form a virion.

18.8.8 Overview of HBV replication cycle

A model for HBV replication is outlined in Figure 18.18. Some aspects of this model have yet to be proven unequivocally and much interesting research remains to be done to unearth more details of the replication of this virus.

18.9 Prevention and treatment of HBV infection

The original HBV vaccines consisted of non-infectious spheres and filaments extracted from blood donations from HBV carriers. Now much vaccine is produced using recombinant yeast cells containing the gene for the S protein (Section 24.7).

Vaccination programmes have been successful in reducing the percentage of HBV carriers in some parts of the world such as Taiwan and Alaska. The requirements to mass produce HBV vaccine cheaply and to vaccinate many more of those at risk are major challenges facing the world.

Alpha-interferon (Section 9.2.1.a) has been used for some years to treat HBV-infected persons. This treatment does not eliminate the infection, but it results in a significant reduction in viraemia in about 20–30 per cent of cases. There is a price to pay for interferon treatment in the form of side-effects such as influenza-like symptoms and weight loss, which may necessitate reducing the dosage or even discontinuing the treatment.

The drug lamivudine, a nucleoside analogue (Section 25.3.1.c), is also used to treat HBV infection. In many ways lamivudine is an improvement over α-interferon as it suppresses virus replication with a low incidence of side-effects, is administered by mouth rather than by injection and is cheaper. Long-term treatment, however, results in the appearance of lamivudine-resistant HBV mutants, though they appear to be susceptible to other nucleoside analogues such as adefovir.

18.10 Other reverse-transcribing DNA viruses

There are some plant viruses with dsDNA genomes that replicate by reverse transcription. These viruses are

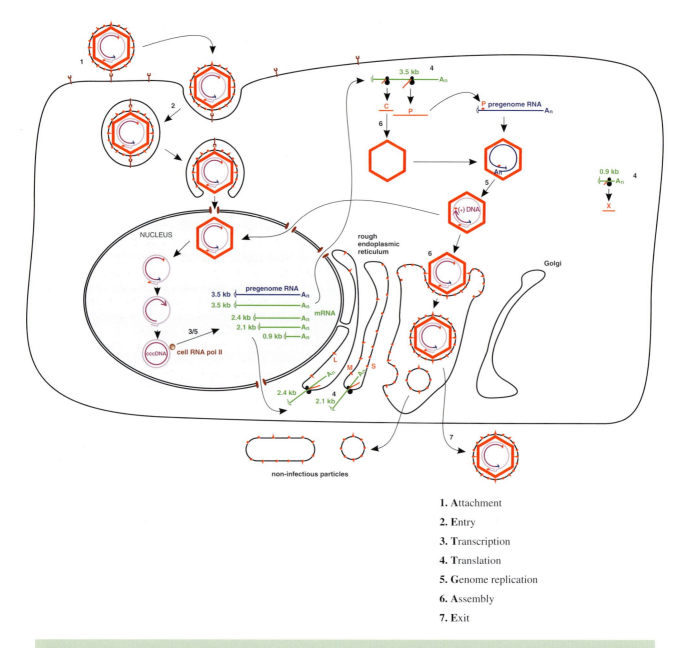

1. Attachment
2. Entry
3. Transcription
4. Translation
5. Genome replication
6. Assembly
7. Exit

Figure 18.18 *Outline of the HBV replication cycle.* Genome replication involves the synthesis of RNA in the nucleus, then copying from RNA to DNA (reverse transcription) in capsids. Stages in virion assembly include capsid construction and acquisition of the envelope by budding. Details of the mode of exit of virions and non-infectious particles from the cell are not shown.

classified in the family *Caulimoviridae*, which includes viruses with isometric virions, such as cauliflower mosaic virus, and viruses with rod-shaped virions, such as rice tungro bacilliform virus.

Learning outcomes

By the end of this chapter you should be able to

- explain the importance of HBV;

- describe the HBV virion and non-infectious particles;

- outline the main features of the HBV genome;

- describe the HBV replication cycle;

- evaluate means of preventing and treating HBV infection.

Sources of further information

Books

Knipe D. M. and Howley P. M. (2001) Chapter 36 in *Fundamental Virology*, 4th edn, Lippincott, Williams and Wilkins

Schultz U. *et al.* (2004) Duck hepatitis B virus: an invaluable model system for HBV infection. *Advances In Virus Research*, Vol. 63, pp. 1–70, Elsevier

Steven A. C. *et al.* (2005) Structure, assembly, and antigenicity of hepatitis B virus capsid proteins *Virus Structure and Assembly*, editor Roy P., *Advances in Virus Research*, Vol. 64, pp. 125–164, Elsevier

Journals

Hilleman M. R. (2003) Critical overview and outlook: pathogenesis, prevention, and treatment of hepatitis and hepatocarcinoma caused by hepatitis B virus *Vaccine*, **21**, 4626–4649

Kidd-Ljunggren K., Miyakawa Y. and Kidd A. H. (2002) Genetic variability in hepatitis B viruses *Journal of General Virology,* **83**, 1267–1280

Kramvisa A., Kewa M. and François G. (2005) Hepatitis B virus genotypes *Vaccine*, **23**, 2409–2423

Lambert C. and Prange R. (2003) Chaperone action in the posttranslational topological reorientation of the hepatitis B virus large envelope protein: implications for translocational regulation *Proceedings of the National Academy of Sciences of the USA*, **100**, 5199–5204

Liu N. *et al.* (2004) *cis*-acting sequences that contribute to the synthesis of relaxed-circular DNA of human hepatitis B virus *Journal of Virology*, **78**, 642–649

Paran N., Cooper A. and Shaul Y. (2003) Interaction of hepatitis B virus with cells *Reviews in Medical Virology*, **13**, 137–143

Seeger C. and Mason W. S. (2000) Hepatitis B virus biology *Microbiology and Molecular Biology Reviews*, **64**, 51–68

Watts N. R. *et al.* (2002) The morphogenic linker peptide of HBV capsid protein forms a mobile array on the interior surface *The EMBO Journal*, **21**, 876–884

Wynne S. A., Crowther R. A. and Leslie A. G. (1999) The crystal structure of the human hepatitis B virus capsid *Molecular Cell*, **3**, 771–780

19

Bacterial viruses

At a glance

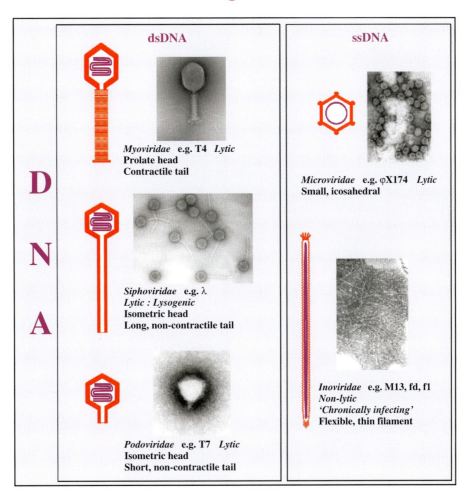

dsDNA

Myoviridae e.g. T4 *Lytic*
Prolate head
Contractile tail

Siphoviridae e.g. λ
Lytic : Lysogenic
Isometric head
Long, non-contractile tail

Podoviridae e.g. T7 *Lytic*
Isometric head
Short, non-contractile tail

D N A

ssDNA

Microviridae e.g. φX174 *Lytic*
Small, icosahedral

Inoviridae e.g. M13, fd, f1
Non-lytic
'Chronically infecting'
Flexible, thin filament

Virology: Principles and Applications John B. Carter and Venetia A. Saunders
© 2007 John Wiley & Sons, Ltd ISBNs: 978-0-470-02386-0 (HB); 978-0-470-02387-7 (PB)

At a glance (continued)

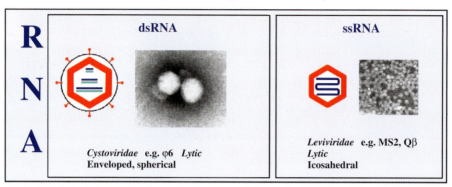

Electron micrographs reproduced by permission of Professor H.-W. Ackermann, Professor R. Duda and Professor R. Inman

19.1 Introduction to bacterial viruses (bacteriophages)

Bacterial viruses, known as bacteriophages or phages (from the Greek *phagein*, 'to eat'), were discovered independently by Frederick Twort (1915) and Felix d'Herelle (1917). A diversity of phages has subsequently been identified and grouped into a number of families. Phage diversity is reflected in both morphological and genetic characteristics. The genome may be DNA or RNA, single- or double-stranded, circular or linear, and is generally present as a single copy. Morphology varies from simple, icosahedral and filamentous phages to more complex tailed phages with an icosahedral head. The majority of phages are tailed.

Phages are common in most environments where bacteria are found and are important in regulating their abundance and distribution. The host controlled modification and restriction systems of bacteria are presumed to protect against phage infection: restriction is levelled against invading double-stranded phage DNA, whilst self-DNA is protected by the modification system. However, in response, certain phages have evolved anti-restriction mechanisms to avoid degradation of their DNA by restriction systems.

Broadly, phages can be classified as either virulent or temperate. A virulent phage subverts the cellular apparatus of its bacterial host for multiplication, typically culminating in cell lysis (for obligately lytic phages) and release of progeny virions. In rare cases, for example the filamentous ssDNA phage M13, progeny are continuously extruded from the host without cell lysis. Accordingly, M13 has been referred to as a 'chronically infecting' phage. Temperate phages have alternative replication cycles: a productive, lytic infection or a reductive infection, in which the phage remains latent in the host, establishing lysogeny. The latter generally occurs when environmental conditions are poor, allowing survival as a prophage in the host (which is referred to as a lysogen). During lysogeny the phage genome is repressed for lytic functions and often integrates into the bacterial chromosome, as is the case for phage lambda (λ), but it can exist extrachromosomally, for example phage P1. The prophage replicates along with the host and remains dormant until induction of the lytic cycle. This occurs under conditions that result in damage to the host DNA. The phage repressor is inactivated and the lytic process ensues. Such a mechanism allows propagation of phages when host survival is compromised. A resident prophage can protect the host from superinfection by the same or similar strains of phages by repressing the incoming phage genome (a phenomenon known as superinfection immunity).

Some temperate phages contribute 'lysogenic conversion genes', for example diphtheria or cholera toxin genes, when they establish lysogeny, thereby converting the host to virulence. Phages can also mediate bacterial genome rearrangements and transfer non-viral genes horizontally by transduction. Tailed phages are the most efficient particles for horizontal (lateral) gene transfer, with the tail effectively guiding injection of DNA into the bacterial cell. Such phage activities generate variability and are a driving force for bacterial evolution.

Bacteriophages have had key roles in developments in molecular biology and biotechnology. They have been used as model systems for animal and plant viruses, and have provided tools for understanding aspects of DNA replication and recombination, transcription, translation, gene regulation and so on. The first genomes to be sequenced were those of phages. Restriction enzymes were discovered following studies on phage infection of different hosts and laid the basis for the development of gene cloning. Phage-encoded enzymes and other products are exploited in molecular biology. Certain phages have been adapted for use as cloning and sequencing vectors and for phage display. Phages are utilized in the typing of bacteria, in diagnostic systems, as biological tracers, as pollution indicators and in food and hospital sanitation (See also Section 1.2.2). There is also renewed interest in the therapeutic potential of phages, due largely to the rapid emergence of antibiotic-resistant bacteria and of new infectious diseases. However, realization of this potential will depend on a number of factors including improved methods of large-scale production and purification of phages, appropriate protocols for administering phages and modifications to enhance therapeutic properties, to remove phage-encoded toxins, to avoid clearance of phages by the host defence system and so on.

This chapter considers the biology of RNA and DNA phages. Properties and applications of a selection of phages that employ different strategies for phage development will be discussed, with emphasis on tailless single-stranded RNA and DNA phages.

RNA PHAGES

19.2 Single-stranded RNA phages

Single-stranded RNA phages are small, icosahedral viruses of the family *Leviviridae* (from the Latin *levis* 'light'), discovered by Tim Loeb and Norton Zinder in 1961. Phages in this family have high mutation rates and some of the smallest RNA genomes known. They are plus strand viruses (with the genome acting as mRNA), containing only a few genes, and infect various Gram-negative bacteria, including *E. coli*, *Pseudomonas* spp. and *Caulobacter* spp. Those infecting the enterobacteria do so by way of the sex pilus. Representative ssRNA phages of the genus *Levivirus* serogroup I (e.g. MS2 and f2) and *Allolevivirus* serogroup III (e.g. Qbeta, Qβ) are considered here.

19.2.1 Virion structure of ssRNA phages

RNA phages typically comprise 180 molecules of major coat (capsid) protein (CP) per virion, one molecule of maturation (A) protein, required for infectivity and maturation, and a linear ssRNA genome of about 3500–4200 nucleotides that displays considerable secondary structure (Figure 19.1).

19.2.2 Genome of ssRNA phages

MS2 was the first virus identified that carried its genome as RNA and the genetic map is shown in

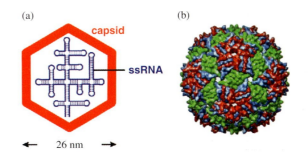

Figure 19.1 *Structure of ssRNA phage MS2.* (a) Virion components. (b) Reconstruction of capsid showing subunit organization. Image rendered by Chimera, from Virus Particle Explorer (VIPER) database (Shepherd C.M. *et al.* (2006) VIPERdb: a relational database for structural virology *Nucleic Acids Research*, **34** (database issue), D386–D389). The icosahedral capsid comprises about 180 copies of major coat protein (as dimers). There is also one copy of maturation protein (minor virion protein), for host infection by recognition of the sex pilus, and the ssRNA genome.

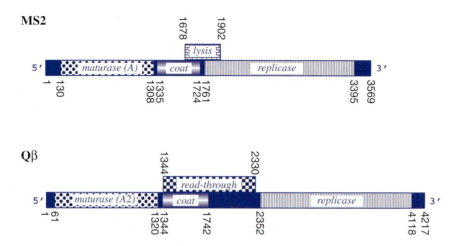

Figure 19.2   *Genetic maps of ssRNA phages MS2 and Qβ.* Nucleotide coordinates are shown along the maps. The solid blue boxes are intergenic regions. MS2: genes for maturase (maturation protein); coat (capsid) protein; replicase protein for RNA-dependent RNA replication and lysis protein (from overlapping coat and replicase genes). The Min Jou interaction involves nucleotides 1427–1433 and 1738–1744. Qβ: genes for maturase; coat protein; replicase protein and read-through protein. The Qβ genome is larger than that of MS2, but there is no separate lysis gene. The maturation protein additionally mediates lysis.

Figure 19.2. The genome is 3569 nucleotides, with four open reading frames (ORFs) for major coat protein, maturation (A) protein, replicase (subunit II) and lysis protein, and is characterized by intergenic spaces.

Lysis protein is encoded by an overlapping ORF between the distal part of the coat protein gene, which appears to be required to regulate expression of the lysis gene, and the proximal part of the replicase gene, which encodes the essential residues for functioning of the lysis protein. Occasionally, ribosomes traversing the coat protein gene fail to maintain the correct reading frame so that once translation of coat protein terminates they can reinitiate at the start of the lysis protein ORF, by shuffling a short distance. Efficiency of translation of the lysis protein ORF by such reinitiation is low. Thus only small amounts of lysis protein are synthesized, in turn guaranteeing that sufficient copies of coat protein will be available for assembly of virus particles before cell lysis occurs.

The genome of Qβ, which is about 4000 nucleotides, is larger than that of MS2. It encodes four proteins: coat protein, replicase, maturation (A2) protein, which also mediates cell lysis, and a minor coat protein (read-through protein, A1), which is generated as a result of

reading through the coat protein gene into the intergenic region (Figure 19.2). Read-through is due to a leaky coat protein terminator (UGA). This allows a low level of misincorporation of tryptophan at the terminator signal and subsequent ribosome read-through. The read-through protein constitutes three to seven per cent of the virion protein and has a role in host infection. Such inefficient termination of translation is not found in the group I phages, where there are tandem stop codons (UAA, UAG) at the end of the coat protein gene and no read-through protein is produced.

Not only does the ssRNA genome encode the phage proteins, it also determines specific secondary and tertiary structure that regulates translation, replication and other functions.

19.2.3. Replication cycle of ssRNA phages

For coliphage MS2, infection involves attachment to pilin (the pilus subunit) along the length of a sex (F) pilus of a susceptible host (see Figure 19.3) via the A protein. Such binding of phages can block conjugation in *E. coli*.

Initially the RBS for the coat protein gene is accessible, occupying the loop position of a hairpin structure, whilst those for the maturation (*A*) and the replicase gene are not, being embedded in secondary structure (see Figure 19.4). The replicase gene is expressed only after the coat protein gene has been translated. Such translational coupling between coat protein and replicase genes involves the Min Jou (MJ) interaction. This is a long distance interaction (LDI), in which a sequence just upstream of the replicase gene RBS is base-paired to an internal sequence of the coat protein gene. The MJ interaction represses translation of the replicase gene. However, passage of ribosomes beyond the first half of the coat protein gene temporarily melts this interaction, opening up the RBS for translation of the replicase gene. The rate of re-folding of MJ in part determines the level of expression of replicase. Another LDI, the van Duin (VD) interaction, which borders the MJ, also appears to have a role in repressing replicase translation.

After about 10 to 20 minutes, translation of the replicase gene stops, due to repression by coat protein, ensuring that only a catalytic amount of replicase is translated. Too much replicase could poison the host. The latter half of the infection cycle is then devoted to synthesis of coat protein, which is required in large quantities for virion assembly. A small amount of lysis protein is synthesized during translation of the coat protein gene, due to an occasional reading frame error.

Once replicase is synthesized, replication of the genome can occur; the genome thus switches from a

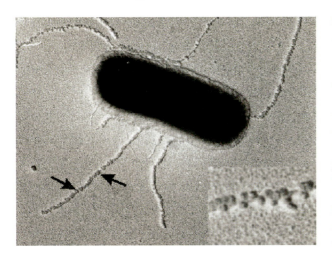

Figure 19.3 *Electron micrograph of ssRNA phage MS2 attached to F pili*. Virions attach along the length of the sex pilus via the maturation protein. The inset shows an enlargement as indicated by the arrows. A. B. Jacobson. Taken from *Genexpress Molecular Biology of Bacteriophages* http://wwwchem.leidenuniv.nl/genexpress/ie/vanduin.htm

The plus strand RNA genome directs protein synthesis immediately upon infection. It adopts a complex secondary structure and this largely controls gene expression via differential access of the ribosome binding sites (RBSs) to host ribosomes during translation.

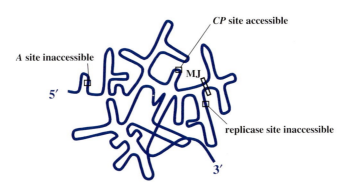

Figure 19.4 *Gene expression in ssRNA phage*. Diagrammatic representation of secondary structure of the genome with ribosome binding site of coat protein (*CP*) gene accessible and those of *A* and replicase genes blocked. MJ: Min Jou, long distance interaction between sequences upstream of the replicase ribosome binding site and sequences near the beginning of the *CP* gene.

template for translation to a template for replication. There would be a topological problem if replication and translation occurred simultaneously, since they proceed in opposite directions along the template ($3' \rightarrow 5'$ and $5' \rightarrow 3'$ respectively). However, it appears that translation of coat protein is repressed by replicase, which may block the coat protein ribosome entry site, and then recommences when sufficient ($-$) RNA has been synthesized. The replicase associates with a number of host proteins to form a phage RNA-specific polymerase (Figure 19.5), which makes both ($+$) and ($-$) strands of RNA. Replication occurs in two stages

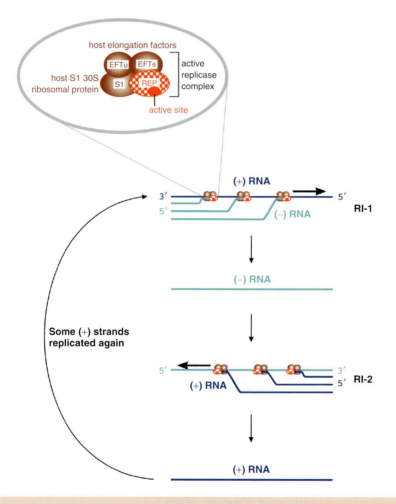

Figure 19.5 *Replication of ssRNA phage genome.* Replication of the single-stranded phage genome occurs through production of two replicative intermediates RI-1 and RI-2. First, the ($-$) strand RNA is synthesized 5′ to 3′ antiparallel and complementary to the ($+$) sense template strand by replicase, in a multi-branched structure. The ($-$) strand RNA then serves as template for formation of new ($+$) strand genomic RNA. This is the only role for the ($-$) strands. Newly replicated ($+$) strands can be re-cycled in replication, translated to yield the capsid proteins or encapsidated in the formation of progeny phages. The inset shows the activation of the replicase protein (REP) by associating with host proteins: for MS2 two elongation factors (EFTs and EFTu) and the ribosomal S1 protein. Replicase is an RNA-templated RNA polymerase, synthesizing both plus ($+$) and minus ($-$) strands of phage RNA 5′ to 3′ through specificity of the replicase for the 3′ end of both template strands. S1 protein is not required for new ($+$) strand synthesis.

and involves replicative intermediates (RIs) composed of multiple, nascent RNA strands synthesized from a single template, and replicase: RI-1 with (+) RNA as template, and RI-2 with (−) RNA as template. Thus minus strands act as templates for synthesis of plus strands for use as mRNA. Secondary structure, which forms as each new strand is synthesized from the template strand, serves to keep the two strands apart, in turn preventing formation of a double-stranded intermediate. It is noteworthy that the error rate for such RNA replication is quite high compared with that for DNA replication. Replicase is error prone and the lack of a double-stranded intermediate means there is no template for error correction.

Expression of the *A* gene occurs independently of the others and is limited to periods of nascent plus strand synthesis. The RBS of the *A* gene is masked by folding of the RNA genome through an LDI in which the Shine-Dalgarno sequence of *A* pairs with a complementary sequence upstream in the 5′ untranslated leader region. The initiator site of *A* becomes exposed transiently when plus strand synthesis begins, due to a delay in folding of the newly synthesized strand. This allows ribosomes to translate the *A* gene until the RBS is sequestered

(Figure 19.6). The number of molecules of A protein is thus maintained in line with the number of new RNA plus strands. There appears to be up-regulation of the A protein in Qβ, consistent with its role in cell lysis as well as infectivity. The LDI is further downstream than in MS2, so that the RBS is exposed for longer, resulting in higher levels of *A* gene expression.

Phage assembly involves spontaneous aggregation, in which the (+) RNA associates with the A protein and is encapsidated through specific recognition of the phage RNA by coat protein dimers. Progeny virions are normally released by cell lysis (see Figure 19.7 for a summary of the replication cycle), with a burst size of about 10^4, compared with a few hundred for DNA phages.

Thus the minimalist genome of the ssRNA phages is utilized very efficiently. Coding capacity is expanded through the use of overlapping genes. Host proteins are utilized to activate phage proteins and some phage proteins have multiple functions. The phage genome serves as a model system for investigating such mechanisms as translational coupling, translational repression and the role of secondary structure of mRNA in the control of gene expression.

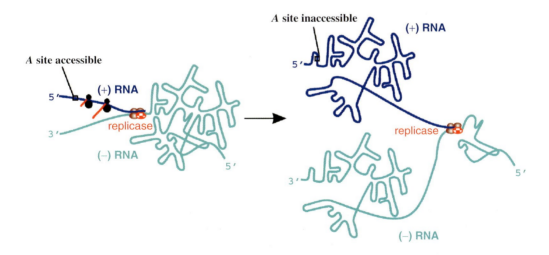

Figure 19.6 *Expression of A gene of ssRNA phage. A* gene expression occurs when plus (+) strand RNA is newly synthesized. The start site for *A* is then exposed, but only transiently, until the ribosome binding site is sequestered by folding of the (+) strand through the long distance interaction involving the Shine-Dalgarno sequence and its upstream complement.

1. Attachment along length of sex (F) pilus

2. Entry of RNA genome

3. Translation:
Host ribosomes attach to phage RNA (+) strand for synthesis of some coat protein and replicase.

4/5. Replication of RNA:
Replicase binds to (+) strand and synthesises (−) strands. (−) strand acts as template for (+) strand synthesis. *A* gene translated during nascent (+) strand synthesis.

Translation of coat protein gene continues from (+) strands.

6. Assembly of progeny virions: 'A' protein associates with RNA (+) strand and coat protein assembles around it to form an icosahedral shell.

7. Exit:
Lysis occurs after ~10,000 virions have formed.

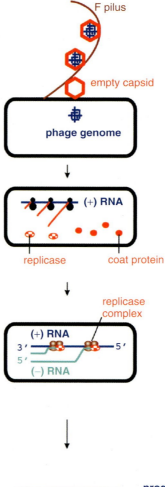
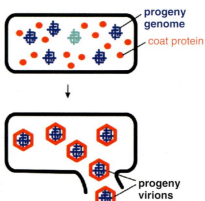

Figure 19.7   *Replication cycle of ssRNA phages.*

19.3 Double-stranded RNA phages

Phages in the family *Cystoviridae* (from the Greek *kystis* 'bladder, sack') contain a dsRNA genome, which is segmented and packaged in a polyhedral inner core with a lipid-containing envelope. Phi6 (φ6) was the first member of the family to be isolated and has been extensively studied. The genome comprises three linear segments: RNA L (large) of about 6400 nucleotides, RNA M (medium) of about 4000 nucleotides and RNA S (small) of about 3000 nucleotides.

Transcription of the double-stranded genome (for synthesis of new plus strands) involves a phage RNA-dependent RNA polymerase. The plus strand transcripts serve as replication templates and mRNAs. Translation of the L segment produces the early proteins that assemble to form the polymerase complex; the M segment produces structural proteins for membranes and spikes in particular, and the S segment produces structural proteins for the capsid, membrane assembly, lysis and entry, and a non-structural protein for envelopment of the capsid.

Phage φ6 infects its host, *Pseudomonas syringae* (pv. phaseolicola), by way of the pilus, which retracts to bring the virion into contact with the cell, and the nucleocapsid enters. Uncoating occurs inside the cell and the polymerase is released. Transcription of the genome is temporally controlled and virions assemble in the cytoplasm, with their envelope deriving from the host. Packaging of the plus strands occurs in the order S–M–L. These single-stranded precursors are then replicated into mature double-stranded genomes inside the capsid. About 100 virions are released following cell lysis.

DNA PHAGES

19.4 Single-stranded DNA phages

There are two groups of ssDNA phages, icosahedral and filamentous. Representative icosahedral phages, φX174 and S-13, were amongst the first studied. Later in the 1960s filamentous phages were isolated by Don Marvin and Hartmut Hoffmann-Berling and by Norton Zinder and co-workers.

19.4.1 Icosahedral ssDNA phages

Icosahedral ssDNA phages belong to the family *Microviridae* (from the Greek *micros*, 'small'). Such phages provided the first evidence for overlapping genes and revealed the economy of genetic coding, which is a feature of several small viruses, including hepatitis B virus (Section 18.6) and the ssRNA phages (Section 19.2). Studies on replication of these phages led to the discovery of rolling circle replication and to the identity of various genes encoding proteins for host DNA replication. φX174 has been most extensively studied.

19.4.1.a Virion structure of phage φX174

The virion of φX174 contains proteins F, G, H and J; F forms the main shell and the spike protein G protrudes on the icosahedral five-fold axes. A molecule of H is found on each spike and acts as a pilot protein. Part of H may lie outside the shell and part within, spanning the capsid through the channels formed by the G proteins. The projections are involved in attachment to the host and delivery of the genome into the host cell. Highly positively charged DNA binding protein J is associated with the ssDNA genome (Figure 19.8).

19.4.1.b Genome of phage φX174

The genome is a circular ssDNA molecule of 5386 nucleotides, coding for 11 proteins. It was the first entire DNA genome to be sequenced (Section 2.9.1). The genes are tightly clustered with little non-coding sequence. The length of the genome is smaller than its coding capacity. However, the different proteins are generated through extensive use of overlapping genes translated in alternative reading frames or employing different start codons. Gene *A* contains an internal translation initiation site to encode protein A*, which corresponds to the C-terminal region of protein A; *B* is encoded within *A* in a different reading frame; *K* is at the end of gene *A* and extends into gene *C* and is translated in a different frame to *A* and *C*; *E* is totally within *D*, but in a different reading frame; the termination codon for *D* overlaps initiation codon for *J* (see Figure 19.9). All genes are transcribed in the same direction.

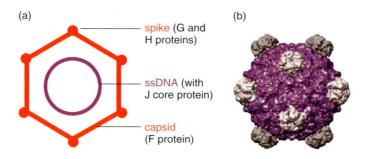

spike (G and
H proteins)

ssDNA (with
J core protein)

capsid
(F protein)

Figure 19.8 *Structure of phage φX174.* (a) Virion components. (b) Reconstruction of capsid showing subunit organization. Image rendered by Chimera, from Virus Particle Explorer (VIPER) database (Shepherd C.M. *et al.* (2006) VIPERdb: a relational database for structural virology *Nucleic Acids Research*, **34** (database issue), D386–D389). The virion contains 60 molecules of major coat protein F (48.4 kD). The spike at each vertex of the icosahedron, composed of five molecules of G protein (19.0 kD) and one molecule of H protein (35.8 kD), has a role in host recognition and attachment. Protein J (4.0 kD) binds to the phage genome for condensation of DNA during packaging. This protein subsequently binds to the internal surface of the capsid to displace the internal scaffolding protein B (13.8 kD).

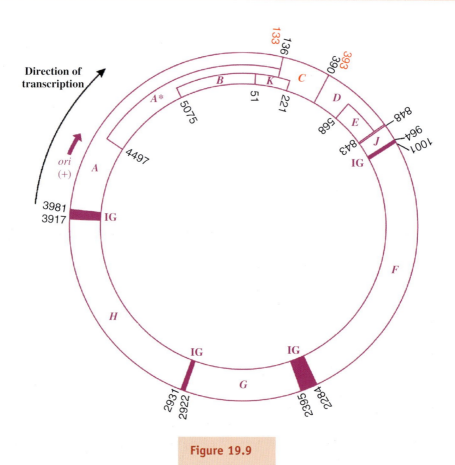

Figure 19.9

19.4.1.c Replication cycle of phage φX174

Phage φX174 recognizes the receptor lipopolysaccharide in the outer membrane of rough strains of *Enterobacteriaceae*, such as *E. coli* and *Salmonella typhimurium*, by way of protein H. This protein then acts as a pilot involved with delivery of the phage DNA into the host cell for replication.

Like M13 (Section 19.4.2.c), replication of the genome of φX174 occurs in three stages and involves a double-stranded intermediate, the replicative form (RF). The viral plus strand is first converted to the RF by host enzymes. However, this process is more complex than for M13, involving the formation of a primosome, that includes a number of proteins to open a hairpin found at the origin (*ori*) of replication of φX174 and to load the replication apparatus on to the DNA. The minus strand is transcribed for synthesis of the phage-encoded proteins. Replication of the RF involves rolling circle replication and requires phage-encoded protein A to synthesize new plus strands. These then serve as templates for minus strand synthesis to generate the new RFs. The third stage is asymmetric replication of progeny ssDNA plus strands. RF synthesis continues until sufficient structural proteins have been synthesized and assembled into empty precursor particles. This is mediated by scaffolding proteins (D and B), which associate transiently with the coat proteins and induce conformational switches to drive the assembly process. A procapsid forms, with D forming the external shell and B occupying an internal location. Viral DNA plus strands are then captured, before they can be used as templates for further minus strands, and are packaged with basic protein J (which neutralizes the negatively charged genome). J protein binds both to the DNA and to the internal surface of the capsid, displacing the B protein, to generate the provirion. Subsequent shedding of the D protein yields the mature infectious virion. Virions accumulate in the cytoplasm and are released by cell lysis, which requires expression of the *E* gene for endolysin production.

19.4.2 Filamentous single-stranded DNA phages

Filamentous ssDNA phages are in the family *Inoviridae* (from the Greek *ina* 'fibre, filament'). The F-specific filamentous (Ff) phages, notably M13, fd and f1, have been most extensively studied. They are plus strand phages and are 'male-specific', infecting *E. coli* strains containing the conjugative plasmid F, by adsorbing to the tip of the F pilus (encoded by the plasmid). Unlike many other DNA phages, filamentous phages do not inject their DNA into the host cell; rather entire phage particles are ingested. Furthermore, these phages do not lyse infected cells, but progeny are continuously extruded through the cell membrane.

Figure 19.9 *Genetic organization of phage φX174.* The numbers are coordinates of the genes. The direction of transcription and approximate location of origin (*ori*) of replication for viral (+) strand are shown. Functions of the gene products (with coordinates of the genes including termination codons) are as follows:

gene *A* (3981–136) protein, viral strand synthesis and RF replication;
gene *A*∗ (4497–136) protein, shutting down host DNA synthesis;
gene *B* (5075–51) protein, capsid morphogenesis;
gene *C* (133–393) protein, DNA maturation;
gene *D* (390–848) protein, capsid morphogenesis;
gene *E* (568–843) protein, host cell lysis;
gene *F* (1001–2284) protein, capsid morphogenesis – major coat protein;
gene *G* (2395–2922) protein, capsid morphogenesis – major spike protein;
gene *H* (2931–3917) protein, capsid morphogenesis – minor spike protein;
gene *J* (848–964) protein, capsid morphogenesis – core protein (DNA condensing protein);
gene *K* (51–221) protein, function not clear-appears to enhance phage yield (burst size).
IG: intergenic region at borders of genes *A, J, F, G, H* contains a ribosome binding site and other features.

Despite this, plaque-like zones are formed in a lawn of sensitive bacteria, since infected cells grow more slowly than uninfected cells.

19.4.2.a Genome of Ff phages

The genome of the Ff phages is a circular ssDNA molecule of about 6400 nucleotides. M13, fd and f1 are 98 per cent identical and generally no distinction will be made between them here. The genes are tightly packed on the genome, with a number of intergenic (IG) regions (Figure 19.10). The major IG region of

507 nucleotides (between genes *IV* and *II*) contains a number of regulatory regions: the origin of replication for (+) and (−) DNA synthesis, a strong rho-dependent transcription terminator, after gene *IV*, and a 78-nucleotide hairpin region, which is the packaging signal (PS) for efficient phage assembly. There is also a strong rho-independent terminator, after gene *VIII*, and a weaker rho-dependent terminator within gene *I*.

The genome encodes 11 proteins. Genes are grouped by function: genes *II*, *V* and *X* for phage DNA synthesis; genes *III*, *VI*, *VII*, *VIII* and *IX* for capsid structure and genes *I*, *IV* and *XI* for assembly.

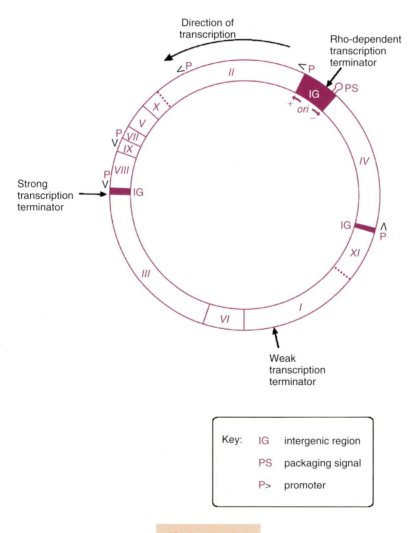

Figure 19.10

Transcription is complex, involving several strong promoters. The two strong transcription terminators divide the genome into two transcription regions: a frequently transcribed region from gene *II* to *VIII* and an infrequently transcribed region from gene *III* to *IV*. All the genes are transcribed in the same direction and there is a gradient of transcription with genes near to a terminator being expressed more frequently, e.g. genes *VIII* and *V*. The weak terminator in gene *I* down-regulates expression of the gene, ensuring that large amounts of gene *I* protein (pI), which would be deleterious to the host, are not synthesized. Gene *XI* protein (pXI) results from an in-frame translation initiator site at residue 241 of gene *I*, and corresponds to the C terminus of pI. Likewise gene *X* protein (pX) is encoded within gene *II*. Such mechanisms ensure that the quantity of each phage protein synthesized is in line with requirements.

19.4.2.b Virion structure of Ff phages

The capsid is a flexible protein filament of variable length, depending on genome size; the wild-type length is approximately 900 nm. The filament comprises about 2700 copies of the 50 amino acid major coat protein (gene *VIII* protein, pVIII), arranged in an overlapping array to form a helix (Figure 19.11). The basic positively charged C-terminal domain of pVIII forms the inside wall of the filament and interacts with the DNA. The N-terminal negatively charged domain is exposed on the outside, giving a negatively charged surface. Hydrophobic interactions hold adjacent coat protein subunits together. The minor coat proteins are present at the ends of the filament. At one terminus (where assembly terminates) are two of these proteins: gene *III* protein (pIII) and gene *VI* protein (pVI), that together form an adsorption complex for recognition of the sex pilus. At the other terminus (where assembly initiates) are gene *VII* protein (pVII) and gene *IX* protein (pIX). The single-stranded circular supercoiled genome is housed lengthwise in the centre of the filament, orientated so that the hairpin packaging loop is located at the pVII/pIX end.

19.4.2.c Replication cycle of Ff phages

The receptor for Ff phages comprises the tip of the F pilus together with TolA, TolQ and TolR membrane proteins of *E. coli*. The phage binds first to the tip of the F pilus through domain N2 of pIII. The pilus retracts bringing pIII to the periplasm and domain N1 of pIII can then interact with TolA (Figure 19.12).

The phage particle enters the cell and disassembles. TolA, TolQ and TolR proteins mediate depolymerization of the coat proteins; the major coat protein and possibly other capsid proteins being deposited in the cytoplasmic membrane to be re-cycled for new particle formation, as the phage DNA is

---

Figure 19.10 *Genetic map of F-specific filamentous phage M13.* The main promoters and terminators are shown. *ori*: origin of replication for viral (+) and complementary (−) DNA strands in main intergenic region. The direction of transcription from promoters (P) is indicated by arrowheads (>). Modified from Webster (2001) In *Phage Display. A Laboratory Manual,* Cold Spring Harbor. Gene product, approximate molecular size and number of copies, and function are as follows:

pI, 35–40 kD, few copies per cell, membrane protein for assembly;
pII, 46 kD, 10^3 copies per cell, endonuclease/topoisomerase for replication of RF, viral strand synthesis;
pIII, 42 kD, ∼ 5 copies per virion, minor capsid protein for morphogenesis and adhesion;
pIV, 44 kD, few copies per cell, membrane protein for assembly;
pV, 10 kD, 10^5 copies per cell, ssDNA binding protein controls switch from RF to viral (+) strand synthesis;
pVI, 12 kD, ∼ 5 copies per virion, minor capsid protein for morphogenesis and attachment;
pVII, 3.5 kD, ∼ 5 copies per virion, minor capsid protein for morphogenesis;
pVIII, 5 kD, 2700–3000 copies per virion, major capsid protein for morphogenesis;
pIX, 3.5 kD, ∼ 5 copies per virion, minor capsid protein for morphogenesis;
pX, 12 kD, 500 copies per cell, replication of RF, viral strand synthesis;
pXI, 12 kD, few copies per cell, membrane protein for assembly.

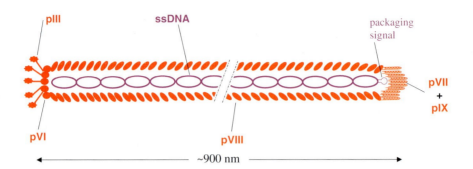

Figure 19.11 Diagrammatic representation of F-specific filamentous phage. The capsid comprises five proteins: pVIII, the major coat protein along the length of the filament; pIII and pVI, minor coat proteins that form an adsorption knob at one end of the filament; pVII and pIX, minor coat proteins that form at the other end. Arrangement of these two proteins is not clear: pVII may interact with pVIII and be shielded from the environment, whilst pIX may interact with pVII and be exposed. The single-stranded genome is oriented with the packaging signal at the pVII/pIX end, which is the first end of the filament to be assembled.

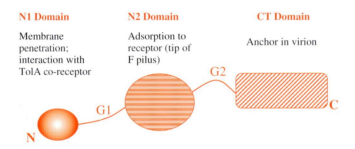

Figure 19.12 *Domains of gene III protein of F-specific filamentous phage.* The gene III protein comprises three domains: N-terminal domains N1 (or D1) and N2 (or D2) required for host cell infection, and C-terminal CT (or D3) required for phage release from the membrane and structural stability; CT anchors pIII in the phage coat by interacting with pVI. The domains are separated by glycine-rich linker regions: G1 and G2. Domains N1 and N2 (exposed on the surface of phage particles) interact in a horseshoe shaped knob-like structure at one end of the filament. Domain N2 binds to the tip of the F pilus and domain N1 dissociates from N2. N1 binds to the TolA membrane protein. Pilus retraction brings the pIII end of the filament to the periplasm. TolQ, R and A proteins are required for translocation of phage DNA to the cytoplasm and of coat proteins to the cytoplasmic membrane. Domain CT contains two sub-domains: one involved in stabilizing the particle once assembled, the other for release of the assembled particle from the membrane.

translocated into the cytoplasm. pIII appears to enter the cell and have a role in the initiation of replication. The circular ssDNA is coated with an *E. coli* ssDNA binding (ssb) protein to protect against degradation by host nucleases. Phage DNA is plus strand and DNA replication occurs immediately upon infection in order to provide the (−) DNA for transcription.

Replication occurs in three stages (Figure 19.13). The first stage is (−) DNA synthesis to generate the double-stranded RF; this depends on host functions. The minus strand can then be transcribed and translated into phage proteins. A specific endonuclease (gene *II* protein, pII) synthesized from the first RF is required for the second stage of replication, which involves multiplication of the RF. The plus strand of the RF is nicked in the major intergenic region, providing a primer (3′ end) for synthesis of a new plus strand, using the minus strand as template, by rolling circle replication. As DNA polymerase III extends the 3′ end, the old plus strand is displaced, cleaved and circularized by pII and converted to a dsDNA (RF), which accumulates as a pool in the cell. As DNA replication proceeds there is concomitant synthesis of phage proteins. Gene *V* protein (pV) accumulates in the cytoplasm and at a critical concentration binds cooperatively to displaced plus strands as they roll off the circle, diverting them from their template role. This initiates stage 3 of replication, in which there is a switch from RF to ssDNA synthesis. The 78-nucleotide hairpin region of the plus strand remains free of pV and initiates the phage assembly reaction. pV additionally serves as a translational repressor of pII, in turn regulating DNA synthesis. Another protein, pX, also located in the cytoplasm, has a role in synthesis of the single-stranded viral DNA. The nucleoprotein complex of pV dimers and ssDNA forms a filamentous structure and is targeted to the cell membrane, where assembly takes place. Capsid proteins are inserted into the cytoplasmic membrane as they form.

Assembly proteins pIV, pI and pXI are integral membrane proteins. pIV forms an extrusion pore in the outer membrane and interacts with pI and pX1 of the inner membrane to form a channel across both membranes, large enough to accommodate extruding phage particles. Assembly occurs where inner and outer membranes appear in close contact at an extrusion site and requires ATP, proton motive force and the bacterial protein, thioredoxin. pV is displaced by coat proteins, which are packaged around the DNA as the phage is extruded through the membrane. The process is initiated by interaction of the packaging signal with pVIII, pVII and pIX, possibly directed by pI. A conformational change induced in pI may allow interaction with pIV to open the gate in the pIV exit pore. The pVII–pIX end of the particle emerges first.

Elongation proceeds as pV is displaced by pVIII, whose positively charged C-terminal region interacts with the negatively charged sugar–phosphate backbone of the DNA. Termination of assembly involves pIII and pVI added to the other end of the filament as it is released, the length of the particle being dictated by the size of the viral DNA. Without pIII, elongation would continue with pVIII encapsidating another viral DNA to generate a polyphage. Thus pIII and pVI are important in maintaining structural integrity and stability of the virion. Particles are continuously extruded without killing the host cell (Figure 19.14).

This assembly process, occurring in the cell membrane and involving polymerization of coat proteins around the Ff phage genome, is in marked contrast to that for dsDNA phages, such as T4, which occurs in the cytoplasm and uses pre-formed heads for packaging DNA (Section 19.5.1).

The filamentous phage CTXφ, which encodes the cholera toxin (the main virulence factor of *Vibrio cholerae*), and a phage associated with invasive meningococci are related to M13. However, unlike the coliphage, these phages can establish lysogeny by site-specific integration into the host chromosome in a similar manner to that of the lambdoid phages (Section 19.5.3).

19.4.2.d Cloning vectors and phage display

The relatively small size and non-lytic infection cycle make the filamentous phages easy to manipulate and use as cloning vectors. Furthermore, since filament size is governed by size of the genomic DNA, insert DNA of variable length can be cloned. Vectors are based on the RF, which is easily obtained from infected cells. Cloned ssDNA can be recovered from the phage particles for use in, for example, DNA sequencing, designing DNA probes and site-directed mutagenesis.

The intergenic (IG) region has been manipulated for the development of cloning vectors. Joachim Messing and co-workers designed a set of vectors, the M13 mp series, by inserting the α-fragment of the *E. coli lacZ* gene containing different polylinkers (multiple cloning sites) into the IG region. Foreign DNA inserted into one of the cloning sites can be identified by insertional inactivation of *lacZ*. Other vectors derived from filamentous phages include phagemids, which combine useful features of plasmids and ssDNA phages. Phagemids enable larger inserts than those used in the M13 mp series to be cloned, double-stranded as plasmids, and amplified and recovered as ssDNA in virions. A phagemid typically contains a selectable marker, an origin of plasmid replication, a cloning

Stage 1 : Second strand synthesis; ssDNA → RF

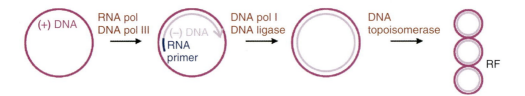

Stage 2 : Multiplication of RF; rolling circle replication

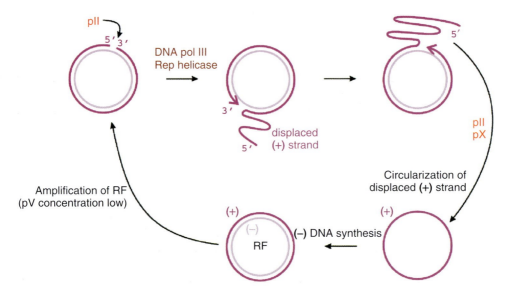

Stage 3 : Amplification of ssDNA; RF → ssDNA
(pV concentration high)

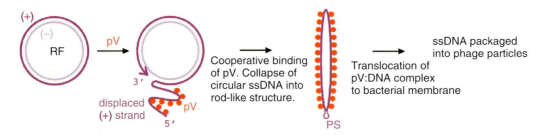

Figure 19.13

region (including polylinker and site for insertional inactivation) and the IG of an ssDNA phage for initiation and termination of DNA replication. Superinfection of the host with an ssDNA helper phage provides phage proteins *in trans* for single-stranded phagemid DNA replication and those for phage capsid and assembly. Defective helper phages can be used to reduce helper phage titre. Various phagemids have been developed, including those for phage display.

In phage display a peptide/protein is fused to one of the phage coat proteins so that it is displayed on the surface of the virion. The DNA sequence encoding the peptide is cloned in the phage/phagemid vector. Display libraries are produced by cloning large numbers of DNA sequences. Displayed peptides of interest can then be selected on the basis of their ability to bind to specific targets by a process of biopanning. This allows high throughput analysis of protein–protein interactions. After several rounds of screening, phages displaying peptides with the tightest binding capacity will be enriched. There is a direct link between the nucleic acid insert (genotype) and the displayed peptide (phenotype).

Early display systems used phage vectors with peptides fused to pVIII or pIII. Only small peptides with six to eight residues could be efficiently displayed; larger peptides compromised phage infectivity. The problem of fusion protein size could be overcome by using phagemid vectors with a single copy of the chosen coat protein gene for fusion, the wild-type copy of that gene being on the helper phage. Both wild-type and fusion coat protein contribute to the phage particles so that the surface is phenotypically mixed, and the phages still infective.

A notable application of phage display technology has been in the engineering of antibodies, in which antibody variable domains have been used to construct combinatorial antibody libraries in phagemid vectors. Affinity binding is employed for rapid selection of antibody fragments binding targets of interest.

Phage display was first developed with filamentous phage M13. However, alternative phage systems, such

---

Figure 19.13 *Replication of F-specific filamentous phage DNA.* Replication occurs in three stages:

1. Second strand synthesis. Plus (+) strand viral genome, introduced into the cell upon infection, is the template for synthesis of the complementary minus (−) strand. Host RNA polymerase (RNA pol) synthesizes an RNA primer in the main intergenic region of the (+) strand. The 3′ end of the primer is extended by host DNA polymerase III (DNA pol III) synthesizing the (−) strand. The primer is removed, the gap filled and a double-stranded, circular, supercoiled replicative form (RF) generated. RF acts as template for production of phage-encoded proteins.

2. Multiplication of RF. RF is replicated by a rolling circle mechanism. Phage-encoded pII, a 'nicking–closing' enzyme, nicks RF at a specific site in the main intergenic region of the viral (+) strand. This allows elongation of the (+) strand by DNA pol III, the original (+) strand being simultaneously displaced by Rep helicase. When a round of replication is completed, pII cleaves and circularizes the displaced (+) strand, which is rapidly converted to another RF. Synthesis of the (−) strand requires an RNA primer and host enzymes. During the early stage of infection pV concentration is low and RF amplification occurs rapidly together with expression of phage proteins.

3. Amplification of ssDNA. Conversion of RF to viral (+) strand DNA for packaging is initiated by ssDNA binding protein pV, which represses translation of a number of phage proteins including pII. Inhibition of pII results in a switch from RF synthesis to production of phage particles. Late in the infection process when pV is present in high concentration it sequesters the viral (+) strands as they are displaced and prevents their conversion to RF. pV dimers bind cooperatively to (+) strand and convert the circular ssDNA into a filament-like pV:DNA complex. The DNA is oriented in the complex so that the packaging signal (PS) hairpin protrudes from one end (pV appears to bind weakly to dsDNA). pV:DNA complex is translocated to the inner membrane for assembly of phage particles, with pVII and pIX assembling on to the PS region of the genome. pX has a role in replication, but it remains unclear.

as those of T phages, lambda (λ) and ssRNA phages, have also been employed. The technology promises to revolutionize the study of protein–protein interactions.

19.5   Double-stranded DNA phages

There is a large variety of dsDNA phages, of which the T phages and λ have been particularly well characterized. T2, T4 and T6 belong to the family *Myoviridae* (from the Greek *mys, myos*, 'muscle', referring to phages with contractile tails), T1 and T5 together with λ are members of the *Siphoviridae* (from

the Greek *siphon*, 'tube', referring to phages with long, flexible, non-contractile tails) and T3 and T7 are in the *Podoviridae* (from the Greek *pous, podos*, 'foot', referring to phages with short, non-contractile tails). A representative of each family is considered.

19.5.1 Phage T4

T4, a virulent T-even phage, is one of the largest phages and was the first prokaryotic 'organism' providing evidence of gene splicing through the presence of introns in the genome. It has a complex morphology,

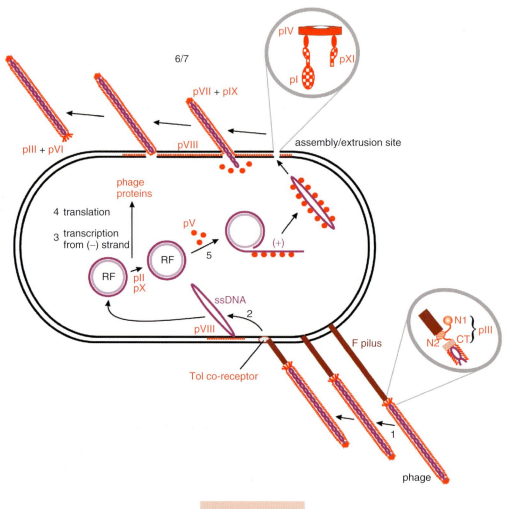

Figure 19.14

consisting of an elongated icosahedral head with a number of different types of protein, including two accessory proteins, Hoc (*h*ighly antigenic *o*uter *c*apsid protein) and Soc (*s*mall *o*uter *c*apsid protein) decorating the external surface, a contractile tail, baseplate, tail fibres and pins (see Figure 19.15).

Binding of Soc stabilizes the capsid and may account for the osmotic shock resistance phenotype. The two accessory proteins have been utilized in display fusions (see also Section 19.4.2.d) with e.g. HIV capsid proteins, to provide vaccine vectors. In addition, the tail fibres can be exploited in phage display and may be used as protein struts for nanodevices.

T4 tail fibres, pins and baseplate are involved in binding to the lipopolysaccharide receptor of the *E. coli* host. The tail sheath contracts, driving the internal tail tube into the cell; tail lysozyme (gene product 5, gp5) facilitates digestion of the peptidoglycan layer to reach the inner membrane. The trans-membrane electrochemical potential is required for transfer of the T4 DNA to the cytoplasm. This occurs rapidly, in contrast to the situation with T7 infection (Section 19.5.2). The empty capsid remains extracellular. The phage shuts down the cell; host RNA polymerase stops recognizing host promoters and instead uses phage promoters for T4 gene expression, and the bacterial genome is degraded and recycled in phage DNA synthesis.

The T4 genome is linear dsDNA of about 169 kbp, which is circularly permuted and terminally redundant (a sequence of about 1.6 kbp at one end is directly repeated at the other) due to the mode of replication and packaging by the head-full mechanism. The genome is AT-rich and contains modified bases in the form of 5-hydroxy-methyl-cytosine, rather than cytosine, which protect the phage DNA from many host restriction systems and from phage-encoded nucleases

---

Figure 19.14 *Replication cycle of F-specific filamentous phage.* Host cell infection is a coordinated, multistep process.

1. Attachment. The phage binds to the tip of the F pilus *via* pIII, with domain N2 attaching to the tip and dissociating from its normal interaction with domain N1, and CT anchored in the virion with pVI.

2. Entry. Retraction of the pilus brings the phage particle to the periplasm to interact with Tol proteins. Phage coat proteins disassemble into the cytoplasmic membrane; the phage ssDNA is translocated into the cytoplasm.

3. Transcription: using (−) strand of RF as template (from stage (i) of genome replication).

4. Translation: for synthesis of all phage proteins, most of which insert in the cytoplasmic membrane. pII, pV, pX reside in the cytoplasm.

5. Genome replication occurs in three stages:

   (i) ssDNA plus (+) strand is converted to the dsDNA replicative form (RF).

   (ii) Amplification of RF by rolling circle replication involving pII and pX. Amplification of RF continues until pV reaches sufficient concentration to bind to (+) strand DNA and prevent its conversion to RF.

   (iii) pV switches most of the DNA replication from dsDNA RF to ssDNA (+) strand.

6. Assembly. The pV:DNA complex migrates to the cell membrane. pV is removed and replaced by the capsid proteins.

7. Exit. The assembly proteins pIV, pI and pXI interact, forming a channel through both cell membranes for extrusion of the progeny phages. The pVII/pIX end of the particle emerges first. The phage is then elongated by replacement of pV by pVIII. pVI and pIII are added to the other end of the filament, the C terminus of pIII being anchored in the particle itself. Particles are extruded continuously without cell lysis.

that degrade cytosine-containing host DNA during phage infection. Furthermore, the majority of the hydroxy-methyl groups are glucosylated after DNA synthesis to counteract the host Mcr (modified cytosine restriction) systems, to which the hydroxy-methyl-cytosine residues confer susceptibility. The genome comprises about 300 probable genes assigned by the timing of their expression, with three different types of promoter: early (Pe), middle (Pm) and late (Pl). The early and middle genes encode functions for DNA replication and for regulating expression of the late genes, which encode head and tail components for the phage particles, and functions for cell lysis (Figure 19.16).

The regulatory circuit is complex and involves transcriptional termination and anti-termination, transcriptional activators and replication-coupled transcription. Early genes rely on the transcription apparatus of the host, being transcribed from normal σ^{70} promoters. Early gene products alter the host RNA polymerase in two ways for expression of middle genes. The enzyme is thereby able to read through a transcription terminator by an anti-termination mechanism (see Section 19.5.3 for an explanation) to express genes downstream of early genes and is modified to recognize middle promoters. These promoters differ from σ^{70} promoters in the −35 sequence, and a transcriptional activator, encoded by the T4 *motA* early gene, allows the host RNA polymerase to transcribe from such promoters.

Transcription of the late genes is coupled to replication. Promoters for these genes again differ from σ^{70} promoters and require an alternative sigma factor, encoded by T4 regulatory gene *55*, to activate transcription. Binding of gene *55* product to host RNA polymerase allows the enzyme to recognize specifically T4 late gene promoters. Other T4 gene products, including those of genes *44*, *45* and *62*, which are involved in replication, are additionally required for late gene transcription. RNA polymerase complexed with gp55 seems to need an actively replicating T4 DNA for effective transcription of the late genes. Such coupling between replication and transcription ensures that phage DNA becomes available for packaging as the capsid proteins are synthesized from the late genes. The DNA is packaged into the pre-made heads. These require a protein scaffold to assemble. Packaging is

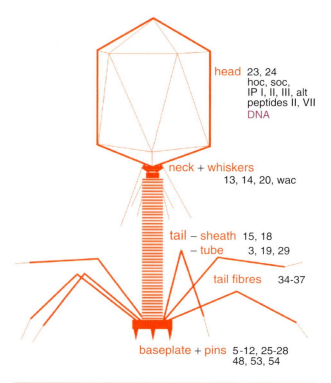

head   23, 24
hoc, soc,
IP I, II, III, alt
peptides II, VII
DNA

neck + whiskers
     13, 14, 20, wac

tail – sheath   15, 18
     – tube   3, 19, 29

tail fibres   34-37

baseplate + pins   5-12, 25-28
     48, 53, 54

Figure 19.15 *Diagrammatic representation of T4 phage.* The capsid comprises a multiple-protein, prolate icosahedral head separated from the tail by the neck. The tail consists of two helical arrangements of protein: one forming the sheath, the other the tube, with a collar and a baseplate with pins and tail fibres. The pins at the baseplate vertices anchor the phage to the host cell. Proteins are labelled with their corresponding gene name or number. Virus assembly requires a protein scaffold. The dsDNA is densely packed as concentric layers in the head. See Leiman *et al.* (2003).

size specific, depending on the size of the head, and is accompanied by a structural change in the procapsid. Tail and baseplate assemble separately. Heads attach to tails and tail fibres are added last to produce mature progeny phages. T4 holin (gene *t* protein) creates a hole in the inner membrane so that lysozyme (gene *e* protein for endolysin) can degrade the peptidoglycan for cell lysis and release of phages.

Upon superinfection of T4-infected bacteria, the phage exhibits the phenomenon of 'lysis inhibition', whereby the latent period is prolonged, lysis is delayed

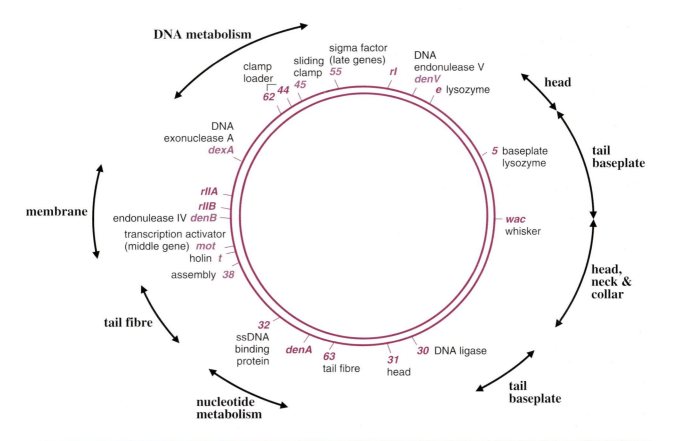

Figure 19.16 *T4 genetic map showing approximate location of some genes and main functional groups.* The map is presented as a circle. For nucleotide positions and functions of genes see Miller *et al.* (2003).

and an increased burst size (about 1000) results. It is noteworthy that T4-like phages have been suggested as possible therapeutic agents in, for example, the treatment of *E. coli* diarrhoea. Lysis of the majority of diarrhoea-causing *E. coli* strains would probably require a cocktail of several phages and/or genetic manipulation to extend host range sufficiently.

19.5.2 Phage T7

T7 is a virulent T-odd phage, with icosahedral head and short, non-contractile tail. The linear genome (about 39 kbp) is smaller than that of T4. It is terminally redundant and comprises about 50 genes, which are closely packed, divided into three groups and numbered from the genetic left end, the first end to enter the host cell (Figure 19.17). Integral numbered genes are essential; most non-integral numbered are non-essential.

T7 early genes (up to *1.3*) are transcribed by the bacterial RNA polymerase, which is afterwards inactivated. One of these early gene products is T7 RNA polymerase, a very active enzyme that takes over transcription of T7 genes. Late genes are transcribed from strong T7 promoters. T7 transcription is regulated both by the slow delivery of the DNA into the host and by the synthesis of the T7 RNA polymerase, which recognizes only T7 promoters. This specificity of T7 RNA polymerase has been exploited in the development of a number of expression

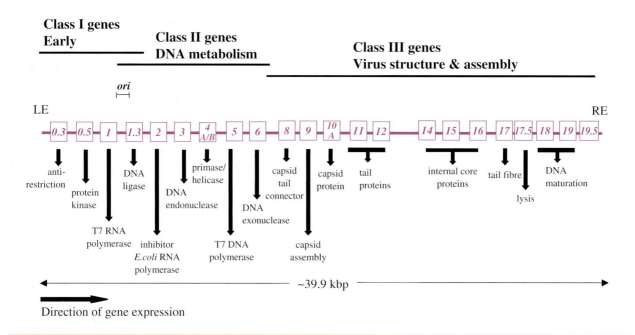

Figure 19.17 *Simplified genetic map of phage T7.* The T7 genome contains about 50 genes, of which a number are shown. Genes are transcribed from the left end (LE) to the right end (RE), with gene *0.3* expressed first and gene *19.5* last. Three classes (I,II,III) of genes are identified by function and timing of their expression. Gene *0.3* encodes an anti-restriction protein to avoid restriction of T7 DNA by type I restriction systems of the host immediately upon infection, allowing the phage to propagate. *E. coli* RNA polymerase transcribes class I genes, one of which encodes T7 RNA polymerase, which subsequently transcribes class II (middle) and class III (late) genes from T7-specific promoters. T7 proteins encoded by genes *0.7* and *2* inactivate the host RNA polymerase, switching transcription to the T7 rather than the bacterial genome. T7 proteins encoded by genes *3* and *6* degrade host DNA. T7 DNA is replicated by T7 DNA polymerase encoded by gene *5*. *ori* is the primary origin of DNA replication. Functions derived from Studier and Dunn (1983) Cold Spring Harbor Quantitative Biology, **47**, 999–1007, and Steven and Trus (1986) In *Electron Microscopy of Proteins Volume 5: Viral Structure* (editors Harris J.R and Horne R.W.) pp. 1–35, Academic Press.

vectors that incorporate these promoters. Degradation of the bacterial chromosome during infection provides the nucleotides for T7 replication, which uses T7 DNA polymerase. This polymerase has been modified and marketed commercially as the DNA sequencing enzyme, Sequenase™.

19.5.3 Phage lambda (λ)

Morphologically, phage λ comprises an isometric head containing a dsDNA genome of about 48.5 kbp and a long, flexible tail with central tail tip fibre and side tail fibres. Genes are clustered according to function

on the genome (see Figure 19.18), with many encoding head and tail components of the phage, but only two (*O* and *P*) for replication, unlike phages T4 and T7, where there are many phage-encoded replication proteins. The genome in the virion is linear, with 12 b single-stranded complementary 5′ ends (cohesive ends) that mediate circularization of the DNA after infection of the host. Circularization of the λ genome protects it from degradation. The cohesive ends are generated due to the mode of packaging of the DNA from concatemers, (containing multiple λ genomes), which are cut asymmetrically at λ *cos* sites. The specificity of this packaging mechanism has been exploited for efficient introduction of insert DNA into recombinant hosts using λ cloning vectors. To subvert bacterial restriction–modification activities upon infection, phage λ encodes the Ral (restriction alleviation) protein. This enhances methylation of the viral DNA, thereby alleviating restriction by type I restriction enzymes.

As a temperate phage, λ can either establish lysogeny, being maintained as a prophage integrated into the bacterial chromosome between the biotin and galactose operons, or multiply lytically to produce progeny virions upon infection (Figure 19.19).

The decision between the lytic cycle and lysogeny depends on a number of genetic, environmental and physiological factors. Gene expression is negatively regulated by repressor–operator interactions, with termination and anti-termination of transcription playing important roles in temporal control.

Phage λ infection involves the *E. coli lamB* protein receptor. Upon introduction into the host the phage genome circularizes and the bacterial RNA polymerase binds to leftward and rightward promoters (P_L and P_R), transcribing *cro* and *N* genes. N protein, together with host proteins, mediates anti-termination at transcription terminators t_{L1}, t_{R1} and t_{R2}, so that delayed early genes including *cII* and *cIII* are expressed. Anti-terminator protein, N, acts by modifying the RNA polymerase, as it passes through the *nut* (N utilization) sequences (*nutL* and *nutR*), so that it ignores termination. N may additionally impede release of the transcript by the terminator hairpin or by the activity of rho helicase. Where the concentration of CII (transcription activator protein) remains below a critical level, which is the case in the majority of cells, transcription from P_L and P_R continues until Q anti-terminator switches on late genes governing synthesis of phage components for heads and tails, and *R*, *Rz* and *S* for lysis. The *S* gene protein allows transport of λ endolysin, encoded by *R* and *Rz*, to the periplasm. The *R* gene product degrades the peptidoglycan of the cell wall, with the *Rz* product having a minor role in cell lysis. The effectiveness and specificity of phage lytic enzymes make them promising candidates for use as antibacterial agents ('enzybiotics') in the control of pathogenic bacteria, including antibiotic-resistant bacteria.

Under conditions where CII protein prevails, it activates transcription of *cI* from P_{RE} (promoter for repressor establishment) and of *int* from P_I (promoter for integrase). This results in integration of the λ genome into the bacterial chromosome and repression of lytic functions, through CI repressor binding to O_L and O_R, for establishment of lysogeny. CII also activates P_{aQ} for transcription of antisense RNA that reduces expression of *Q*. Lysogeny is maintained by continued synthesis of CI repressor from P_{RM} (promoter for repressor maintenance). Bacterial proteins, such as HflA protease, and phage-encoded CIII protein affect stability and hence concentration of CII. These proteins are themselves affected by environmental stimuli, e.g. temperature and nutrients.

Physiological stress, effected by e.g. exposure of a λ lysogen to UV irradiation, results in damage to DNA and induces the SOS response, in turn activating RecA co-protease. This stimulates a latent protease activity of CI with subsequent cleavage of the repressor, so that it no longer binds to O_R and O_L. Anti-repressor Cro in turn binds to O_R and O_L to prevent transcription of *cI* and stimulate its own transcription. The phage is thereby induced to enter the lytic cycle. Genes *xis* (for excisionase) and *int* (for integrase) are expressed and the λ genome excises to direct the lytic cycle.

The λ lysis–lysogeny decision represents a paradigm for control of gene expression in bacteria and lysogenic phages. Relatives of phage λ may carry verocytotoxin (VT) genes to convert their hosts into pathogens upon lysogenization, as is the case with *E. coli* O157. Phage λ, which itself carries *bor*, conferring serum resistance on the lysogen, provides a useful model system for studying such lambdoid phages.

Phages of the lambda family and certain filamentous phages (Section 19.4.2.c) are important vectors for spread of virulence genes. Thus, despite potential lethality to their hosts, such phages may confer selective advantages, through e.g. lysogenic conversion and superinfection immunity, to contribute to bacterial fitness in the environment.

(a)

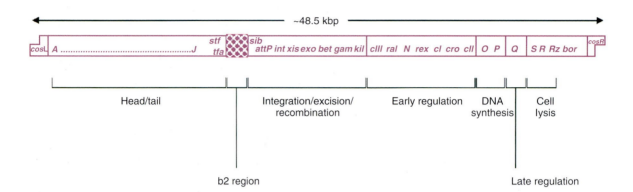

(b)

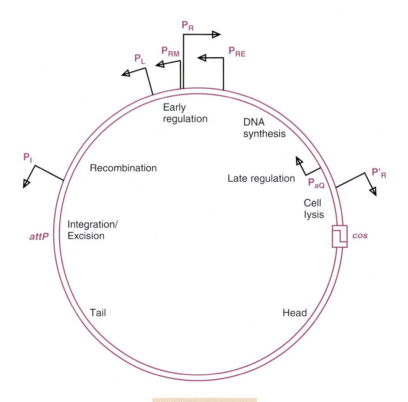

Figure 19.18

Figure 19.18 *Genetic organization in phage lambda.* (a) Simplified genetic map (non-integrated). Some of the essential λ genes are shown. Gene products/functions are as follows: *A* to *J*, head and tail proteins and assembly; *stf* and *tfa*, side tail fibres; b2 region, non-essential genes, part of the region replaced in λ cloning vectors; *sib*, regulatory sequence for *int* and *xis* expression – *int* can be transcribed from P_I and P_L. Expression from P_L is inhibited when *sib* is present. P_L transcript terminates beyond *sib* and the site forms a secondary structure (hairpin) that is recognized for cleavage by RNAse III, removing the *int* portion (= a form of retroregulation). P_I transcript, which terminates at t_I (upstream of *sib*), is stable and *int* expressed. P_L transcript from integrated prophage DNA is also stable, since *sib* is removed, and *int* and *xis* expressed. *attP* (P.P' or a.a'), λ attachment site – location of site-specific recombination (15 b homology) with host chromosome integration site *attB* (B.B' or b.b') between *gal* and *bio* operons; *int*, integrase for integration and with *xis* λ prophage excision; *xis*, excisionase for λ prophage excision; *exo*, 5' to 3' exonuclease for recombination; *bet*, promotes annealing of complementary ssDNA strands for recombination; *gam*, inhibits host RecBCD nuclease; *kil*, host killing by blocking cell division; *cIII*, required for establishment of lysogeny, CIII blocks degradation of CII by host proteases (HflA and HflB); *ral* (restriction alleviation), for partial alleviation of host B and K restriction systems; *N*, anti-terminator protein promotes expression of delayed early genes, causes RNA polymerase to overcome transcription terminators and not stop transcription, requires *nut* (*N* utilization) site and host factors; *rex* [*rexB/rexA*] (restriction of exogenous phage), blocks superinfection by phage such as P22, expressed in the lysogen; *cI*, λ repressor – binds to O_L and O_R to repress transcription from P_L and P_R, activates transcription from P_{RM} when bound to O_{R2}, *cI* expressed from P_{RE} to establish lysogeny and from P_{RM} to maintain lysogeny; *cro*, anti-repressor – binds to O_L and O_R to repress early transcription, block activation of P_{RM} and promote the lytic cycle; *cII*, regulatory protein for establishment of lysogeny, activates transcription from P_{RE} for CI synthesis, P_I for integrase synthesis, P_{aQ} for antisense *Q* resulting in reduced expression of *Q*. Stability of CII controlled by host proteases and CIII. *O*, *P*, proteins for phage DNA replication; *Q*, anti-terminator protein promotes expression of late genes from P'_R, requires *qut* site; *S*, holin protein, creates holes in cytoplasmic membrane permitting endolysin (*R* product) to gain access to periplasm for cell lysis – two forms of S protein are translated (from two in-frame translation start sites) and the ratio determines the timing of lysis to ensure progeny phages are ready for release before lysis occurs. *R*, an endolysin – a transglycosylase that digests peptidoglycan for cell lysis; *Rz*, for cell lysis; *bor*, lipoprotein in the membrane of λ lysogen conferring serum resistance on host. *cos*L,*cos*R, cohesive ends(left and right respectively) – 12 b single-stranded complementary 5' extensions, anneal to form circular genome in the host. Note that the integrated prophage DNA is a linear permutation of the virion DNA, such that *sib* is no longer downstream of *att*. (b) Simplified circular map. Circularization of genome by annealing of cohesive ends to form the *cos* site brings all late genes (cell lysis, head, tail genes) together. Main promoters are shown:

P_L	promoter leftward
P_R	promoter rightward
P'_R	late promoter
P_{RE}	promoter for λ repressor establishment
P_{RM}	promoter for λ repressor maintenance
P_I	promoter for *int*
P_{aQ}	promoter for anti-*Q*.

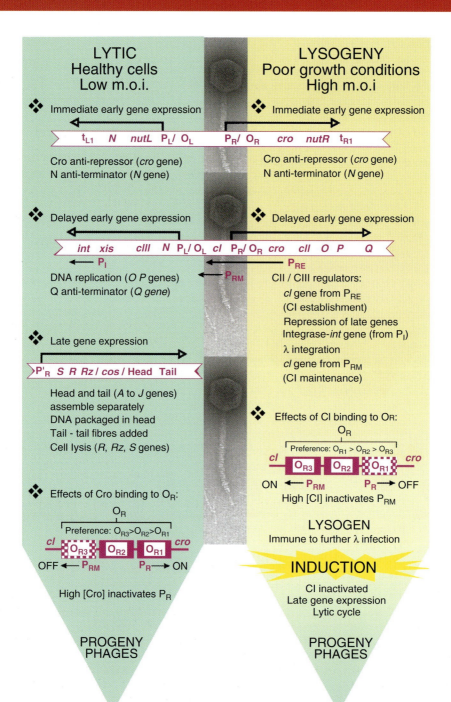

Figure 19.19 *Requirements for lytic cycle and lysogeny in phage lambda.* Important genes, gene products/functions required for the lytic cycle and lysogeny are shown. For description of gene functions see Figure 19.18. Electron micrograph of phage lambda reproduced by permission of Professor R. Duda.

Learning outcomes

By the end of this chapter you should be able to

- discuss the replication cycle and control of gene expression in ssRNA coliphages;

- outline the infection process of dsRNA phages;

- review the biology of the Ff and icosahedral ssDNA phages;

- describe the structure and replication cycle of dsDNA phages, T4 and T7;

- discuss the molecular events that govern the decision between the lytic cycle and lysogeny in phage λ;

- explain some uses of phages in modern biotechnology;

- appreciate the role of phages in the evolution of bacterial diversity.

Sources of further information

General

Calendar R., editor (2006) *The Bacteriophages,* 2nd edition, Oxford University Press

Kutter E. and Sulakvelidze A., editors (2005) *Bacteriophages – Biology and Applications*, CRC Press

Wackett L. P. (2005) Bacteriophage. An annotated selection of world wide web sites relevant to the topics in environmental microbiology *Environmental Microbiology*, **17**, 299–300

Single-stranded RNA phages

van Duin J. and Tsareva N. (2006) Single-stranded RNA bacteriophages. In *The Bacteriophages*, 2nd edition, pp. 175–196 (Calendar R., editor), Oxford University Press

Double-stranded RNA phages

Bamford D.H. and Wickner R.B. (1994) Assembly of double-stranded RNA viruses: bacteriophage φ6 and yeast virus L-A *Seminars in Virology*, **5**, 61–69

Phage φX174

LeClerc J. E. (2002) Single-stranded DNA phages. In *Modern Microbial Genetics*, 2nd edition, pp. 145–176 (Streips U. N. and Yasbin R. E., editors) Wiley-Liss

Filamentous single-stranded DNA phages

Russel M., Lowman H. B. and Clackson T. (2004) Introduction to phage biology and phage display. In *Phage Display: a Practical Approach*, pp. 1–26 (Clackson T. and Lowman H. B., editors), Oxford University Press

Webster R. (2001) Filamentous phage biology. In *Phage Display. A Laboratory Manual*, pp. 1.1–1.37 (Barbas C. F. III, Burton D. R. and Scott, J. K., editors), Cold Spring Harbor Laboratory Press

Phage T4

Leiman P.G. *et al.* (2003) Structure and morphogenesis of bacteriophage T4 *Cellular and Molecular Life Sciences*, **60**, 2356–2370

Miller E.S. *et al.* (2003) Bacteriophage T4 genome *Microbiology and Molecular Biology Reviews*, **67**, 86–156

Phage T7

Dunn J.J. and Studier F.W. (1983) Complete nucleotide sequence of bacteriophage T7 DNA and locations of T7 genetic elements *Journal Molecular Biology*, **166**, 477–535

Molineux I.J. (2001) No syringes please, ejection of phage T7 DNA from the virion is enzyme driven *Molecular Microbiology*, **40**, 1–8

Phage lambda

Gottesman M E. and Weisberg R. A. (2004) Little lambda, who made thee? *Microbiology and Molecular Biology Reviews*, **68**, 796–813

Oppenheim A. B. *et al.* (2005) Switches in bacteriophage lambda development *Annual Review of Genetics*, **39**, 409–429

Ptashne M. (2004) *A Genetic Switch. Phage Lambda Revisited*, 3rd edn, Blackwell

20

Origins and evolution of viruses

At a glance

Possible Virus Origins:
- RNA molecules that existed before cells
- cell components
- micro-organisms.

Viruses evolve as a result of:
- errors during nucleic acid replication;
- recombination between virus strains;
- reassortment between virus strains;
- acquisition of cell genes.

Evolution of viruses can be monitored by sequencing their genomes and creating phylogenetic trees:

New viruses may evolve as a result of viruses infecting new host species, e.g. HIV-1 and HIV-2.

Virology: Principles and Applications   John B. Carter and Venetia A. Saunders
© 2007 John Wiley & Sons, Ltd   ISBNs: 978-0-470-02386-0 (HB); 978-0-470-02387-7 (PB)

20.1 Introduction to origins and evolution of viruses

In this chapter we shall speculate on how viruses may have originated and consider the mechanisms by which viruses evolved in the past and continue to evolve today. The origins and evolution of many cellular organisms can be inferred from fossils, but there is very little fossil record of viruses! The discussion of the possible origins of viruses will be highly speculative, though when we come to mechanisms of virus evolution we will be on firmer ground as sequencing of virus genomes enables the construction of phylogenetic trees and the monitoring of virus evolution as it occurs.

20.2 Origins of viruses

Viruses by definition are parasites of cells, so there could be no viruses until cells had evolved. There is evidence for the presence of living cells about 3.9 billion years ago (Figure 20.1), though the first cells probably came into existence several hundred million years earlier. It is believed that, prior to the development of these primitive prokaryotes, there was a phase of evolution involving organic molecules. These molecules probably included proteins and RNA molecules, and some of the latter may have evolved capacities for self-replication.

The archaea and the bacteria that inhabit the earth today are the prokaryotic descendants of the early cells. It is likely that viruses developed at an early stage in these primitive prokaryotes, but the extent to which the viruses of the modern prokaryotes (Chapter 19) resemble these early viruses is unknown. Eukaryotic cells appeared much later in evolution (Figure 20.1), so on the evolutionary timescale the viruses that infect eukaryotes are much more recent than those of the prokaryotes. The viruses of modern mammals and birds are presumably descended from those that infected their dinosaur ancestors.

The answer to the question 'Where did viruses come from?' is 'Basically, we do not know!'. We can, however, speculate about possible ancestors of viruses; these include molecular precursors of cellular organisms when the earth was young, components of cells and intracellular micro-organisms. We shall consider each of these in turn.

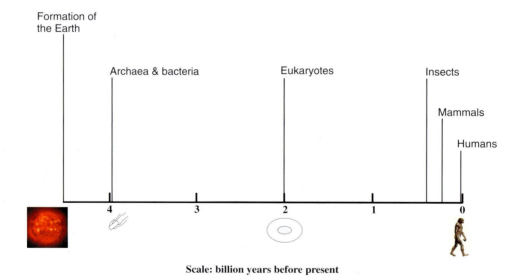

Scale: billion years before present

Figure 20.1 *Timeline of the Earth's history.* The approximate times when various groups of organisms first appeared are indicated.

20.2.1 Molecular precursors of cellular organisms

It is thought that on the early Earth, before cellular organisms had come into existence, RNA molecules evolved and developed enzyme activities (ribozymes) and the ability to replicate themselves. Once cells had evolved perhaps some were parasitized by some of these RNA molecules, which somehow acquired capsid protein genes. If this happened then these were the first viruses.

20.2.2 Components of cells

Perhaps some cellular components evolved abilities to replicate themselves, independent of host cell control, and thus became parasites of those cells. Potential candidates for precursors of viruses include mRNA molecules and DNA molecules such as plasmids and transposons (Figure 20.2).

The genomes of many plus-strand RNA viruses of eukaryotic hosts have characteristics of eukaryotic mRNA. For example, the 5' end of the genome is capped in tobamoviruses, the 3' end is polyadenylated in picornaviruses, while the genomes of coronaviruses are both capped and polyadenylated. The plus strands in the dsRNA genomes of reoviruses are also capped and polyadenylated. Could some RNA viruses be descendants of cell mRNAs?

Covalently closed circular DNA molecules known as plasmids are found in prokaryotic and eukaryotic cells. Could some DNA viruses be descendants of such molecules? Some bacterial plasmids carry genes that specify the production of a protein tube (a sex pilus) that can attach to other bacterial cells and allow the transfer of a copy of the plasmid from one cell to the other. Could some phages, such as the filamentous phages or some of the tailed phages (Chapter 19), be descended from ancient plasmids and their sex pili? On the other hand it has been suggested that the sex pili of modern bacteria may have originated from filamentous phages!

Transposons are sequences of mobile DNA in the genomes of prokaryotes and eukaryotes. They are described as mobile because they can move from one part of the genome to another by 'cut and paste' and 'copy and paste' mechanisms. Perhaps some DNA viruses are descendants of transposons.

For any of these DNA or RNA molecules to evolve into a virus the putative virus genome would somehow have to acquire a range of genes, including those for capsid protein(s) and, in many cases, for a polymerase to replicate the virus genome.

Another class of cellular components that are strong contenders as precursors of viruses are retrotransposons, which are retrovirus-like sequences in eukaryotic cell genomes. A retrotransposon consists of two long terminal repeats (LTRs) flanking a *gag* gene, a *pol* gene (encodes a reverse transcriptase and an integrase) and sometimes an *env* gene. Retrotransposons thus resemble the proviruses of retroviruses (Chapter 16); they can replicate by transcribing mRNAs, which are translated into Gag and Gag–Pol proteins. Reverse transcription produces a new copy of the retrotransposon, which can be inserted into another site in the cell genome. Retrotransposons are found in the genomes of humans (Figure 20.3), other vertebrates, invertebrates, plants and fungi.

It is likely that retroviruses originated from retrotransposons, though this is a chicken-and-egg situation and it is also likely that some retrotransposons are descended from retroviral proviruses that became

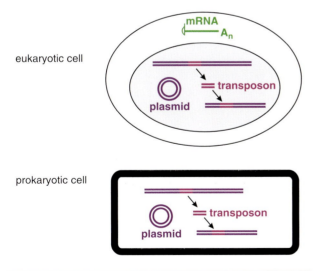

Figure 20.2 *Cell components that are potential candidates as precursors of viruses.*

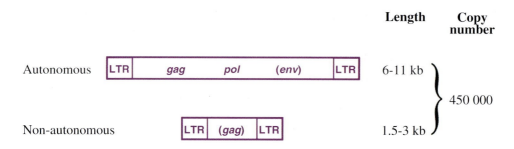

Figure 20.3 *Retrovirus-like elements in the human genome.* The autonomous elements (retrotransposons) resemble the retroviral provirus (Section 16.3.3).

inserted into the germ line of their hosts. Retrotransposons are described as autonomous elements because they have a *pol* gene that enables them to replicate.

The human genome also contains many non-autonomous elements flanked by LTRs (Figure 20.3). Some of these elements consist only of *gag* sequences, which could be remnants of either retrotransposons or retroviral proviruses.

Results from the human genome sequencing project indicate that about eight per cent of our genome is composed of LTR-flanked sequences, which are often referred to as endogenous retroviruses (Section 16.6). David Baltimore (one of the discoverers of reverse transcriptase) has said 'in places, the genome looks like a sea of reverse-transcribed DNA with a small admixture of genes'.

Other reverse transcribing viruses, such as hepadnaviruses (Chapter 18), might have similar origins to retroviruses.

20.2.3 Intracellular micro-organisms

There is strong evidence that the mitochondria and chloroplasts of eukaryotic cells are derived from prokaryotic cells that adopted new modes of life within the cells of host organisms. It is assumed that the ancestors of these organelles adopted parasitic or mutually beneficial modes of life in host cells, and that over time they became increasingly dependent on their hosts, losing the ability to perform various functions, and losing the genes that encode those functions. Perhaps a similar evolutionary process continued further, leading to greater degeneracy and

loss of functions such as protein synthesis, until the intracellular intruder was no longer a cell or an organelle, but had become a virus.

A virus that may have arisen in this way is the mimivirus (Section 1.3.1). The 1.2 Mb genome of this virus encodes a wide variety of proteins, including enzymes for polysaccharide synthesis and proteins involved in translation; it also encodes six tRNAs.

20.2.4 How did viruses originate?

The answer to this question is that we do not know! The great variety of virion structures and of virus genome types and replication strategies indicates that viruses had multiple origins. Small simple viruses, such as parvoviruses and picornaviruses, perhaps evolved from molecular precursors, while some large complex viruses, such as the mimivirus, perhaps evolved from cellular precursors.

We noted in Chapter 3 that the viruses found in certain categories of organism have a predominance of certain types of genome: dsDNA genomes in prokaryote viruses, ssRNA genomes in plant viruses and dsRNA genomes in fungal viruses. We can now note that certain structural types of virus are restricted to particular categories of host: naked, rod-shaped ssRNA viruses are restricted to plant hosts, while viruses with a head-plus-tail structure are rarely found outside the prokaryotes. The reasons for these distributions presumably concern diverse origins of the viruses in these very different hosts.

There is one feature, however, that is universal throughout the virus world: icosahedral symmetry. There are icosahedral viruses with dsDNA, ssDNA, dsRNA and ssRNA genomes, and icosahedral viruses infect bacterial, fungal, plant and animal hosts. These viruses have evolved from a variety of origins, but evolution has chosen the icosahedron for capsid design because it is a particularly stable structure.

One final thought on virus origins: we must entertain the possibility that today new viruses are continuing to evolve from molecular and cellar precursors.

20.3 Evolution of viruses

> '...endless forms most beautiful and most wonderful have been, and are being, evolved' Charles Darwin (1859).

Charles Darwin's concepts of over-reproduction and survival of the fittest that have been found to apply to cellular organisms also apply to viruses. Like a cellular organism, a virus has genes that strive to perpetuate themselves. The expression of the combination of genes that constitutes the genome permits the virus to replicate itself. Changes to individual genes and new gene combinations continually throw up new genotypes, most of which are less successful than the parental genotypes and do not survive. Occasionally, however, a new genotype is more successful than the parental genotype and might even supplant it. Sometimes a new virus genotype enables the infection of a new host species.

20.3.1 Viruses from the past

The ancestors of modern viruses ('virus fossils') are not readily available, but the reverse transcription–polymerase chain reaction (RT-PCR) has been used to recover and amplify genomes of at least two RNA viruses from the past: influenza A virus from the 1918–19 pandemic (see Section 20.3.3.c) and tomato mosaic virus. The latter virus is very stable and it can be detected in the atmosphere. Its presence in Arctic ice was predicted, and when cores of ice between 500 and 140 000 years old were examined the RNAs of 15 strains of tomato mosaic virus were detected.

20.3.2 Monitoring virus evolution

Comparing sequences of virus genomes allows the degrees of evolutionary relationship between viruses to be estimated, and even allows changes in rapidly evolving viruses (e.g. HIV-1) to be followed. Sequence data can be used to produce phylogenetic trees (Section 10.2.1), which indicate how modern virus strains may have arisen from common ancestors (Figure 20.4).

20.3.3 Mechanisms of virus evolution

In many respects, the underlying processes that drive virus evolution are the same as those that drive the evolution of cellular organisms. These processes involve the generation of genome variants, the vast majority of which are deleterious and do not survive, but a few provide an advantage in a particular niche. For a virus the niche might be a new host species or the presence of an antiviral drug, and a variant may proliferate in that niche as a new virus strain. Virus genome variants arise as a result of mutations, new gene arrangements and acquisition of cell genes; we shall consider each process in turn.

20.3.3.a Mutations

When nucleic acids are copied by polymerases some errors are made. If an error is in a protein-coding sequence and if the error results in a change in the amino acid encoded, then the error results in a mutation. Natural selection operates and the mutations that survive are those that best fit the virus for its continued survival.

There are many selection pressures on viruses. These pressures include the immune response of the host; for example a new antigenic type of an animal virus that the host's immune system has not previously encountered is at an advantage compared with antigenic types against which the host has acquired immunity. There is, therefore, heavy selection pressure on virus proteins (e.g. HIV-1 gp120) that are targets of the

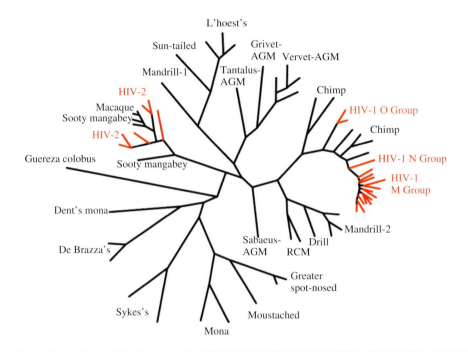

Figure 20.4 *Phylogenetic tree showing relationships between primate lentiviruses based on* pol *sequences.* Simian immunodeficiency viruses (SIVs) are denoted by the name of the host species in black. Human immunodeficiency viruses (HIVs) are indicated in red. AGM: African green monkey. RCM: red-capped mangabey. Adapted from Müller and De Boer (2006) *PLoS Pathogens*, **2** (3), e15.

host's immune response, and these proteins tend to be the least conserved. Those regions of the proteins that are targets for neutralizing antibodies are especially variable. There are, however, constraints on the evolutionary process. Virus attachment proteins must retain the configurations that enable them to bind to cell receptors, and enzymes, such as reverse transcriptases, must retain their catalytic abilities.

For some viruses there are additional constraints. A virus such as potato yellow dwarf virus, which needs to replicate in both a plant host and an insect host, can tolerate a mutation that enhances its replication in the insect host only if the mutation does not compromise its ability to replicate in the plant host, and *vice versa*.

When virus DNA genomes are replicated, the error rates are much lower than when RNA genomes are replicated (Figure 20.5). This is because DNA-dependent DNA polymerases have a proofreading mechanism that corrects most of the errors, whereas RNA polymerases and reverse transcriptases have no

such mechanism. As a result, DNA viruses, such as papillomaviruses, evolve much more slowly than RNA viruses, such as picornaviruses.

> 'RNA viruses deserve their reputation as Nature's swiftest evolvers'
> Michael Worobey and Edward Holmes (1999).

The high error rates during RNA replication and reverse transcription mean that for an RNA virus there is no fixed sequence of bases for the genome. Instead, the virus genome exists as a large number of variants; the term 'quasispecies' has been coined to describe the group of variants that collectively constitute the genome of an RNA virus (Figure 20.6). Many variants have only a fleeting existence, while those best adapted to a particular niche dominate in that niche. The high mutation rate produces some variants that may enable a

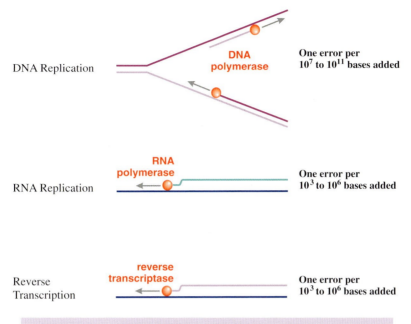

Figure 20.5   Error rates of DNA polymerases and RNA polymerases.

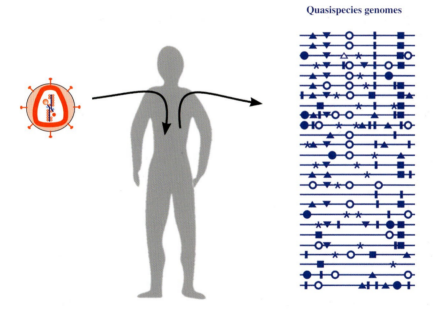

Figure 20.6   *Quasispecies genomes of an RNA virus in an infected human.* Virus genomes are represented as lines, and mutations as symbols on the lines. Mutations result in a spectrum of related genomes, termed viral quasispecies. Modified with the permission of the author and the publisher (Springer Verlag) from Domingo *et al.* (2006) Viruses as quasispecies: biological implications *Current Topics in Microbiology and Immunology*, Vol. 299, pp. 51–82.

virus to adapt to new niches, for example an expansion of its host range, should the opportunities arise.

The first step in the replication of HIV genomes involves reverse transcription, with an estimated error rate around one error per 10^4 bases added. As the HIV genome contains about 10^4 bases, this means that there is an average of one error each time the genome is reverse transcribed. There is evidence that the human immunodeficiency viruses evolved from viruses that jumped from chimpanzees and sooty mangabeys into humans (Section 20.3.4). Since jumping species, the viruses have evolved rapidly, especially the M (main) group of HIV-1, which has radiated into many subtypes and sub-subtypes.

Rapid evolution of HIV can be detected within the body of an infected individual, in which there may be $10^{10}-10^{12}$ new virions produced each day. There is therefore constant production of a huge amount of genetic variability. We mentioned in Chapter 17 that R5 strains of HIV-1 predominate in a newly infected host, but in many individuals as the infection proceeds there is evolution towards X4 and R5X4 strains. The constant production of new variants also means that the virus can evolve rapidly in response to a selection pressure, such as the presence of an anti-retroviral drug.

High diversity is also found in the genomes of other RNA viruses, such as hepatitis C virus, where isolates from a patient can have up to five per cent difference in their nucleotide sequences.

The capacity for rapid evolution of RNA viruses creates problems for the development and maintenance of some vaccination programmes. Vaccination with attenuated poliovirus vaccine initiates infection in the gut with one or more of the vaccine strains, and the infection normally lasts for 1–2 months. Some immunodeficient vaccinees, however, remain infected for much longer periods of time, and in these individuals the vaccine viruses almost always evolve. There are differences of only a few bases between the genomes of neurovirulent poliovirus strains and those of the vaccine strains (Section 24.2), so reversion to neurovirulence is common in these chronically infected individuals, who may develop paralytic poliomyelitis and may transmit virulent poliovirus to others.

Another example of RNA virus variability having practical consequences for vaccination is foot and mouth disease virus. Seven distinct serotypes of this virus have evolved, and vaccination against one serotype does not provide protection against the others. In addition, there are many subtypes of each serotype and immunity to one subtype confers only partial immunity to the related subtypes.

Hepatitis B virus (HBV) is a DNA virus, but, because it replicates its genome by reverse transcription, error rates similar to those of RNA viruses are observed. More than half of the HBV genome is read in two reading frames and regulatory elements are all within protein coding regions (Section 18.6). There are, therefore, severe constraints on the ability of HBV to mutate and remain viable but, in spite of these constraints, variation is much greater than in DNA viruses that replicate their genomes solely with DNA-dependent DNA polymerases.

20.3.3.b Recombination

Recombination is a process that results in the production of a new genome, derived from two parental genomes. For a cellular organism recombination can take place between DNA molecules within a cell. For viruses recombination may occur when a cell is infected with two related viruses; the new virus that is produced is referred to as a recombinant virus. Both DNA and RNA viruses can undergo recombination.

The first evidence for recombination in a virus came from studies with the DNA phage T2, and since then recombination has been documented for other DNA viruses. The formation of recombinants between different baculoviruses (DNA viruses that infect invertebrates such as insects), and between herpes simplex virus 1 and herpes simplex virus 2, has been reported. The recombination mechanism operating for DNA viruses is cleavage of the parental DNA molecules followed by ligation (Figure 20.7). The genomes of RNA viruses may be able to undergo recombination by a similar mechanism.

For ssRNA viruses such as picornaviruses, it is thought that recombinant genomes can also be produced by a template switching mechanism. This involves displacement of an RNA polymerase transcribing a molecule of RNA, followed by association of the enzyme with another RNA molecule and resumption of RNA synthesis using the new template (Figure 20.8).

DNA strain 1

DNA strain 2

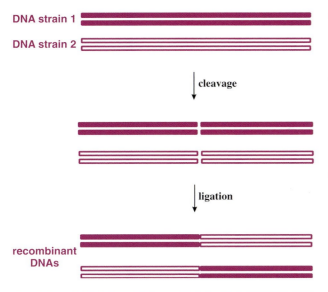

cleavage

ligation

recombinant
DNAs

Figure 20.7 *Recombination between the DNAs of two virus strains.*

Recombination appears to be a common event in retroviruses and pararetroviruses. It probably occurs during reverse transcription (Sections 16.3.2 and 18.8.6) by a template switching mechanism similar to that described above, with the reverse transcriptase switching between two RNA templates. The RNA templates for a retrovirus are in the infecting virion, derived from two virus strains co-infecting the cell in which the virion was assembled. For a pararetrovirus the RNA templates are present in the cytoplasm of a cell co-infected with two strains. Many recombinant strains of HIV-1 have evolved; for example, a recombinant derived from HIV-1 subtypes A and E has become prevalent in Thailand.

Recombination occurs in nature, but recombinants with desired combinations of genes can be generated in the laboratory. Examples are the recombinant vaccinia virus that was produced for vaccination of foxes (Section 15.2.1) and recombinant baculoviruses that are produced for use in gene expression systems.

20.3.3.c Reassortment

Reassortment is a category of recombination that may occur with segmented genome viruses that have all the segments packaged in one virion, for example reoviruses (Chapter 13), bunyaviruses and influenza viruses (Section 15.5). If a cell is co-infected by two strains of one of these viruses the progeny virions may contain mixtures of genome segments from the two parental strains. These virions are known as reassortants.

The formation of influenza A virus reassortants is illustrated in Figure 20.9. The virus genome is composed of eight segments of minus-strand RNA; two of the segments encode the envelope glycoproteins haemagglutinin (H) and neuraminidase (N). The figure shows how reassortment gives rise to a virus strain with a new H gene.

During the 20th century there were three pandemics of influenza caused by new reassortants of influenza A virus (Table 20.1). The most devastating pandemic occurred in 1918–19, when the number of people killed by the virus was greater than the number killed during the First World War. Using RT-PCR, influenza virus RNA sequences have been recovered from the tissues of some victims. In one project lung tissue that had been fixed in formalin and embedded in paraffin was the source, while another project used lung tissue from a body that had been buried in permafrost in Alaska. The influenza virus sequences that were found differ from any H1N1 viruses known today and the origins of the 1918 virus remain obscure.

Using reverse genetics it has been possible to clone the genes of the 1918 virus. This has enabled studies on the roles of individual genes in virus pathogenicity. It was shown that the H gene of the 1918 virus conferred

Table 20.1 Influenza A virus reassortants responsible for pandemics. Each pandemic was caused by a reassortant with a new combination of haemagglutinin (H) and neuraminidase (N) genes

Year	Influenza A virus reassortant	Parental viruses
1918	H1N1	Uncertain
1957	H2N2	Human H1N1, avian H2N2
1968	H3N2	Human H2N2, avian H3

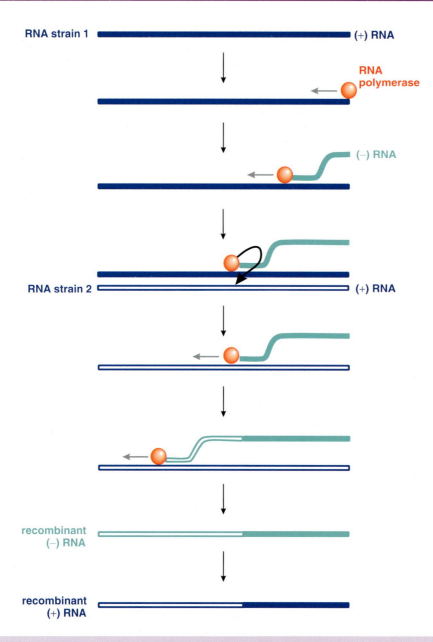

Figure 20.8 *Template switching mechanism of recombination in ssRNA viruses.* A cell is infected with two strains of a virus. During replication of strain 1 RNA the RNA polymerase, together with the incomplete complementary strand of RNA, dissociates from the template RNA and associates with a new template (strain 2 RNA). RNA synthesis is completed on the new template, resulting in a hybrid complementary strand with part of its sequence derived from strain 1 and part from strain 2. When this strand is used as a template to generate new genome strands, the resulting RNA molecules are recombinant.

pathogenicity to human influenza A virus strains that were non-pathogenic in mice.

The origins of the viruses responsible for the two more recent influenza pandemics have been traced. In each case the virus was a reassortant between a human virus and an avian virus; the 1957 reassortant derived the H and N genes from an avian virus, while the 1968 reassortant derived the H gene from an avian virus (Table 20.1).

20.3.3.d Acquisition of cell genes

The discovery of similarities between some virus proteins and cell proteins has led to the suggestion that some viruses have 'captured' cell genes. The genomes of some retroviruses contain oncogenes that they probably acquired from their host cells, which contain similar genes (proto-oncogenes). An example is the *src* gene of Rous sarcoma virus (Section 16.4); a very similar gene (c-*src*) is present in the cells of all vertebrates. A retrovirus might acquire an oncogene by recombination between the proviral DNA and host DNA, or by RNA synthesis during transcription continuing from the provirus through into a proto-oncogene adjacent in the cell chromosome.

Some virus proteins modulate the immune system of the host, often by mimicking cell proteins. For example, some of the human herpesviruses encode proteins similar to cytokines and major histocompatibility complex proteins. Several large DNA viruses encode proteins similar to interleukin-10 and some of these proteins have been shown to suppress cell-mediated immune responses. Are some of the virus genes that encode these proteins derived from cell genes and, if so, how did the viruses acquire the genes?

One possible mechanism is recombination, though in order for this to occur the genomes of the virus and the cell must come into close proximity. This may happen with DNA viruses that replicate in the nucleus (e.g. herpesviruses), but there are many DNA viruses (e.g. poxviruses) that do not enter the nucleus, so cellular and viral genomes are never in close proximity.

A second possible mechanism for viral acquisition of a cell gene might be through synthesis of a DNA copy of a cell mRNA, followed by insertion of the DNA into the virus genome. The DNA synthesis would require a reverse transcriptase, possibly supplied by the cell or by a retrovirus co-infecting the cell.

20.3.4 Evolution of new viruses

From time to time a new virus appears, and mystery surrounds its origin. Often there are suspects for the origin of the virus and these suspects may be viruses in other host species. The genomes of the new virus and the suspects can be sequenced, and if a close match is found between the new virus and one of the suspects then the mystery has probably been solved. It is commonly found that a new virus is one that has infected a new host species in which some further evolution has then taken place.

> 'Crossing the species barrier – one small step to man, one giant leap to mankind'
> Mark Klempner and Daniel Shapiro (2004).

Many virus infections across species barriers are dead ends for the viruses as no transmissions to further individuals occur, but sometimes a virus adapts to enable transmission from host to host. A virus may have to undergo one or more adaptations in order for it to

- replicate in the new host;

- evade immune responses of the new host;

- transmit to other individuals of the new host.

In 1978 a new parvovirus appeared in dogs and rapidly spread around the world. The genome sequence of this canine parvovirus was found to be more than 99 per cent identical to that of a parvovirus in cats (feline panleukopenia virus). Both viruses can infect feline cells, but only the canine virus can infect canine cells. The feline virus is thought to be the precursor of the canine virus.

Similarly, the genome sequences of HIV-1 and HIV-2 are very similar to those of simian immunodeficiency viruses found in chimpanzees (SIV_{cpz}) and sooty

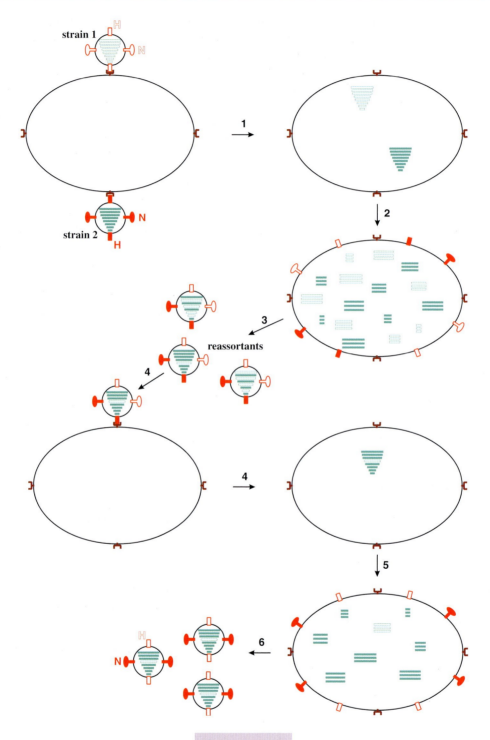

Figure 20.9

mangabeys (SIV_{sm}), respectively (Figure 20.4). The human viruses are believed to have originated when these simian viruses jumped species, perhaps when humans came into contact with the blood of infected animals. Other examples of viruses that have emerged in this way are discussed in Chapter 21.

20.3.5 Co-evolution of viruses and their hosts

A virus–host association that has existed for a long period is likely to have evolved a relationship in which the host suffers little or no harm. Examples of such viruses are the dependoviruses (Chapter 12), and some reoviruses, which acquired the 'o' in their name because they were found to be 'orphan' viruses (not associated with any disease; Chapter 13). The members of these virus groups that infect *Homo sapiens* have probably been with us since we diverged as a separate species.

When a virus extends its host range into a new species it is often much more virulent in the new host than in the old. SIV_{cpz} and SIV_{sm} are both avirulent in their simian hosts, but their derivatives, HIV-1 and HIV-2, are lethal to humans. Of course the potential for evolution is usually much greater for a virus than it is for its host. As noted earlier, the reproductive capacity of HIV (10^{10}–10^{12} new virions per day) dwarfs that of the host.

A virulent virus may affect the evolution of its host by causing genes for susceptibility to infection with the virus to become extinct, whilst selecting in favour of genes for resistance. HIV-1 is currently selecting for the 32-nucleotide deletion in the *CCR5* gene that confers resistance to infection (Section 17.4.1).

A good example of co-evolution of a virus–host relationship is myxoma virus in the European rabbit. In the natural host of the virus, a rabbit species in South America, infection results in the development of skin tumours, which eventually heal. In contrast, when myxoma virus infects the European rabbit it causes myxomatosis, characterized by acute conjunctivitis (Figure 20.10), lumps on the skin and loss of appetite. The disease is almost always fatal, and for this reason the virus was introduced into Australia as a biological control agent against the European rabbit. Within a year of its introduction there was evidence of virus evolution. Myxoma virus strains with attenuated virulence had evolved, apparently selected because they were transmitted more effectively. Subsequently, selection for rabbits with resistance to myxomatosis occurred and both the virus and its new host have continued to co-evolve.

Figure 20.9 *Influenza virus reassortment*

1. A cell is co-infected with two strains of influenza virus. Each strain depicted has a unique haemagglutinin (H) and a unique neuraminidase (N).

2. The RNAs of both strains are replicated.

3. Progeny virions with genomes made up from mixtures of RNA segments from the parental strains are called reassortants. H and N proteins of both virus strains are present in the virion envelopes as the infected cell expresses the genes of both strains.

4. A reassortant infects another cell. This reassortant has the H gene of strain 1 (the fourth genome segment), while the other genes are derived from strain 2.

5. The RNAs of the reassortant are replicated.

6. Virions of the reassortant bud from the infected cell.

Figure 20.10 *Rabbit with myxomatosis.*

The ultimate in virus–host co-evolution occurs if a virus genome becomes permanently integrated into the genome of its host. This is probably what happened to create at least some of the endogenous retrovirus/retrotransposon sequences in the genomes of eukaryotes (Sections 16.6 and 20.2.2).

The genomes of prokaryotes too contain sequences that originated as a result of integration of viral sequences through the process of lysogeny (Section 19.1). Phage sequences (prophages) have been found in the majority of the bacterial genomes that have been sequenced. Some of these prophages are inducible, meaning that they can be activated, virus replication can ensue and the host can be lysed. Other prophages are defective and replication cannot be induced. The genes for some characteristics of bacteria are encoded by prophage sequences, for example, the cholera toxin gene of *Vibrio cholerae* and toxin genes in *E. coli* O157.

Sources of further information

Books

Domingo E., editor (2006) *Quasispecies: Concept and Implications for Virology, Current Topics in Microbiology and Immunology* Vol. 299, Springer

Flint S. J. *et al.* (2004) Chapter 20 in *Principles of Virology: Molecular Biology, Pathogenesis and Control of Animal Viruses*, 2nd edn, ASM Press

Pevsner J. (2003) Chapter 13 in *Bioinformatics and Functional Genomics*, Wiley

Journals

Forterre P. (2006) The origin of viruses and their possible roles in major evolutionary transitions *Virus Research*; **117**, 5–16

Froissart R. *et al.* (2005) Recombination every day: abundant recombination in a virus during a single multi-cellular host infection *PLoS Biology*, **3**, e89

Ghedin E. *et al.* (2005) Large-scale sequencing of human influenza reveals the dynamic nature of viral genome evolution *Nature*, **437**, 1162–1166

Holmes E. C. (2003) Error thresholds and the constraints to RNA virus evolution *Trends In Microbiology*, **11**, 543–546

Hueffer K. *et al.* (2003) The natural host range shift and subsequent evolution of canine parvovirus resulted from virus-specific binding to the canine transferrin receptor *Journal of Virology*, **77**, 1718–1726

Klempner M. S. and Shapiro D. S. (2004) Crossing the species barrier – one small step to man, one giant leap to mankind *New England Journal of Medicine*, **350**, 1171–1172

Lukashev A. N. (2005) Role of recombination in evolution of enteroviruses *Reviews in Medical Virology*; **15**, 157–167

Shackelton L. A. and Holmes E. C. (2004) The evolution of large DNA viruses: combining genomic information of viruses and their hosts *Trends in Microbiology*, **12**, 458–465

Worobey M. and Holmes E. C. (1999) Evolutionary aspects of recombination in RNA viruses *Journal of General Virology*, **80**, 2535–2543

21

Emerging viruses

At a glance

Emerging Viruses	Examples
• Viruses in new hosts	Nipah virus in pigs and humans
• Viruses in new areas	West Nile virus in North America
• Newly evolved viruses	Influenza virus reassortants
• Recently discovered viruses	Human metapneumovirus
• Re-emerging viruses	Mumps virus

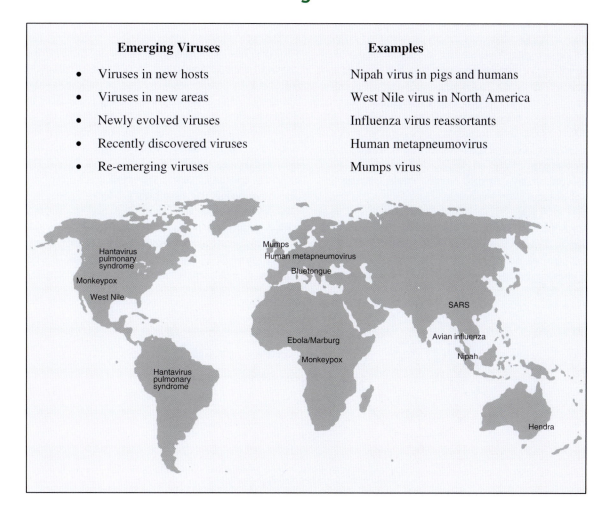

Virology: Principles and Applications   John B. Carter and Venetia A. Saunders
© 2007 John Wiley & Sons, Ltd   ISBNs: 978-0-470-02386-0 (HB); 978-0-470-02387-7 (PB)

21.1 Introduction to emerging viruses

The term 'emerging virus' is used in a number of contexts: it may refer to a virus that has recently made its presence felt by infecting a new host species, by appearing in a new area of the world or by both. Sometimes a virus is described as a 're-emerging virus' if it has started to become more common after it was becoming rare. Foot and mouth disease virus (Section 14.2.5) re-emerges in the UK from time to time.

Human activities may increase the likelihood of virus emergence and re-emergence. Travel and trade provide opportunities for viruses to spread to new areas of the planet, and the introduction of animal species into new areas, e.g. horses into Australia, may provide new hosts for viruses in those areas.

Other activities that may result in virus emergence involve close contact with animals, including the hunting and killing of non-human primates for bushmeat. It has been shown that viruses such as simian immunodeficiency viruses (SIVs) can be present in the meat and there is a risk of acquiring an infection when the meat is handled. Simian-to-human transmission of SIVs is thought to have occurred several times, resulting in the major groups of HIV-1 and HIV-2 (Figure 20.4), and there is concern that further viruses might emerge as a result of contact with bushmeat.

If a virus jumps into a new host species it may undergo some evolutionary changes in the new host, resulting in a new virus. This is how HIV-1 and HIV-2 (Section 20.3.4) were derived from their SIV precursors. Other new viruses emerge when recombination and reassortment result in new viable combinations of genes (Section 20.3.3); new strains of influenza A virus come into this category. Some 'new' viruses that are reported are actually old viruses of which mankind has recently become aware, such as Kaposi's sarcoma-associated herpesvirus (Chapters 11 and 22).

This chapter considers examples of viruses that have 'emerged' and 're-emerged' in the late 20th century and the 21st century. Most of the examples discussed are human viruses.

21.2 Viruses in new host species

21.2.1 Bunyaviruses

In 1993 in a region of the south-western US known as Four Corners some residents became afflicted by an illness resembling influenza; many of them developed severe lung disease and died. The region is normally very dry, but there had been unusually high rainfall and snowfall that had resulted in a burst of plant growth followed by explosions in the populations of small mammals. One of those mammals was the deer mouse, a species that is attracted to human habitation.

On investigation, it was found that many of the deer mice were persistently infected with a virus that was excreted in their urine, droppings and saliva. Humans exposed to these materials were becoming infected and developing the respiratory illness. The virus was characterized as a new hantavirus and the disease was called hantavirus pulmonary syndrome.

Hantaviruses are members of the family *Bunyaviridae* (Figure 21.1) and are named after the River Hantaan in Korea, where the first of these viruses was

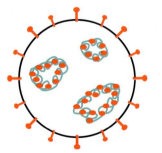

- Genome: ssRNA (three segments, which are either all minus strand, or a mixture of minus strand and ambisense). Each segment is circular as a result of base pairing at its ends.

- Capsid symmetry: helical

- Enveloped virion

- Includes genus *Hantavirus*, e.g. Sin Nombre virus

Figure 21.1 Family *Bunyaviridae*.

isolated during the Korean War from soldiers who had developed haemorrhagic fever with renal syndrome. Similar viruses are known elsewhere in Asia and in Europe.

There was much heated debate as to what name should be given to the virus that had appeared in Four Corners. Local people did not want it named after a place as they did not want potential tourists to be discouraged by publicity surrounding a virus disease. Eventually the Spanish name Sin Nombre virus (the virus without a name) was agreed.

Since 1993 hantavirus pulmonary syndrome, caused by Sin Nombre virus and other hantaviruses, has been reported in many parts of North America and South America, with mortality rates around 50 per cent.

21.2.2 Paramyxoviruses

21.2.2.a Hendra virus

In 1994 at Hendra in south-east Australia there was an outbreak of pneumonia in horses, then a trainer and a stable hand who worked with the horses developed severe respiratory disease. The horses and the two humans were found to be infected with a virus that had characteristics of the family *Paramyxoviridae* (Figure 21.2). This family contains well known viruses such as measles and mumps viruses, but the virus that was isolated was previously unknown. Thirteen of the horses and the trainer died as a result of infection with the virus, which was named Hendra virus.

A couple of years later a man who lived about 800 km north of Hendra developed seizures and paralysis. He was found to be infected with Hendra virus and he died from meningoencephalitis. It turned out that he had helped with post mortems of two horses 13 months earlier. Tissue from the horses had been preserved and when it was examined it was found to contain Hendra virus.

It looked as though the infected humans had acquired the virus from horses, but the source of infection for horses was a mystery. An investigation was initiated, and evidence was obtained that the virus is present in populations of fruit-eating bats; Hendra virus-specific antibodies were found in all four local species of the bat genus *Pteropus*, and infectious virus was isolated from one bat. Bats appear to be the normal hosts of Hendra virus, and experimental infection of several bat species has provided no evidence that infection causes disease in these animals. It appears that occasional transmission to horses occurs, with very occasional transmission from horses to humans (Figure 21.3). In these species the outcome of infection is very different, with the virus demonstrating a high degree of virulence.

21.2.2.b Nipah virus

In 1997 a new disease of pigs appeared in Malaysia. Affected animals developed respiratory disease and encephalitis. Soon afterwards workers on the affected pig farms began to develop encephalitis; over a two-year period there were several hundred human cases, with over 100 deaths. In 1999 a new virus was isolated from the brain of one of the patients who had died. The

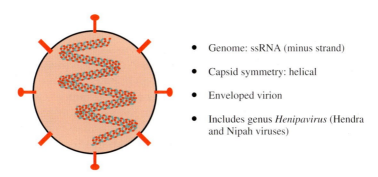

- Genome: ssRNA (minus strand)
- Capsid symmetry: helical
- Enveloped virion
- Includes genus *Henipavirus* (Hendra and Nipah viruses)

Figure 21.2 Family *Paramyxoviridae*.

Figure 21.3 *Hendra virus transmission routes.*

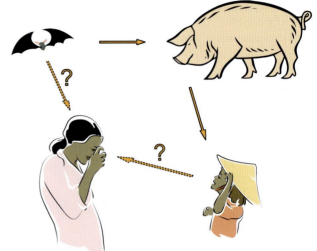

Figure 21.4 *Nipah virus transmission routes.* Bat-to-pig and pig-to-human transmission occur. Bat-to-human and human-to-human transmission may also occur.

virus turned out to be a paramyxovirus with similar characteristics to Hendra virus (Figure 21.2) and it was named Nipah virus. In an attempt to contain the outbreak over one million pigs were slaughtered.

A search was initiated to find the reservoir of infection. As with Hendra virus, bats were investigated and evidence of Nipah virus infection was found in fruit-eating bats of the genus *Pteropus*. Infectious virus was isolated from bats' urine and virus-specific antibodies were found in their blood. Thus the situation parallels that of Hendra virus, with a reservoir of infection in bats and transmission to humans via an intermediate mammalian host (Figure 21.4). More recently there have been outbreaks of Nipah virus encephalitis in humans in Bangladesh and India. In these outbreaks there was no evidence of transmission from pigs and it is believed that transmission may have occurred from bats to humans. There was also evidence of human-to-human transmission of the virus.

21.3 Viruses in new areas

21.3.1 West Nile virus

In 1999 some people in New York became ill with apparent viral encephalitis; several of them died. At about the same time several birds in the zoo became ill, and deaths of wild birds, especially crows, were reported. Diagnostic tests revealed that these human patients and the birds were infected with West Nile virus (WNV).

This virus has been known since 1937 when it was first isolated from the blood of a woman in the West Nile region of Uganda. In addition to Africa the virus is also known in Eastern Europe, West Asia and the Middle East, but until 1999 it had never been reported in North America. Most infections of humans and animals with WNV are asymptomatic, but about 20 per cent of human infections result in signs and symptoms such as fever and aching muscles. Less than one per cent of infections spread to the central nervous system, causing severe illness (such as encephalitis), which may result in death; these cases are mainly in those over 60 years old.

WNV is classified in the family *Flaviviridae* (Figure 21.5). The virus has a wide host range and is common in birds, humans and other vertebrates; it also replicates in mosquitoes, which act as vectors between the vertebrate hosts (Chapter 4). Protective measures include use of insect repellents, wearing 'mosquito jackets' and treating mosquito breeding areas with insecticides.

It is not known how WNV was introduced into North America, but once there it spread rapidly throughout the US (Figure 21.6); within five years

- Genome: ssRNA (plus strand)

- Capsid symmetry: icosahedral

- Enveloped virion

- Includes West Nile virus and hepatitis C virus

Figure 21.5 Family *Flaviviridae*.

it was also present in Canada, Central America and islands in the Caribbean. The virus has caused hundreds of human deaths in North America.

The spread of WNV to North America prompted an investigation to determine whether the virus is present in birds in the UK. WNV-specific antibodies were found in a number of species and, using RT-PCR, WNV RNA was detected in magpies and a blackbird, though all the birds appeared healthy.

21.3.2 Bluetongue virus

Bluetongue virus (BTV), a member of the genus *Orbivirus* in the family *Reoviridae* (Chapter 13), causes disease in ruminant animals, including sheep,

goats and cattle. Like WNV it has a wide host range in vertebrate animals and replicates in insects, which act as vectors.

All ruminant species are susceptible to infection with BTV, but infection results in severe disease and death only in certain hosts, mainly some breeds of sheep and some species of deer. In some affected animals the tongue may become swollen and blue. Infected animals that remain healthy can act as reservoirs of infection. The virus is transmitted between its ruminant hosts by certain species of biting midge in the genus *Culicoides*, so the virus is restricted to areas of the world where these midge species occur, and transmission is restricted to those times of the year when the midges are active.

Bluetongue has been known in Africa and the Middle East for a long time. Occasionally it has made brief incursions into Europe, but in 1998 six serotypes of BTV began to spread north. By 2004 there were cases 800 km further north than ever before (Figure 21.7) and over a million sheep had been killed by the disease. It has been found that the midge species that is the main BTV vector has extended its range north in Europe, presumably because of the climate becoming warmer, and this is a major factor responsible for the spread of the virus.

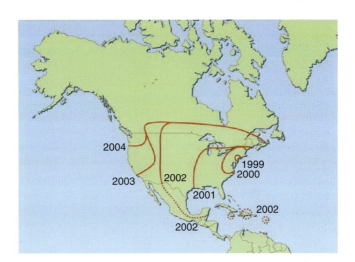

Figure 21.6 *Approximate distribution of West Nile virus in the Americas 1999–2004.* From Mackenzie *et al.* (2004) *Nature Medicine*, **10**, S98. Reproduced by permission of Nature Publishing Group and the author.

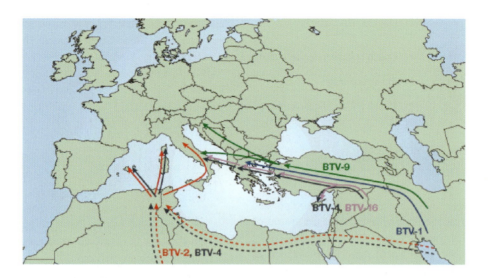

Figure 21.7 *Spread of bluetongue virus serotypes in Europe 1998–2004.* From Purse *et al.* (2005) *Nature Reviews Microbiology*, **3**, 171. Reproduced by permission of Nature Publishing Group and the authors.

21.4   Viruses in new host species and in new areas

21.4.1 Filoviruses

In 1967 in the town of Marburg in Germany some laboratory workers became ill with a haemorrhagic fever (a terrible disease characterized by diarrhoea and vomiting, as well as by haemorrhaging and fever). These people had been in contact with blood, organs and cell cultures from African green monkeys caught in Uganda. Seven of those affected died, and there were five cases of the illness in hospital staff that had been in contact with patients' blood. Investigations revealed that the monkeys had been infected with a virus that had been transmitted to the laboratory staff and from them to the hospital staff.

This virus was of a type never previously encountered, with elongated virions, some of which were straight and some were curved. It was named Marburg virus and is now classified in the family *Filoviridae*, named from the Latin *filum*, meaning a thread (Figure 21.8).

In 1976 there were outbreaks of a similar disease in Africa near the River Ebola in the Democratic Republic of Congo (then Zaire) and in Sudan. A virus

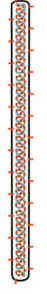

- Genome: ssRNA (minus strand)

- Capsid symmetry: helical

- Enveloped virion

- Members are Marburg and Ebola viruses

Figure 21.8   Family *Filoviridae*.

similar to Marburg virus was isolated from patients and was named Ebola virus. Since then there have been a number of outbreaks of disease caused by Ebola and Marburg viruses across central Africa, from Côte d'Ivoire in the west to Kenya in the east and to Angola in the south.

The way in which the outbreaks start has long been a mystery. There is increasing evidence that some outbreaks start when a human becomes infected as a result of contact with the blood of an infected non-human primate. It was known from the original outbreak in Germany that African green monkeys can be infected with Marburg virus. It has since been found that gorillas, chimpanzees and duikers can be infected with Ebola virus, which may be responsible for significant mortality of these species. Ebola and Marburg viruses are present in the blood of infected hosts and transmission to humans can occur through contact with the flesh of infected animals after they have been hunted and killed. Human-to-human transmission readily occurs through contact with the blood of infected individuals.

There are several species of Ebola virus in Africa. A further species has appeared in the US and Italy in monkeys imported from the Philippines; there is serological evidence that animal handlers have been infected, but there have been no reports of this virus causing disease.

It is thought likely that additional animal species act as 'reservoirs' of Marburg and Ebola viruses, and that infection in these species is likely to cause few or no signs of disease. There have been many expeditions to search for reservoir hosts, but very few produced positive results; Ebola virus RNA has been detected in some small mammals including fruit bats, and anti-Ebola virus antibodies have been detected in some fruit bats. There remain unanswered questions concerning reservoir species and host ranges of these viruses.

21.4.2 Monkeypox virus

Monkeypox is a disease of monkeys that is similar to smallpox in man. The viruses that cause these diseases are members of the family *Poxviridae* (Figure 21.9). Since 1970 there have been cases in Africa of human disease caused by infection with monkeypox virus; some cases have been fatal.

In 2003 some people in the US developed a mystery illness that involved a rash and a fever. Investigations revealed that they were infected with monkeypox virus, which had entered the country in small mammals imported from Ghana. The virus had infected pet prairie dogs and had then been transmitted to humans.

Sequencing the genomes of monkeypox virus isolates has revealed the existence of two strains that appear to differ in virulence. The virus that was imported into the US was the less virulent strain.

21.5 New viruses

21.5.1 SARS coronavirus

In 2002 a new human respiratory disease emerged in southern China. The following year one of the doctors who had been treating patients travelled to Hong Kong, where he became ill and died. Subsequently, people who had stayed in the same hotel as the doctor travelled to Singapore, Vietnam, Canada and the US, taking the infectious agent with them. The epidemic of severe acute respiratory syndrome (SARS) was under way.

The signs and symptoms of SARS resemble those of influenza and include fever, aching muscles, sore throat, cough and shortness of breath. About 90 per

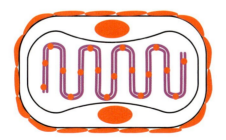

- Genome: dsDNA

- Large, complex virion

- Members include smallpox and vaccinia viruses

Figure 21.9 Family *Poxviridae*.

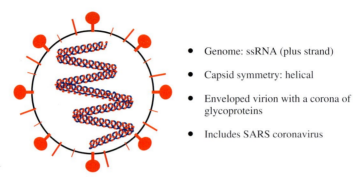

- Genome: ssRNA (plus strand)
- Capsid symmetry: helical
- Enveloped virion with a corona of glycoproteins
- Includes SARS coronavirus

Figure 21.10 Family *Coronaviridae*.

cent of patients recovered, but for the remainder the infection proved to be lethal. These were mainly individuals who had an underlying condition such as diabetes, heart disease or a weakened immune system. On the face of it SARS is a respiratory tract disease, but in many patients the infection spread to other parts of the body. Diarrhoea developed in some patients and the virus was shed in the faeces and urine for several weeks.

The causative agent was found to be a new coronavirus (Figure 21.10). No natural reservoir for the SARS coronavirus has been found, but coronaviruses with very similar genome sequences were isolated from animals sold in markets in the region of China where the first SARS cases appeared. Antibodies to these viruses were found in many workers in the markets, but none of the workers had any history of SARS. A similar virus has also been isolated from several bat species. It is likely that coronaviruses have repeatedly crossed into humans from other mammalian species, but in the majority of cases the virus has lacked the ability for efficient human-to-human transmission. It would seem that the SARS coronavirus evolved from a rare virus that had this ability.

The SARS outbreak was brought under control by quarantine measures, but only after there had been over 8000 cases with nearly 800 deaths.

21.5.2 New strains of influenza virus A

The family *Orthomyxoviridae* includes the influenza A viruses (Figure 3.20). The virions have two species of surface glycoprotein: a haemagglutinin (H) and a neuraminidase (N). There are 16 types of H antigen and nine types of N antigen, and there are many subtypes of each type. From time to time a virus emerges with a new combination of H and N genes formed by reassortment, and causes a pandemic (see Chapter 20).

The hosts of influenza A viruses are principally birds that frequent aquatic habitats. The birds (e.g. ducks, geese, gulls) acquire infections by ingestion or inhalation and the viruses infect their intestinal and/or respiratory tracts. Infection with most virus strains results in few or no signs of disease, but some strains are highly pathogenic and can kill their avian hosts. The viruses can be spread to new areas when the birds migrate.

Some influenza A viruses infect mammalian species including pigs, horses and humans; the respiratory tract is the main site of virus replication. Normally humans are infected only with viruses that have H type 1, 2 or 3 and N type 1 or 2. Patients are commonly very ill and some die, either as a direct result of the virus infection, or indirectly from secondary pathogens, which are able to infect as a result of damage to the respiratory epithelium.

The previous two paragraphs describe the normal situation, but exceptions occur: some avian strains of influenza A virus can be highly pathogenic in birds, can be transmitted from wild birds to domestic poultry and can be transmitted to humans (Figure 21.11). This was the situation in Hong Kong in 1997, when an H5N1 virus caused an outbreak of serious disease in poultry. Eighteen people also became infected with the virus and six of them died. In order to bring the outbreak to an end all poultry in Hong Kong were slaughtered.

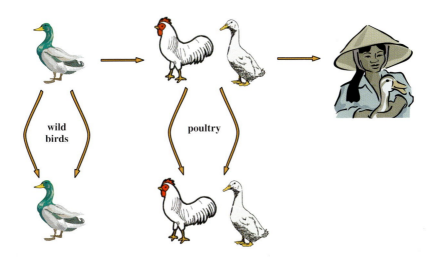

Figure 21.11 *Transmission of avian influenza A viruses.* The viruses are present in wild birds from which they may be transmitted to poultry. Sometimes there is transmission to humans who have close contact with poultry.

H5N1 viruses appeared in a number of Asian countries in 2003, in Europe in 2005 and in Africa in 2006. Millions of ducks, chickens and turkeys died from the disease or were slaughtered; H5N1 viruses also infected humans, causing severe respiratory disease, culminating in death in many cases. At about the same time H9N2 viruses emerged in Asia, causing disease in poultry with some transmission to humans.

Most, if not all, of the human cases of 'bird flu' caused by H5N1 and H9N2 viruses were in people who work with poultry and who presumably became infected as a result of direct contact with virus on the birds themselves, or in their faeces. Like many of the animal coronaviruses that have infected humans (Section 21.5.1), these avian influenza viruses appeared to have little or no propensity for human-to-human transmission. H5N1 viruses have also infected other mammalian species including domestic cats; tigers and leopards at a zoo in Thailand died as result of infections that they acquired from eating virus-contaminated chickens.

21.6 Recently discovered viruses

21.6.1 Human metapneumovirus

In 2001 a new member of the family *Paramyxoviridae* (Figure 21.2) was isolated from young children in the Netherlands. The virus was named human metapneumovirus. Infection can cause mild respiratory problems, but can also cause bronchiolitis and pneumonia, with symptoms similar to those caused by infection with respiratory syncytial virus, another member of the family *Paramyxoviridae*. There is serological evidence that the metapneumovirus had been present in humans for at least 50 years, and that in the Netherlands most children have become infected with the virus by the age of five.

Infants and children with acute respiratory tract infection due to human metapneumovirus have since been reported in a number of countries, including the UK.

21.7 Re-emerging viruses

21.7.1 Measles and mumps viruses

Further members of the family *Paramyxoviridae* (Figure 21.2) are measles and mumps viruses. Both are important human pathogens, especially measles, which is a major cause of mortality in developing countries.

Until the late 1990s cases of measles and mumps had been declining in the UK as a result of widespread uptake of the measles, mumps and rubella (MMR) vaccine, but after fears were raised surrounding the safety of the vaccine fewer children were vaccinated. Measles, and especially mumps, began to 're-emerge', with hundreds of measles cases and thousands of mumps cases each year.

21.7.2 Bioterrorism

There are concerns that the world may witness the re-emergence of a virus as a result of a deliberate release by terrorists. There is a long list of potential agents, including viruses, that terrorists might consider using. Smallpox virus, a member of the family *Poxviridae* (Figure 21.9), is high on the list. There have been no smallpox cases since the 1970s, smallpox vaccination programmes have been discontinued and most humans lack immunity to this virus. Some governments have stockpiled vaccine to be used in the event of a smallpox release.

21.8   Virus surveillance

In order that the threats posed by emerging viruses and other agents can be dealt with effectively, it is important that world-wide surveillance systems are in place. There need to be effective systems that warn when agents such as the SARS coronavirus emerge.

The scheme for monitoring influenza virus strains provides a good model for a surveillance system. In this scheme, which is co-ordinated by the United Nations World Health Organisation (WHO), isolates of influenza virus from laboratories around the world are sent to Collaborating Centres for Influenza Reference and Research in London, Atlanta, Melbourne and Tokyo. Antigenic variation of the virus in humans and in animals (especially birds and pigs) is monitored, and twice a year the WHO recommends the virus strains to be mass-produced for incorporation into influenza vaccines.

21.9   Dealing with outbreaks

It is important that resources for diagnosis, research and treatment of infectious diseases are available when the need arises. A useful case study is provided by the measures taken to deal with the emergence of the SARS coronavirus (Section 21.5.1).

There was some delay in alerting the world to SARS but, once it was apparent that this virus posed a major threat, work got under way in a number of virology laboratories. The virus was isolated in February 2003 and three months later its genome had been sequenced. The following year a paper was published reporting compounds that inhibit replication of the virus, while other papers reported the cell receptor of the virus and the structure of its replicase protein.

Diagnostic laboratory methods to detect evidence of SARS coronavirus in samples from patients rapidly became available; tests based on immunofluorescence and RT-PCR were developed. The health of international travellers was monitored (Figure 21.12) and those found to be infected with the virus were nursed in isolation. These measures brought the SARS outbreak under control.

When there is an outbreak of a highly infectious virus, such as the SARS coronavirus or Ebola virus, infected patients and their contacts are quarantined. Control measures for some outbreaks include slaughter of animals that are infected and those that have been in contact with infected animals. This has been a key measure in dealing with outbreaks of foot and mouth disease (cattle and sheep), Nipah virus (pigs) and avian influenza (poultry).

Figure 21.12 Checkpoint for monitoring the health of people crossing the Thailand–Burma border during the SARS outbreak, June 2003. Courtesy of Professor A.J.S. Whalley, Liverpool John Moores University.

Medical staff, veterinary staff and other personnel dealing with humans and animals infected with high-risk viruses must take precautions to protect themselves and to avoid spreading infection. Precautions include wearing protective clothing and breathing filtered air.

Learning outcomes

By the end of this chapter you should be able to

- evaluate the term 'emerging virus';

- discuss examples of viruses that have recently appeared in new host species;

- discuss examples of viruses that have recently appeared in new parts of the world;

- discuss examples of new viruses;

- discuss examples of re-emerging viruses;

- assess measures that can be taken to prevent and contain outbreaks of infectious disease.

Sources of further information

Books

Hartman A. L., Towner J. S. and Nichol S. (2005) Pathogenesis of Ebola and Marburg viruses, pp. 109–124 in *Molecular Pathogenesis of Virus Infections*, editors Digard P., Nash A. A. and Randall R. E., 64th Symposium of the Society for General Microbiology, Cambridge University Press

Komar N. (2003) West Nile virus: epidemiology and ecology in North America, pp. 185–234 in *Advances in Virus Research*, Vol. 61, Elsevier

Parrish C. R. and Kawaoka Y. (2005) The origins of new pandemic viruses: the acquisition of new host ranges by canine parvovirus and influenza A viruses, pp. 553–586 in *Annual Review of Microbiology*, Vol. 59, Annual Reviews

Schmidt A., Wolff M. H. and Weber O., editors (2005) *Coronaviruses with Special Emphasis on First Insights Concerning SARS*, Birkhauser

Smolinski M. S., Hamburg M. A. and Lederberg J., editors (2003) *Microbial Threats to Health: Emergence, Detection and Response*, National Academies Press

Journals

Belshe R. B. (2005) The origins of pandemic influenza – lessons from the 1918 virus *New England Journal of Medicine*, **353**, 2209–2211

Eaton B. T. *et al.* (2005) Hendra and Nipah viruses: different and dangerous *Nature Reviews Microbiology*, **4**, 23–35

Horimoto T. and Kawaoka Y. (2005) Influenza: lessons from past pandemics, warnings from current incidents *Nature Reviews Microbiology*, **3**, 591–600

Leroy E. M. *et al.* (2005) Fruit bats as reservoirs of Ebola virus *Nature*, **438**, 575–576

Likos A. M. *et al.* (2005) A tale of two clades: monkeypox viruses *Journal of General Virology*, **86**, 2661–2672

Purse B. V. *et al.* (2005) Climate change and the recent emergence of bluetongue in Europe *Nature Reviews Microbiology*, **3**, 171–181

van den Hoogen B. G. *et al.* (2001) A newly discovered human pneumovirus isolated from young children with respiratory tract disease *Nature Medicine*, **7**, 719–724

Walsh P. D., Biek R. and Real L. A. (2005) Wave-like spread of Ebola Zaire *PLoS Biology*, **3** (11), e371

22
Viruses and cancer

At a glance

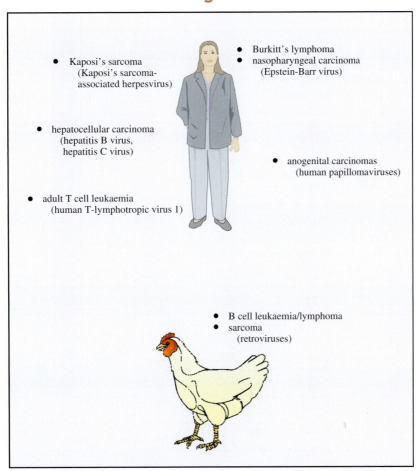

Virology: Principles and Applications   John B. Carter and Venetia A. Saunders
© 2007 John Wiley & Sons, Ltd   ISBNs: 978-0-470-02386-0 (HB); 978-0-470-02387-7 (PB)

22.1 Introduction to viruses and cancer

A cancer in a human or an animal is a malignant tumour and involves continuous proliferation of a clone of cells derived from one of the body's normal cells. The cell undergoes changes known as transformation as a result of events that include mutation, activation of oncogenes (tumour genes) and inactivation of tumour suppressors. The types of cancer that will be discussed in this chapter, and the cell types from which they are derived, are listed in Table 22.1.

For most cancers the full sequence of events that turns a normal cell into a tumour cell (the process of oncogenesis) is not fully understood, but it is thought that between four and six steps are involved. Some of the steps in this transformation can be triggered by environmental factors including some chemicals, some forms of irradiation and some viruses.

A virus that is able to cause cancer is known as an oncogenic virus. Evidence that a virus is oncogenic includes the regular presence in the tumour cells of virus DNA, which might be all or a part of the virus genome. In some types of tumour the virus DNA is integrated into a cell chromosome, while in other types it is present as multiple copies of covalently closed circular DNA (cccDNA), as discussed in the context of latent virus infections (see Figure 9.7). In many cases one or more of the virus genes are expressed in the tumour cell and virus proteins can be detected.

For some types of cancer there is evidence for involvement of a virus in most, if not all, cases of the cancer. For other types virus DNA and/or proteins are detectable in only a minority of cases and it

is possible that the virus is just one of a number of carcinogenic factors that can give rise to these cancers. Some human adenoviruses are oncogenic in that they can transform cells in culture and can cause tumours when inoculated into animals, though there is no evidence that adenoviruses cause cancer in humans. Studies with these viruses have, however, contributed to understanding of oncogenic mechanisms.

This chapter is mainly concerned with viruses that cause human cancers, of which about 20 per cent of cases in females and about eight per cent in males are thought to be caused by viruses. The relative incidences of the main virus-associated cancers are shown in Figure 22.1.

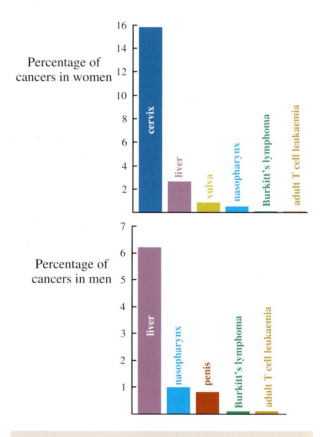

Figure 22.1 *Virus-associated cancers in women and men.* Modified from *Microbiology Today*, August 2005, with the permission of Professor D.H. Crawford (University of Edinburgh) and the Society for General Microbiology.

Table 22.1 Types of cancer and the cell types from which they are derived

Cancer	Cell type
Carcinoma	Epithelium
Leukaemia	Blood-forming
Lymphoma	Lymphocyte
Mesothelioma	Mesothelium
Sarcoma	Connective tissue

The viruses involved are found world-wide and some are very common, though the prevalence of some varies between regions. We shall also briefly consider some animal cancers caused by viruses.

22.2 Papillomavirus-linked cancers

Cervical carcinoma is the third most common cancer in women, with approximately half a million new cases and 280 000 deaths in the world each year. Most, if not all, of these cancers result from infection with a papillomavirus.

The papillomaviruses are small DNA viruses of mammals and birds (Figure 22.2). There are well over 100 human papillomavirus (HPV) types, differentiated by their DNA sequences. They enter the body through small abrasions and infect keratin-making cells (keratinocytes) in skin or a mucous membrane. Each HPV type infects a preferred site, such as the hands or the genitals, and infection may result in a benign wart (papilloma) or a carcinoma.

The papillomaviruses that infect the genitals are transmitted between individuals during sexual contact. Most papillomavirus infections do not become persistent, but in a minority of hosts the infection is not cleared by the host's immune response. In individuals who harbour a persistent infection there is a small risk of cancer developing. This risk is associated with about 15 of the HPV types; these 'high-risk' types include HPV-16 and 18. Infection with other HPV types that infect the genitals carries little or no risk of cancer; these 'low-risk' HPV types include HPV-6 and 11, and are associated almost exclusively with benign genital warts (Figure 22.3).

Keratinocytes, the host cells of papillomaviruses, stop dividing as they differentiate, but a papillomavirus needs much of the DNA-replicating machinery of the host cell, so the virus induces an infected keratinocyte into the S phase of the cell cycle (Figure 4.5). The cell then undergoes cycles of cell division, which generally have a finite number, but occasionally division of a cell infected with a high-risk HPV continues unchecked as a cancer.

The stages from normal cervical cell to cancer cell involve a number of morphological changes that can be detected by observation of a cervical smear. If precancer cells are detected they can be killed or removed to prevent the development of the tumour. The presence of all or part of the genome of a high-risk HPV type can nearly always be demonstrated in precancer cells and in cancer cells; the virus DNA is integrated into a chromosome (Figure 9.7). The virus replication cycle is not completed in these cells and no progeny virions are formed.

There is evidence that high-risk HPVs are also causative agents of some carcinomas at other body sites, including the vulva, penis, anus, head and neck. Again the evidence is the presence in the tumour cells of the virus DNA.

HPVs are also involved in a very rare form of skin cancer, which has a genetic basis: epidermodysplasia verruciformis. In this disease the patient is highly susceptible to infection with HPVs, especially a number of types that are rare in the normal population, mainly types 5 and 8 (Figure 22.3). Warts spread over the entire body during childhood and 25–33 per cent of patients develop cancer (squamous cell carcinoma) in areas of the skin exposed to ultraviolet light (sunlight). HPV DNA can be detected in more than 80 per cent of these cancers but, in contrast to cervical cancers, the DNA is rarely integrated into a cell chromosome.

- Genome: dsDNA (covalently closed circular)

- Capsid symmetry: icosahedral

Figure 22.2 *Virion characteristics of the families* Papillomaviridae *and* Polyomaviridae. *The virions of both families consist of icosahedral capsids containing dsDNA; papillomavirus virions are larger (50–55 nm diameter) than those of polyomaviruses (40–45 nm diameter).*

22.3 Polyomavirus-linked cancers

Polyomavirus virions have the same characteristics as those of the papillomaviruses, except that they are slightly smaller (Figure 22.2). They are found in mammals and birds, and most infections are subclinical.

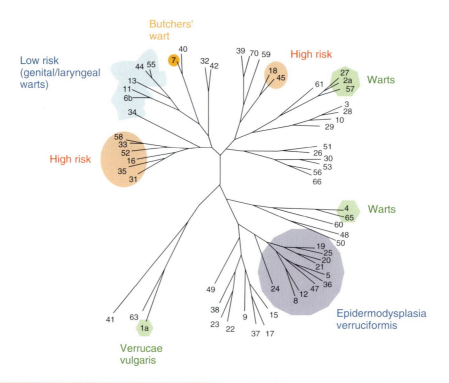

Figure 22.3 *Phylogenetic tree showing relationships between some HPVs.* The high risk HPVs cluster in two regions of the tree. Modified from *Microbiology Today*, August 2005, with the permission of Professor N.J. Maitland (University of York) and the Society for General Microbiology. Data from Los Alamos National Laboratory HPV website (http://hpv-web.lanl.gov).

Injection into certain animal species, however, can induce the development of many types of tumour ('poly-oma'). There is mounting evidence that polyomaviruses may have roles in some cases of human cancer.

Two human polyomaviruses are known: JC and BK viruses, named after the initials of the persons from whom they were first isolated. Both viruses are extremely common, with over 80 per cent of adults seropositive for them. These viruses can induce tumours in newborn hamsters and they can transform cells in culture. There is evidence that they may have roles in some cases of certain types of tumour in humans, especially brain tumours. The evidence is the presence of JC or BK virus DNA and protein in the tumour cells.

A monkey polyomavirus, simian virus 40 (SV40), was first isolated from primary cell cultures derived from rhesus monkey kidneys. The cell cultures were being used to produce poliovirus for Salk vaccine. The quality control procedures included injection of vaccine into hamsters, and some of these animals developed tumours. It was subsequently found that injection of SV40 into a newborn hamster could cause a carcinoma, sarcoma, lymphoma or leukaemia, depending on the injection site. Millions of humans became infected with SV40 when they received polio vaccine that was unknowingly contaminated with the virus. There have been suggestions that SV40 is linked with some types of human cancer, including primary brain cancer and malignant mesothelioma, but these suggestions are disputed.

22.4 Epstein-Barr virus-linked cancers

Burkitt's lymphoma (BL) is a B cell tumour that occurs with a high frequency in children in central Africa. Shortly after it was first described (by Denis Burkitt) Anthony Epstein established cell lines from the tumour of a patient. The cells were found to be persistently infected with a herpesvirus, which was named Epstein-Barr virus (EBV). Please see Chapter 11 for general characteristics of herpesviruses. Cases of BL also occur in other parts of the world, but generally without an association with EBV.

A consistent abnormality in BL tumour cells is a chromosomal rearrangement that results in the c-*myc* gene being placed next to an enhancer of an immunoglobulin gene. This results in the expression of c-*myc* at abnormally high levels. The chromosomal rearrangement is present in all cases of BL, irrespective of whether EBV is present or not, so perhaps EBV is one of several agents that can trigger the rearrangement.

Another tumour that has an association with EBV is nasopharyngeal carcinoma (NPC). This tumour, like BL, is more prevalent in particular regions (Figure 22.4). In both tumours the EBV genome is present in the tumour cells as cccDNA molecules. You will recall that when a herpesvirus infects a cell the linear virus genome becomes circularized (Section 11.5.1).

EBV is found world-wide and the majority of humans are infected, but the factors that restrict high incidences of BL and NPC to particular geographical locations are not understood. The zone of high incidence of BL in Africa corresponds closely with the zone of malaria caused by *Plasmodium falciparum*, so this parasite may play a role, as well as EBV. In southern China there are clusters of NPC in families, so there may be a human gene that increases the risk. Components of the diet, such as salted fish, are also suspected as cofactors in NPC.

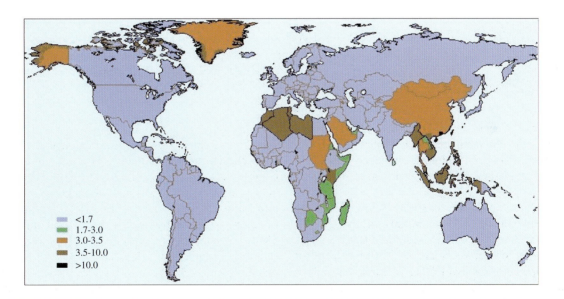

Figure 22.4 *Incidence rates of nasopharyngeal carcinoma in males.* The data (age standardized incidence rates per 100 000) were published by Busson *et al.* (2004) *Trends in Microbiology*, **12**, 356. The map was drawn by V. Gaborieau and M. Corbex (Genetic Epidemiology Unit, International Agency for Research on Cancer). Reproduced by permission of Elsevier Limited and the authors.

A number of other cancers have been found to have associations with EBV, including

- some cases of Hodgkin's lymphoma

- non-Hodgkin's lymphoma in AIDS

- post-transplant lymphoproliferative disorder.

In contrast to BL and NPC, Hodgkin's lymphoma is not restricted to particular geographical areas. The other two cancers listed develop in patients who are immunocompromised as a result of AIDS or immunosuppressive treatment.

The oncogenic potential of EBV can be demonstrated when cultures of B cells are infected with the virus. Infected cells synthesize a number of virus proteins that push the cells permanently into the cell cycle (Figure 4.5), resulting in the formation of lymphoblastoid cell lines. In the body this B cell proliferation is normally controlled by T cells, but in individuals who are T cell immunocompromised the control may be inadequate and a B cell tumour may result.

22.5　Kaposi's sarcoma

Kaposi's sarcoma was first described in the 19th century as a rare skin cancer that affected elderly men in the Mediterranean region. Since the arrival of AIDS this picture has changed. It is one of the most common cancers in people with AIDS, and in these patients the cancer disseminates throughout the body and is more aggressive. It is thought that the tumour cells are derived from endothelial cells.

In 1994 it was discovered that the tumour cells contain the DNA of a new herpesvirus. The virus was subsequently isolated and was named Kaposi's-sarcoma-associated herpesvirus (KSHV). The virus can be found in most parts of the world, but serological studies have shown that it is more common in certain regions, such as central Africa, and it is more common in homosexual men. The risk of developing Kaposi's sarcoma parallels the prevalence of the virus in the population. As with EBV-associated tumours, copies of the KSHV genome are present as cccDNA in the nuclei of the tumour cells.

There is also good evidence linking KSHV with two other human cancers: primary effusion lymphomas and multicentric Castleman's disease. Both are tumours derived from B cells.

22.6　Adult T cell leukaemia

Adult T cell leukaemia is associated with human T-lymphotropic virus 1 (HTLV-1) infection. Regions of the world with a high prevalence of the virus, such as south-west Japan, have a high prevalence of these tumours. HTLV-1 is a retrovirus and the tumour cells each have a copy of the proviral DNA integrated into a chromosome (Chapter 16).

22.7　Hepatocellular carcinoma

Hepatocellular carcinoma (liver cancer) accounts for four to five per cent of cancer cases in the world. A number of factors are implicated as causative agents, including consumption of mould toxins in food, and two viruses: hepatitis B virus (HBV) and hepatitis C virus (HCV).

HBV is the most significant agent; the prevalence of liver cancer closely parallels the prevalence of persistent HBV infection (Section 18.2), with the highest incidences in Asia and in central and southern Africa. In most of the tumours, HBV DNA is integrated into the cell genome, and in most cases the virus DNA has undergone rearrangements, including deletions. The P and C ORFs have generally been destroyed, but the S and X ORFs are often intact (please see Figure 18.7 for a reminder of the HBV genes). The only viral gene product that is consistently present in the tumour cells is the X protein.

The other virus associated with liver cancer, HCV, is a flavivirus (Figure 21.5). It is the only class IV virus (plus-strand RNA virus) that is known to be oncogenic. Its genome is not found routinely in the cancer cells, in contrast to the other human oncogenic viruses.

Both HBV and HCV elicit immune responses when they infect the body. In some individuals the immune response successfully eliminates the infection, but in many cases the infection persists for life, as discussed in Section 18.2 for HBV. As far as HCV is concerned, it is estimated that about 80 per cent of individuals who become infected are unable to eliminate the infection and that about 175 million people in the world are infected. In some individuals who are persistently infected with HBV or HCV liver cirrhosis develops, and for some this eventually leads on to cancer.

22.8 Virus-associated cancers in animals

A number of cancers in animal species have been found to be associated with viruses. Outbreaks of some of these diseases can have serious economic consequences in agriculture. A few examples are given here of animal cancers associated with retroviruses and a herpesvirus.

It was demonstrated early in the 20th century that filterable agents cause a leukosis (a B cell leukaemia/lymphoma) and a sarcoma in chickens. These agents were subsequently shown to be retroviruses and the sarcoma-causing virus was named Rous sarcoma virus after its discoverer, Peyton Rous (Section 16.4). Other retroviruses were subsequently found to be causative agents of leukaemia in cats and cattle, and a retrovirus known as Jaagsiekte sheep retrovirus was found to be responsible for adenocarcinoma of the lung in sheep.

Marek's disease in chickens is a lymphoma, similar to that caused by retroviruses, but the causative agent is a herpesvirus and the transformed lymphocytes contain multiple copies of the viral genome. Papillomaviruses, already discussed as causative agents of human cancers (Section 22.2), cause cancers in cattle, horses and rabbits.

22.9 Cell lines derived from virus-associated cancers

Cancer cells, which by definition multiply continuously in the body, will often continue to multiply continuously if transferred into cell culture, when permanent cell lines may be established from them. Cell lines derived from most virus-induced cancers contain part or all of the virus genome. The most famous of these, the HeLa cell line, was established from a cervical cancer in the mid-20th century and is now used in laboratories around the world. Each HeLa cell contains a copy of part of the genome of HPV-18; this will be discussed further in Section 22.10.1.

Cell lines derived from Burkitt's lymphoma contain copies of the EBV genome and some produce infectious EBV; the virus was first discovered in one of these cell lines. Multiple copies of Marek's disease virus genome are present in a cell line derived from a Marek's tumour. A cell line derived from a liver cancer contains HBV DNA integrated at seven sites, and the cells secrete HBsAg. Cell lines derived from Kaposi's sarcoma, however, do not contain KSHV DNA; the virus DNA is lost from the tumour cells when they grow in culture.

22.10 How do viruses cause cancer?

Most virus-induced cancers develop after a long period of persistent infection with an oncogenic virus; for adult T cell leukaemia this period is exceptionally long (around 60 years). The virus infections persist in their hosts in spite of immune responses, such as the production of virus-specific antibodies. Some persistent infections are latent for much of the time (e.g. EBV and KSHV), with only small numbers of virus genes expressed. Others, including HBV and HCV infections, are productive. Both of the latter viruses are able to evolve rapidly (Chapter 20), and this probably allows them to keep one step ahead of acquired immune responses.

Although many humans are persistently infected by viruses that are potentially oncogenic, only small percentages develop virus-linked cancers. Relatively few people develop EBV-related tumours, though over 90 per cent of adults worldwide are infected with the virus. About three per cent of women persistently infected with one of the high-risk strains of HPV develop cervical carcinoma, and similar percentages of individuals persistently infected with HBV, HCV or HTLV-1 develop a virus-linked cancer.

As cancer develops in only small percentages of virus-infected hosts it is clear that the virus infections alone do not cause cancer. Other factors are involved and these include exposure to particular environmental factors, host genetic factors and immunodeficiency.

Immunodeficiency increases the risk of a virus-associated tumour, the nature of the immunodeficiency influencing the types of tumour that may develop. AIDS patients are much more likely to develop Kaposi's sarcoma than immunosuppressed transplant recipients. Non-Hodgkin's lymphoma in AIDS and post-transplant lymphoproliferative disorder (EBV-related tumours; Section 22.4) each occurs in a distinct class of immunodeficient patient. Immunodeficiency does not increase the risk of all tumours with a viral link; for example, there is no increased incidence in AIDS patients of NPC or of liver cancer.

The probability of cancer developing in a host infected with an oncogenic virus therefore depends on a complex interplay between the state of the host, environmental factors to which the host is exposed and cellular changes induced by the virus infection. There is increasing evidence that virus proteins synthesized during persistent infections play roles in the conversion of normal cells to cancer cells, and these are discussed in the next sections.

22.10.1 'Deliberate' interference with control of the cell cycle

In order to complete its replication cycle a virus needs to manipulate the internal environment of its host cell so that all the requirements for virus replication are present. Some animal viruses require the host cell to be in a particular phase of the cell cycle (Figure 4.5).

Many cells in the animal body grow and divide either slowly or not at all; the latter are arrested in the G1 phase of the cell cycle and are said to be in the G0 state. The control of the cell cycle is mediated by many proteins; two that play key roles in humans are p53 and retinoblastoma protein (pRb). The latter protein is so named because the absence of the protein in some individuals, due to mutations in both copies of the gene, leads to the development of retinoblastoma (cancer of the retina). It has been demonstrated that several proteins produced by oncogenic viruses can interact with p53, pRb and other proteins that control cell growth and division, increasing the probability of a cell being pushed into repeated cycles of division.

Virus proteins that can interfere with control of the cell cycle include the HPV early proteins E6 and E7. Papillomaviruses have small genomes (about 8 kbp) and require the cell's DNA synthesizing functions that are available in the S phase of the cell cycle. E6 and E7 make these functions available by localizing to the nucleus and binding to cell proteins. E6 binds to p53, promoting its degradation, and E7 binds to pRb–E2F complexes, causing their dissociation (E2F is a cell transcription factor that activates DNA synthesis). The outcome of these interactions is the transition of the cell to the S phase.

An important feature of the high-risk HPVs is a propensity for integration of all or part of the virus genome into a cell chromosome. In most cervical carcinomas HPV DNA is integrated; often the virus genome is incomplete, but the E6 and E7 genes are always present (Figure 22.5). Expression of these genes in a variety of cell types in culture can transform the cells, and the transformed state can be maintained by the continuous expression of the genes. Part of the genome of HPV-18 is integrated into the genome of HeLa cells, which were derived from a cervical cancer (Section 22.9).

Many other proteins of viruses considered in this chapter can interact with p53 and/or pRB. Examples include

- SV40 large T (= tumour) antigen (binds to p53 and pRB)

- KSHV latency-associated nuclear antigen (binds to p53 and pRB)

- HTLV-1 Tax protein (binds to p53)

- HBV X protein (binds to p53).

22.10.2 'Accidental' activation of cell genes

Some virus proteins are able to bind to cell proteins that may not be the intended targets and may thus trigger events that are of no value to the virus, but may be harmful to the host. A virus protein might inadvertently push a cell towards a cancerous state by activating a cell gene that is switched off, or by enhancing the rate of transcription of a gene that is being expressed at a low level. An example of a virus protein that might act in this way is the Tax protein of HTLV-1. This virus is

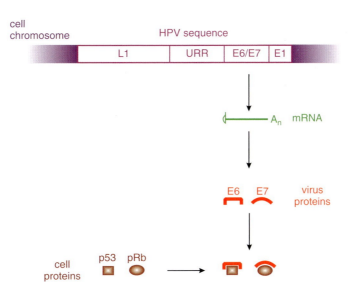

Figure 22.5 *Expression of HPV genes from an integrated virus sequence.* The HPV-18 sequence that is integrated into the HeLa cell genome is shown. The E6 and E7 proteins are synthesized and bind to the cell proteins p53 and pRb, respectively. E: early gene. L: late gene. URR: upstream regulatory region.

a complex retrovirus (Section 16.4) and the *tax* gene is one of its auxiliary genes. The gene is expressed from the integrated provirus and the Tax protein, in conjunction with cell transcription factors, functions as a transcription factor for the virus. The Tax protein, however, can also influence the expression of many cell genes, including cell cycle regulators (Figure 22.6).

Some retroviruses can induce tumour formation as a consequence of integrating their provirus into a cell chromosome at a site that puts a cell gene under the control of the virus promoter. This may result in the activation of a gene that was switched off, or it may result in a boost in the expression of a gene that was already active. Some cell genes that can be activated in this way are known as proto-oncogenes because their activation may result in a cancer. The proteins encoded by proto-oncogenes are generally involved in gene regulation or cell cycle control. An example of a proto-oncogene is c-*myc*. Infection of a chicken with an avian leukosis virus (Section 22.8) can cause a tumour when proviral integration next to c-*myc* results in greatly enhanced transcription of this gene. An unfortunate outcome of retroviral activation of a proto-oncogene has been the development of leukaemia

in some patients in gene therapy trials involving a retroviral vector.

22.10.3 Retroviral oncogenes

Some retroviruses have the ability to cause cancer because of the presence in the viral genome of an oncogene; for example, the Rous sarcoma virus genome contains the v-*src* gene (Figure 16.12). The retroviral oncogenes were derived from cell proto-oncogenes and are closely related to them (Table 22.2), hence v-*src* is closely related to c-*src*.

When a proto-oncogene is mutated or aberrantly expressed it becomes an oncogene. The *src* gene, which has been implicated in the development of many cancers, encodes a protein kinase; overphosphorylation of the enzyme's substrates is a key process in the development of these tumours. C-*myc* encodes a transcription factor and is expressed at abnormally high levels in Burkitt's lymphoma (Section 22.4).

Nearly all of the oncogene-carrying retroviruses are defective as a result of genome deletions. These viruses can replicate only with the help of an endogenous retrovirus, or in a cell co-infected with a helper virus; for example, murine sarcoma viruses are defective

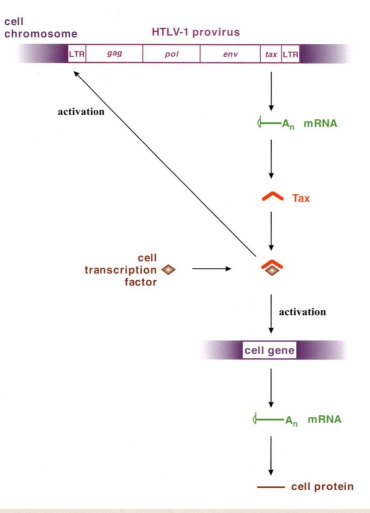

Figure 22.6 *HTLV-1 oncogenesis.* The *tax* gene of the provirus is expressed. The Tax protein, complexed with cell proteins, activates transcription of the provirus and may also activate cell genes.

Table 22.2 Examples of retroviral oncogenes and the cell proto-oncogenes from which they were derived. The names of some genes are derived from the virus names. The names of viral oncogenes are prefixed by 'v-', while the names of cell proto-oncogenes are prefixed by 'c-'

Retrovirus	Viral oncogene	Cell proto-oncogene
Rous *sarc*oma virus	v-*src*	c-*src*
Avian *myc*elocytomatosis virus	v-*myc*	c-*myc*
*Ha*rvey murine sarcoma virus	v-H-*ras*	c-H-*ras*

and murine leukaemia virus may act as a helper. Rous sarcoma virus is one of the few non-defective retroviruses with an oncogene. Retroviral oncogenes lack the sequences that control the expression of cell proto-oncogenes, so when a retroviral oncogene is expressed in an infected cell the protein is often at a higher concentration than the product of the cell proto-oncogene.

The oncogene-carrying retroviruses have the ability to induce rapid formation of tumours, usually 1–6 weeks post-infection, in contrast to tumours induced by other oncogenic viruses, which develop only after years or decades of persistent virus infection.

22.10.4 Damage to immune defences

Interactions between cell proteins and proteins produced by oncogenic viruses can lead to breakdown of immune defences that may allow the development of a cancer. Papillomavirus proteins interfere with apoptosis, and hence prevent the death of virus-infected cells. Some of the proteins produced by EBV and KSHV in latently infected cells can interfere with acquired immune responses.

22.11 Prevention of virus-induced cancers

Logical approaches to preventing virus-induced cancers include attempting to prevent transmission of the viruses, so knowledge of the modes of transmission of the viruses is important (Table 22.3).

Knowledge of the modes of virus transmission allows the development of strategies aimed at reducing the risks of transmission. Most of the oncogenic human viruses can be transmitted sexually and in blood, so measures implemented to prevent transmission of HIV (Section 17.7) also reduce the risk of transmission of these viruses.

Vaccination provides another approach to reducing rates of virus transmission. Most transmission of HBV occurs from mother to baby around the time of birth, so several countries with high incidences of HBV infection have initiated programmes to give babies HBV vaccine within 2 days of birth. There is evidence from Taiwan of a reduced incidence of liver cancer in children as a result of this programme, and in time there should be a reduced incidence in adults too.

Trials in women with vaccines designed to prevent infection with the most common high-risk HPVs have produced promising results, and at the time of writing it is anticipated that HPV vaccines will soon be in widespread use, with the aim of protecting against cervical and other cancers associated with these viruses.

Marek's disease in chickens is economically important and vaccines are used to protect poultry against infection with the causative virus. Some vaccines contain an attenuated strain of the virus, while others contain a related herpesvirus from turkeys.

Table 22.3 The main modes of transmission of oncogenic human viruses

Virus	Transmission mode				
	Mother to baby		Via saliva	Sexual	Via blood
	Before or during birth	Breast feeding			
HPV				✓	
EBV			✓		✓
KSHV			✓	✓	
HTLV-1		✓		✓	✓
HBV	✓			✓	✓
HCV	✓			✓	✓

An alternative approach to the prevention of virus-induced cancers is the attempted elimination from the body of persistent infections with oncogenic viruses. So far this approach has had limited success. HCV infection in some patients has been eliminated by treatment with α-interferon (Section 9.2.1.a), either alone or in combination with ribavirin. Persistent HBV infections may be controlled, though not eliminated, using α-interferon and/or lamivudine. Please see Chapter 25 for more information on these anti-viral drugs.

Learning outcomes

By the end of this chapter you should be able to

- outline the characteristics of viruses that are associated with cancers;

- evaluate the evidence for association of viruses with some cancers;

- discuss possible mechanisms for virus induction of cancer;

- suggest how virus-induced cancers may be prevented.

Sources of further information

Books

Basser R. L., Edwards S. and Frazer I. H. (2005) Development of vaccines against human papillomavirus (HPV), Chapter 12 in Moingeon P., editor *Vaccines: Frontiers in Design and Development*, Horizon

Campo M. S., editor (2006) *Papillomavirus Research: from Natural History to Vaccines and Beyond*, Caister

Gray L., Jolly C. and Herrington C. S. (2005) Human papillomaviruses and their effects on cell cycle control and apoptosis, pp. 235–251 in *Molecular Pathogenesis of Virus Infections*, editors Digard P., Nash A. A. and Randall R. E., 64th Symposium of the Society for General Microbiology, Cambridge University Press

Moore P. S. and Chang Y. (2003) Kaposi's sarcoma-associated herpesvirus immunoevasion and tumorigenesis:

two sides of the same coin? pp. 609–639 in *Annual Review of Microbiology*, Vol. 57, Annual Reviews

Journals

Bekkers R. L. M. *et al.* (2004) Epidemiological and clinical aspects of human papillomavirus detection in the prevention of cervical cancer *Reviews in Medical Virology*, **14**, 95–105

Butel J. S. (2000) Viral carcinogenesis: revelation of molecular mechanisms and etiology of human disease *Carcinogenesis*, **21**, 405–426

Cathomas G. (2003) Kaposi's sarcoma-associated herpesvirus (KSHV)/human herpesvirus 8 (HHV-8) as a tumour virus *Herpes*, **10**, 72–77

Collot-Teixeira S. *et al.* (2004) Human tumor suppressor p53 and DNA viruses *Reviews in Medical Virology*, **14**, 301–319

Guerrero R. B. and Roberts L. R. (2005) The role of hepatitis B virus integrations in the pathogenesis of human hepatocellular carcinoma *Journal of Hepatology*, **42**, 760–777

Hammerschmidt W. and Sugden B. (2004) Epstein-Barr virus sustains Burkitt's lymphomas and Hodgkin's disease *Trends in Molecular Medicine*, **10**, 331–336

Hebner C. M. and Laimins L. A. (2005) Human papillomaviruses: basic mechanisms of pathogenesis and oncogenicity *Reviews in Medical Virology*, **16**, 83–89

Jarrett R. F. (2006) Viruses and lymphoma/leukaemia *Journal of Pathology*, **208**, 176–186

Longworth M. S. and Laimins L. A. (2004) Pathogenesis of human papillomaviruses in differentiating epithelia *Microbiology and Molecular Biology Reviews*, **68**, 362–372

Lopes V., Young L. S. and Murray P. G. (2003) Epstein–Barr virus-associated cancers: aetiology and treatment *Herpes*, **10**, 78–82

Schulz T. F. (2006) The pleiotropic effects of Kaposi's sarcoma herpesvirus *Journal of Pathology*, **208**, 187–198

Snijders P. J. F. *et al.* (2006) HPV-mediated cervical carcinogenesis: concepts and clinical implications *Journal of Pathology*, **208**, 152–164

Thorley-Lawson D. A. (2005) EBV the prototypical human tumor virus – just how bad is it? *Journal of Allergy and Clinical Immunology*, **116**, 251–261

Pagano J. S. *et al.* (2004) Infectious agents and cancer: criteria for a causal relation *Seminars in Cancer Biology*, **14**, 453–471

Vilchez R. A. and Butel J. S. (2004) Emergent human pathogen simian virus 40 and its role in cancer *Clinical Microbiology Reviews*, **17**, 495–508

23

Survival of infectivity

At a glance

- Viruses vary greatly in the rate at which they lose infectivity.

- The rate of virus infectivity loss is affected by temperature, radiation and the chemical environment including pH.

- There may be two rates of infectivity loss.

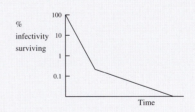

- Understanding the kinetics of infectivity loss is important in practical situations, such as devising disinfection and sterilization procedures, storage of virus preparations and vaccine manufacture.

- In vaccine manufacture it is vital to ensure that maximum infectivity is preserved in live vaccines, and that 100% of infectivity is destroyed in inactivated vaccines.

Virology: Principles and Applications   John B. Carter and Venetia A. Saunders
© 2007 John Wiley & Sons, Ltd   ISBNs: 978-0-470-02386-0 (HB); 978-0-470-02387-7 (PB)

23.1 Preservation of virus infectivity

There are several practical situations in which there is a need to preserve virus infectivity. A prime example is during the storage and distribution of live virus vaccines (Sections 24.2 and 24.5), when there is a need to retain the potency of the vaccine. A second example is during the transport and storage of blood, faeces, saliva and other specimens from human and animal patients prior to testing in a diagnostic virology laboratory. Last, but by no means least, virologists need to preserve the infectivity of the viruses that they work with.

23.2 Destruction of virus infectivity

There are many situations in which there is a requirement to destroy virus infectivity. Procedures to minimize the transmission of pathogenic viruses (and other micro-organisms) in hospitals and other health care facilities include the disinfection of surfaces and the sterilization of equipment and materials. A good example of a situation requiring rigorous action is an outbreak of winter vomiting disease caused by a norovirus; patients' vomit contains significant quantities of virus, which must be inactivated if the outbreak is to be brought under control.

Inactivated (killed) virus vaccines (Section 24.3) are made from preparations of virulent viruses, so it is essential that all infectivity is destroyed. Blood products, such as clotting factors for haemophiliacs and immunoglobulins, are treated to destroy any viruses (e.g. HIV, hepatitis viruses) that might be present in the donated blood. Continuing the theme of human and animal health, water supplies that are free from pathogenic viruses (and other micro-organisms) are of fundamental importance, so water treatment procedures must include processes that destroy these agents. The same applies to the treatment of swimming pool water.

The processes for production of cheese, yogurt and other dairy products utilize lactic acid bacteria to ferment the milk. There are many phages that infect these bacteria, and a phage infection during a production process can result in poor product quality, or in delay or failure of the fermentation. To minimize the risk, procedures are required for treatment of the milk or whey to inactivate any phages present, and for the disinfection of the factory environment.

Finally, virologists need supplies of pipettes, cell culture vessels and other laboratory materials that are guaranteed free from viruses (and other micro-organisms), then after these materials have been used any viruses present must be inactivated before disposal.

23.3 Inactivation targets in virions

The infectivity of a virion can be destroyed by the alteration or removal of its nucleic acid and/or one or more of its proteins (Figure 23.1). A protein molecule may be altered by the induction of a conformational change or, more drastically, by the breaking of covalent bonds, such as peptide bonds and disulphide bonds.

23.4 Inactivation kinetics

The rate at which a virus loses infectivity under defined conditions (for example, at a particular temperature and pH value) can be determined by incubating aliquots of the virus preparation under those conditions. At

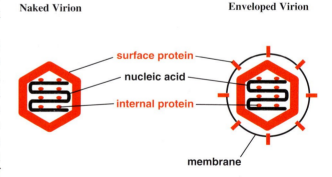

Figure 23.1 *Inactivation targets in virions.* Alteration of a surface protein might prevent a virion from attaching to its host cell and/or from entering the cell. Stripping the envelope from an enveloped virion removes the surface proteins and achieves the same outcome. Alteration of internal virion proteins can destroy properties, such as enzyme activities, essential for the replication of the virus.

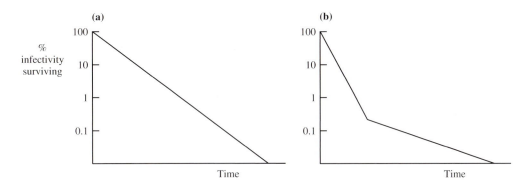

Figure 23.2 *Loss of virus infectivity at one rate (a) and at two rates (b).*

various time intervals the surviving infectivity in an aliquot is determined in an infectivity assay appropriate for the virus (Section 2.8). The logarithm of the percentage infectivity surviving is plotted against time (Figure 23.2).

In some circumstances infectivity is lost at a constant rate until no further infectivity can be detected (Figure 23.2(a)). The slope of the line is a measure of the rate of infectivity loss. Sometimes there is a change in the rate of infectivity loss after most (usually more than 99 per cent) of the infectivity has been destroyed (Figure 23.2(b)). It is not entirely clear why a small fraction of some virus preparations loses infectivity at a slower rate. It has been suggested that this fraction might contain virions that have a higher degree of resistance to the inactivating factor and/or that the preparation might contain virion clumps, with virions at the centres of the clumps having some protection.

23.5 Agents that inactivate virus infectivity

23.5.1 Physical agents

23.5.1.a Temperature

All infective viruses lose infectivity over time, the rate of infectivity loss being highly dependent on temperature. There is a lot of variability between viruses in the rates at which they are inactivated. In general, naked viruses are more stable than enveloped

viruses; most naked viruses lose little infectivity after several hours at 37 °C, whereas most enveloped viruses lose significant infectivity under the same conditions. The generalization is, as stated, a broad one, as there may be significant differences in stability between closely related viruses. Under dry conditions varicella-zoster virus survives well, while the related herpes simplex viruses survive poorly.

At temperatures below −35 °C most viruses lose infectivity very slowly, so for long-term storage of viruses they are placed in freezers or in liquid nitrogen. Preservation of infectivity may be enhanced by storing virus preparations in a freeze-dried form, rather than as aqueous suspensions. Moving to higher temperatures, most viruses are completely inactivated within 30 minutes at 60 °C or within a few seconds at 100 °C.

Some of the bacteriophages of lactic acid bacteria that cause problems in the dairy industry are quite heat resistant, and some infectivity may survive the heat treatments used in the decontamination procedures (Figure 23.3).

The virions of some viruses, such as hepatitis A virus (HAV) and parvoviruses, are exceptionally heat stable. Data for the minute virus of mice (a parvovirus) indicate some infectivity surviving after 60 minutes at 80 °C (Figure 23.4).

Exceptions to the generalization concerning the stability of naked and enveloped virions include those of hepatitis B virus (HBV), which are enveloped but have high stability, and those of poliovirus, which are naked but have low stability. Interestingly, HAV (heat-stable virions) and poliovirus (heat-labile virions) are

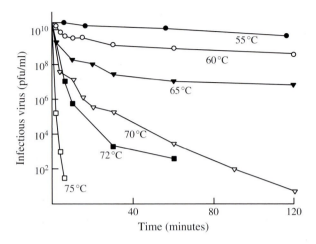

Figure 23.3 *Inactivation of* Lactococcus lactis *phage P008 infectivity in a broth at temperatures between 55 and 75°C. Data from Müller-Merbach* et al. *(2005)* International Dairy Journal, **15**, *777. Reproduced by permission of Elsevier Limited and the author.*

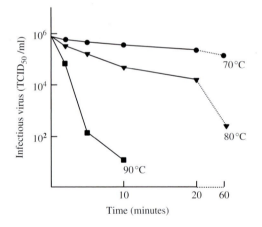

Figure 23.4 *Inactivation of minute virus of mice infectivity in water at 70, 80 and 90°C. Data from Boschetti* et al. *(2003)* Biologicals, **31**, *181. Reproduced by permission of Elsevier Limited and the author.*

both members of the family *Picornaviridae*, so, as with the herpesviruses, there is not always consistency in the degree of stability between viruses with similar structures.

The fact that virus preparations lose infectivity means that special precautions have to be taken to minimize infectivity loss during transport and storage of live vaccines, such as polio, mumps, measles and rubella. These precautions include a 'cold chain' of containers and vehicles that are refrigerated and/or insulated. This adds considerably to the expense of vaccination programmes.

The rate of infectivity loss depends on the nature of the medium containing the virus. The presence of proteins, fats, glycerol or certain salts may protect infectivity. Magnesium ions reduce the rate of inactivation of many RNA viruses, including poliovirus, so magnesium chloride is added to live polio vaccines to extend their shelf life.

The most heat-resistant viruses known are those that infect the thermophilic prokaryotes that inhabit hot springs. The optimum temperature for the growth of many of the hosts of these viruses is around 85 °C. Prions are also remarkably heat resistant (Section 26.3).

23.5.1.b Radiation

Some forms of radiation inactivate virus infectivity because they damage nucleic acids; for example, X rays, gamma rays and ultraviolet rays cause breaks in nucleic acids. Ultraviolet rays also cause other types of damage to nucleic acids, including the formation of thymine dimers in dsDNA (Figure 23.5). The thymine–adenine hydrogen bonds at two adjacent

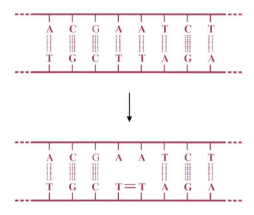

Figure 23.5 *Formation of a thymine dimer in dsDNA.*

base pairs are broken and covalent bonds are formed between the two thymine residues.

Sunlight inactivates virus infectivity because of its ultraviolet component. A practical aspect of this is the rapid inactivation by sunlight of baculoviruses applied to crops and forests as insecticides.

23.5.2 Chemical agents

Chemicals that inactivate virus infectivity do so with similar kinetics to physical agents, with either one or two inactivation rates (Figure 23.2). When some inactivation data are plotted the initial part of the graph has a small shoulder before the straight line. This could be due to a delay while the chemical reaches its target in the virion. Many chemicals can inactivate virus infectivity, but for many of them the inactivation target(s) are uncertain. We shall consider the effects of hydrogen ion concentration (pH) and of a variety of other chemical agents on virus infectivity.

23.5.2.a Acids and alkalis

Most viruses are fairly stable at pH values between 5 and 9, but at more extreme pH values virions may undergo conformational changes and rates of infectivity loss usually increase. At high pH values both nucleic acids and proteins can be damaged. Alkali, in the form of dilute sodium hydroxide (sometimes in combination with steam), is used to disinfect manufacturing vessels and tools in the pharmaceutical industry.

The infectivity of some viruses is inactivated at low pH values as a result of alterations to surface proteins. Viruses that infect humans and other mammals via the gastrointestinal route, however, are relatively stable at low pH values. These viruses, which include rotaviruses (Chapter 13), have to survive in the stomach contents, which are commonly at a pH value between 1 and 2. Furthermore, their virions must be able to withstand the sudden increase in pH when they are moved from the stomach to the intestine (pH about 7.5), and they must be resistant to inactivation by bile salts and proteolytic enzymes.

The enteroviruses and HAV infect their hosts via the gastrointestinal tract. These viruses are picornaviruses (Chapter 14), but, when compared to those picornaviruses (such as rhinoviruses and aphthoviruses) that infect their hosts via other routes, their virions are much more acid resistant. It can be seen in Figure 23.6 that HAV infectivity can still be detected after 6 minutes

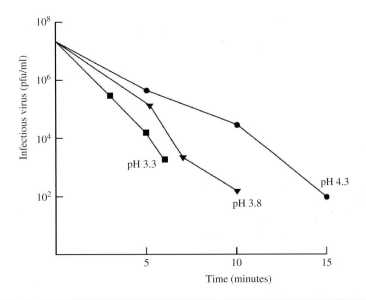

Figure 23.6 *Inactivation of hepatitis A virus infectivity at three pH values at 85 °C.* Data from Deboosere *et al.* (2004) *International Journal of Food Microbiology*, **93**, 73. Reproduced by permission of Elsevier Limited and the author.

at pH 3.3 at a temperature of 85 °C, thus HAV is an example of a virus with high resistance to inactivation by both heat and acid.

Phenol, which is a weak acid, destroys the capsids of virions without altering nucleic acids. It is commonly used in the laboratory to extract nucleic acids from virions.

23.5.2.b Hypochlorite

One of the most widely used disinfectants is hypochlorite. Both the hypochlorite ion (OCl^-) and the undissociated hypochlorous acid (HOCl) are excellent disinfectants; they act by oxidizing organic molecules, such as proteins. Hypochlorite ions are either supplied as sodium hypochlorite or they are generated by dissolving chlorine gas in water. The latter process is used to treat drinking water supplies and swimming pool water.

Sodium hypochlorite is one of the disinfectants used in the dairy industry to control phages of lactic acid bacteria such as *Lactobacillus* spp. The inactivation of a *Lactobacillus* phage at four concentrations of

hypochlorite is shown in Figure 23.7. At the three highest concentrations of hypochlorite there were slower rates of inactivation for a small percentage of the virions, as discussed in Section 23.4.

23.5.2.c Aldehydes and alcohols

Formaldehyde and glutaraldehyde damage both enveloped and non-enveloped virions. Formaldehyde cross-links proteins and reacts with amino groups in nucleic acids; it is used to destroy the infectivity of poliovirus and other viruses for use in inactivated vaccines (Section 24.3). Glutaraldehyde is useful for disinfecting metal surfaces because it is non-corrosive.

Ethanol and isopropyl alcohol are used as general purpose disinfectants and in waterless hand washes.

23.5.2.d Lipid solvents

Detergents and many organic solvents, such as chloroform, destroy the infectivity of enveloped viruses by removing their lipid membranes and associated glycoproteins. Detergents are used in the preparation of clotting factors and other blood products to destroy enveloped viruses (particularly HIV, HBV and HCV) that may be present in pooled blood donations. Lipid solvents may also damage virions with internal lipid, such as the corticoviruses and the tectiviruses (Section 3.5.2).

Quaternary ammonium compounds are cationic and disrupt virion envelopes. A number are used as disinfectants, especially chlorhexidine gluconate, which is widely used in health care institutions. Anionic detergents, such as sodium dodecyl sulphate, disrupt virion envelopes and capsids.

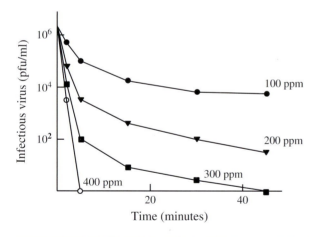

Figure 23.7 *Inactivation of* Lactobacillus *phage BYM by hypochlorite.* The concentrations are shown as parts per million (ppm) of free chlorine. The phage suspensions were at 25 °C and pH 7. Data from Quiberoni *et al.* (2003) *International Journal of Food Microbiology*, **84**, 51. Reproduced by permission of Elsevier Limited and the author.

Learning outcomes

By the end of this chapter you should be able to

- discuss situations in which preservation of virus infectivity is important;

- discuss situations in which destruction of virus infectivity is important;

- give examples of physical and chemical agents that damage virions and cause them to lose infectivity;

- discuss the kinetics of inactivation of virus infectivity.

Sources of further information

Books

Fraise A. P., Lambert P. A. and Maillard J.-Y., editors (2004)

Russell, Hugo and Aycliffe's Principles and Practice of Disinfection, Preservation and Sterilization, 4th edition, Blackwell

Wolff M. H. *et al.* (2005) Environmental survival and microbicide inactivation of coronaviruses, pp. 201–212 in Schmidt A., Wolff M.H. and Weber O., editors, *Coronaviruses with Special Emphasis on First Insights Concerning SARS*, Birkhauser

Journals

Farshid M. *et al.* (2005) The clearance of viruses and transmissible spongiform encephalopathy agents from biologicals *Current Opinion in Biotechnology*, **16**, 561–567

Maillard J.-Y. (2001) Virus susceptibility to biocides: an understanding *Reviews in Medical Microbiology*, **12**, 63–74

McDonnell G. and Russell A. D. (1999) Antiseptics and disinfectants: activity, action, and resistance *Clinical Microbiology Reviews*, **12**, 147–179

Rzeźutka A. and Cook N. (2004) Survival of human enteric viruses in the environment and food *FEMS Microbiology Reviews*, **28**, 441–453

24

Virus vaccines

At a glance

Vaccine Types		Examples
Live attenuated virus		Polio (Sabin)
Inactivated virus		Polio (Salk)
Virion subunit		Influenza (haemagglutinin and neuraminidase)
Live recombinant virus		Rabies (for wildlife vaccination)
Virus-like particles		Hepatitis B
DNA		SARS coronavirus (experimental)

24.1   Introduction to virus vaccines

The term vaccination is derived from the Latin word *vacca*, meaning cow. This is because the original procedure involved the inoculation of material from cowpox lesions into healthy people. Edward Jenner tried the procedure first in 1796 after he noticed that the faces of most milkmaids were unmarked by pocks; this was because milkmaids rarely contracted smallpox. They did, however, commonly contract cowpox, so Jenner inoculated material from a milkmaid's cowpox lesion into the arm of an 8-year-old boy. Six weeks later the boy was inoculated with material from a smallpox scab; he remained healthy. The immune response against cowpox virus had protected against smallpox virus, the protection resulting from related antigens in the two viruses.

A vaccine, therefore, contains material intended to induce an immune response, and this may involve both B cells (which develop into antibody-producing cells) and T cells (responsible for cell-mediated immunity). Both arms of adaptive immunity can be important for providing protection against a virus infection (Section 9.2.2).

The purpose of most viral vaccines is to induce long-term immunity against the virus by establishing immunological memory that will be triggered if the virus ever invades the body. In order to establish strong immunological memory there is a requirement for vaccines that induce vigorous immune responses; in other words, highly immunogenic virus materials are required.

Effective vaccines are in use to protect against diseases caused by many viruses such as polio, rubella, rabies and foot and mouth disease. This chapter will describe the various categories of virus vaccine that are in medical and veterinary use, and will outline some aspects of their manufacture. Effective vaccines have yet to be developed against many other viruses, including HIV-1, hepatitis C, Ebola and the herpes simplex viruses. Those involved in virus vaccine research face many difficulties, such as multiple antigenic variants of target viruses and the requirements for high standards of safety. Some vaccines that have been developed have not been accepted for widespread use because of safety concerns. The great need for new vaccines has spawned attempts to produce novel categories of vaccine, such as replication-defective viruses, peptide vaccines and DNA vaccines, and some of these will be introduced in this chapter.

24.2   Live attenuated virus vaccines

A live attenuated vaccine contains a mutant strain of a virus that has been derived from a wild-type virulent strain. Vaccines of this type have a number of advantages over most other types of vaccine. One advantage is that there are increasing amounts of virus antigen in the body as the virus replicates. Another is that a wide-ranging immune response is induced that involves B cells, CD4 T cells and CD8 T cells.

There are two properties that the vaccine virus must possess. First, its antigens must be identical, or very similar, to those of the wild-type virus so that an immune response against the vaccine virus provides protection from infection with the wild-type virus. Second, the virulence of the wild-type virus must have been attenuated; in other words the vaccine virus must have little or no virulence.

Most attenuated virus strains have been derived by 'hit and miss' procedures such as repeated passage of wild-type virus in cells unrelated to the normal host. The vaccine strains of the three serotypes of poliovirus (Chapter 14), which are attenuated as a result of their loss of ability to infect neurones, were derived from wild-type strains by passage in monkeys and in monkey kidney cell cultures. Albert Sabin did this pioneering work and the procedure he used to derive one of the attenuated strains is depicted in Figure 24.1. The mutations responsible for attenuation of the three serotypes are now known and some of them are shown in Figure 24.2.

Other attenuated vaccines that have been developed using a similar approach include those for mumps, measles, rubella, yellow fever, canine parvovirus and canine distemper. The attenuated strains of the first three of these, now combined in the MMR vaccine, were all developed by Maurice Hilleman. The mumps virus strain is called Jeryl Lynn after his daughter, from whom the wild-type virus was isolated.

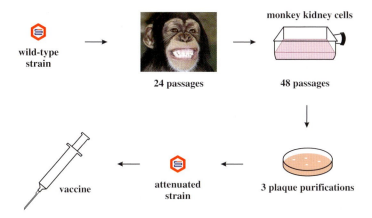

wild-type
strain

24 passages

monkey kidney cells

48 passages

3 plaque purifications

attenuated
strain

vaccine

Figure 24.1 *Derivation of attenuated poliovirus strain (Sabin type 1) from wild-type poliovirus strain (Mahoney 1).*

Other approaches to the development of attenuated virus vaccines include the following.

- *'Cold-adapted' virus strains.* These have been derived by incubating virus-infected cell cultures at temperatures below the optimum for virus replication. 'Cold-adapted' virus strains with reduced virulence have been used in influenza and respiratory syncytial virus vaccines, though the latter were not sufficiently attenuated for use in children.

- *Reassortants (Section 20.3.3.c).* Rotavirus reassortants have been produced, with some genes derived from a human virus and some from an animal virus.

- *Reverse genetics (Sections 14.6.1 and 15.7).* The power of reverse genetics has been used to develop a respiratory syncytial virus vaccine by engineering into the virus genome mutations that attenuate virulence.

One of the risks in using a live vaccine is that, during virus replication, nucleotide substitutions might occur, resulting in reversion to virulence. For poliovirus this can occur when the virus replicates in the gut after vaccination. It is also possible that recombinants between vaccine strains and wild type strains may be formed (Section 20.3.3.b). Because of these risks many countries switched from using attenuated polio vaccine to inactivated vaccine.

Each batch of a live virus vaccine is subjected to extensive quality control tests. These include testing for the presence of every known virus that could be a contaminant, and testing for neurovirulence in animals.

24.3 Inactivated virus vaccines

Inactivated, or killed, virus vaccines are made by mass producing the virulent virus and then inactivating the infectivity, usually by treatment with a chemical such as formaldehyde (Section 23.5.2.c). The trick lies in finding the combination of chemical concentration and reaction time that completely inactivates the virus, but leaves its antigens sufficiently unchanged that they can still stimulate a protective immune response.

Jonas Salk developed a treatment for poliovirus that led to the development of the vaccine that bears his name. The treatment involves suspending virions in formalin (formaldehyde solution) at 37 °C for about 10 days. Other examples of killed virus vaccines are those containing inactivated virions of influenza, hepatitis A and foot and mouth disease viruses.

Because the virus used to produce an inactivated vaccine is a virulent strain, it is vital that 100 per cent of the infectivity is destroyed in the production process. As discussed in Section 23.4, there are some situations in which a small proportion of virions is inactivated at a slower rate. It is essential that the kinetics of virus inactivation are understood and that an inactivation procedure is developed that guarantees 100 per cent inactivation.

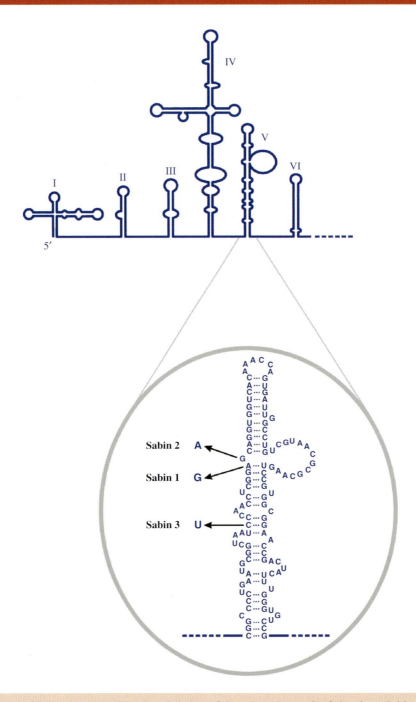

Figure 24.2 *5' end of poliovirus RNA with expanded view of domain V.* For each of the three Sabin strains a mutation in domain V that contributes to the attenuation of neurovirulence is indicated. Modified from Ochs *et al.* (2003) *Journal of Virology*, **77**, 115, with the permission of the authors and the American Society for Microbiology.

The need for complete inactivation was under-lined in the US in 1955 when four million doses of inadequately inactivated Salk vaccine were inoculated into children. Amongst the vaccinees there were 204 cases of paralytic polio and 11 deaths. A similar problem occurred when there were outbreaks of foot and mouth disease in France and the UK in the period 1979–81 after batches of foot and mouth disease vaccine had been inadequately inactivated.

24.4 Virion subunit vaccines

A subunit vaccine contains purified components of virions. In the case of influenza the vaccines contain the haemagglutinin (H) and neuraminidase (N) surface glycoproteins. A typical production method is outlined in Figure 24.3. The infectivity of a batch of influenza virions is inactivated with formaldehyde or β-propiolactone, then the virion envelopes are removed with a detergent, such as Triton X-100. This releases the glycoproteins, which form aggregates of H 'cartwheels' and N 'rosettes'. These structures are purified by centrifugation in a sucrose gradient, and then material from three influenza virus strains is combined to form the vaccine.

For protection against influenza the subunit vaccines are preferred over the inactivated vaccines as they cause fewer side-effects, especially in children. The subunit vaccines, however, induce poorer immune responses, so two doses are necessary to provide adequate immunity.

24.5 Live recombinant virus vaccines

A recombinant vaccinia virus engineered to contain the gene for the rabies virus G protein has been used to vaccinate wild mammals against rabies. The use of this vaccine was discussed in Section 15.2.1.

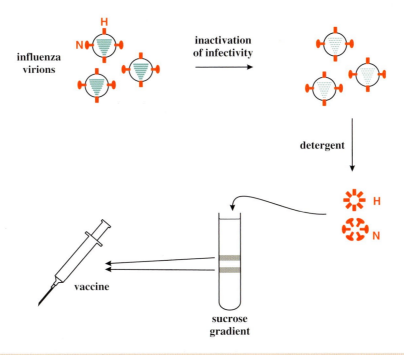

Figure 24.3 *Outline of production method for influenza virus subunit vaccine.* Haemagglutinin (H) and neuraminidase (N) are extracted from inactivated influenza virions and purified by sucrose gradient centrifugation. The bands from the gradient are harvested and incorporated into the vaccine.

24.6   Mass production of viruses for vaccines

The production of the types of vaccine discussed so far (live attenuated, inactivated, subunit and live recombinant) requires large quantities of virions; for most viruses these are produced in cell cultures. John Enders demonstrated in 1949 that poliovirus can be grown in primary monkey cell culture and this led to procedures for the mass production of poliovirus for the Salk and Sabin vaccines. It was subsequently found that some monkey cell cultures used for vaccine production had been harbouring viruses, including simian virus 40 (Section 22.3). It was therefore recommended that human cell cultures should be used for the production of vaccines for human use, though cell lines of monkey origin, such as Vero cells (African green monkey cells), are used for the production of some.

Another preference is for the use of diploid cell lines, rather than heteroploid cell lines that have been derived from tumours, as there are perceived risks associated with tumour-derived cells. Examples of human diploid cell lines used for production of vaccine viruses are MRC-5 (developed by the Medical Research Council in the UK) and WI-38 (developed by the Wistar Institute in the US). Both of these cell lines were derived from human lung fibroblasts.

Most of the cells are grown on surfaces such as parallel plates and small plastic particles. The virus inoculum is added to the cell culture at 0.1–10 pfu per cell and the maximum titre is usually reached 12–24 hours after inoculation. The manufacturer might expect a virus yield up to 100 000 times the inoculum.

Some viruses are mass-produced in chick embryos (Figures 2.1 and 24.4), either because no suitable cell culture system is available, or because such a system has been developed only recently. Influenza and yellow fever vaccines are produced in chick embryos that are inoculated by automatic systems. The yield of virus from each embryo provides sufficient virus for a few doses of vaccine.

Figure 24.4   *Production of vaccine in chick embryos.* Courtesy of the World Health Organisation.

For influenza virus vaccines, decisions have to be made each year as to which H and N subtypes of the virus to include, as new influenza strains with altered subtypes are constantly arising through antigenic drift (Chapter 20). New strains are monitored and characterized by Collaborating Centres for Influenza Reference and Research around the world, and the World Health Organisation recommends the strains that should be produced for the next batches of vaccine.

Newly isolated strains of influenza virus rarely grow well in chick embryos, so a reassortant (Section 20.3.3.c) between each new strain and a high-yielding strain is produced. A high-yielding strain that is commonly used is strain A/Puerto Rico/8/34 (an influenza A virus isolated in Puerto Rico in 1934). This strain evolved during passage in the laboratory; it has roughly spherical virions and it replicates well in chick embryos, in contrast to many wild-type strains, which have filamentous virions and replicate poorly in chick embryos. For vaccine production reassortants are selected that have the H and N genes of the new strain and the remaining genes of the high-yielding strain.

Table 24.1 Examples of viruses that are mass produced for vaccine manufacture

Vaccines for humans	Vaccines for animals
Polio	Foot and mouth disease
Influenza	Marek's disease
Measles	Newcastle disease
Mumps	Canine distemper
Rubella	Rabies
Rabies	
Yellow fever	

Although influenza virus has traditionally been produced in chick embryos for incorporation into inactivated and subunit vaccines, cell culture systems for virus production have now been developed. Furthermore, live attenuated influenza virus vaccines have been licensed in some countries, so the types of influenza virus vaccine and the ways in which they are mass produced may change in the future.

There are major enterprises around the world producing vaccines for human and veterinary use. Some examples of the vaccines produced are given in Table 24.1.

24.7 Virus-like particles

Virus-like particles are structures assembled from virus proteins. The particles resemble virions, but they are devoid of any nucleic acid and can therefore be deemed safer than vaccines containing attenuated or inactivated virions.

Hepatitis B virus (HBV) vaccine is produced in recombinant yeast cells that have the gene for the HBsAg inserted into the genome. The cells are grown in bulk and then broken to release the virus protein. After purification the HBsAg molecules receive a chemical treatment that causes them to aggregate into spherical structures similar to the non-infectious particles produced in the HBV-infected human host (Section 18.4).

The major capsid protein of papillomaviruses can self-assemble into virus-like particles that bear the epitopes required for generating neutralizing antibodies. (An epitope is a short sequence of amino acids critical for inducing an immune response.) Human papillomavirus vaccines (Section 22.11) contain virus-like particles that are produced either in insect cells using a baculovirus vector or in yeast cells.

24.8 Synthetic peptide vaccines

Each protein antigen has one or more epitopes. These short amino acid sequences can be synthesized in a machine and it was suggested that the resulting peptides might be used as vaccines. Compared with traditional vaccines it would be easier to ensure the absence of contaminants such as viruses and proteins.

A lot of work has been done to try to develop peptide vaccines against foot and mouth disease virus. In this virus there is an important epitope within the virion protein VP1 (Section 14.3.1). Synthetic peptides of this sequence induced reasonable levels of neutralizing and protective antibodies in laboratory animals, but when vaccine trials were done in farm animals the results were disappointing.

24.9 DNA vaccines

The most revolutionary approach to vaccination is the introduction into the vaccinee of DNA encoding an antigen, with the aim of inducing cells of the vaccinee to synthesize the antigen. One advantage of this approach is that there is a steady supply of new antigen to stimulate the immune system, as with live virus vaccines. Because the antigen (a virus protein in this case) is produced within the cells of the vaccinee, it is likely to stimulate efficient T-cell-mediated responses.

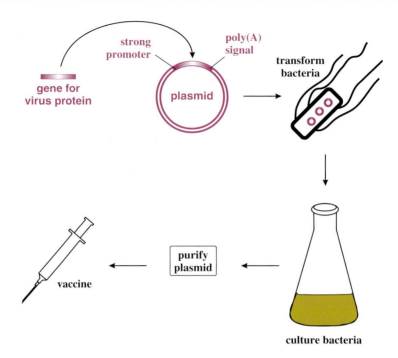

Figure 24.5 *Production of a DNA vaccine.* The virus protein gene is inserted into a plasmid, which is then cloned in bacteria. The plasmid is extracted from the bacterial cells, purified and incorporated into a vaccine.

A procedure for the production of a DNA vaccine is outlined in Figure 24.5. The antigen-encoding sequence, obtained directly from a DNA virus or by reverse transcription from an RNA virus, is inserted into a plasmid between a strong promoter and a poly(A) signal. The plasmid is replicated in bacterial cells and is then purified for use as a vaccine. Administration of the vaccine can be by injection into a muscle or by using a gene gun that delivers DNA-coated gold beads directly into skin cells.

Experimental DNA vaccines have been produced for a number of viruses, including HIV-1, SARS coronavirus, West Nile virus and foot and mouth disease virus; trials have been carried out in mice, pigs, horses and humans. Before any DNA vaccine goes into clinical use some important questions about safety must be answered. There must be confidence that injection of the DNA will not trigger an anti-DNA autoimmune disease, and that the DNA will not create cancer-causing mutations by insertion into host genomes.

24.10   Storage and transport of vaccines

Once a vaccine has been manufactured there is a need to preserve its efficacy until it is used. For live vaccines this means preserving virus infectivity; for vaccines containing inactivated virions, subunits and virus-like particles it means preserving immunogenicity. The major physical and chemical factors that can reduce the infectivity of a live vaccine were considered in Chapter 23; some of these factors can also reduce the immunogenicity of vaccines.

Most vaccines are stored and transported at low temperatures (Figure 24.6); this minimizes losses of infectivity and immunogenicity. Substances that reduce the rate of infectivity loss are included in some vaccines, an example being magnesium chloride in live polio vaccines.

Figure 24.6 *Low temperature storage of virus vaccines.* Photograph courtesy of Novartis Vaccines.

Sources of further information

Books

Dale J. W. and von Schantz M. (2002) Vaccines, pp. 307–314 in *From Genes to Genomes*, Wiley

Flint S. J. *et al*. (2004) Chapter 19 in *Principles of Virology: Molecular Biology, Pathogenesis and Control of Animal Viruses*, 2nd edn, ASM Press

Haaheim L. R. *et al*., editors (2002) Virus vaccines. Chapter 5 in *A Practical Guide to Clinical Virology*, 2nd edn, Wiley

Jiskoot W. *et al*. (2002) Vaccines. Chapter 12 in *Pharmaceutical Biotechnology*, editors Crommelin D. J. A. *et al*., 2nd edn, Taylor and Francis

Kew O. M. *et al*. (2005) Vaccine-derived polioviruses and the endgame strategy for global polio eradication, pp. 587–635 in *Annual Review of Microbiology*, Vol. 59, Annual Reviews

Mims C. A. *et al*. (2004) Chapter 34 in *Medical Microbiology*, 3rd edn, Mosby

Journals

Forde G. M. (2005) Rapid-response vaccines – does DNA offer a solution? *Nature Biotechnology*, **23**, 1059–1062

Gerdil C. (2003) The annual production cycle for influenza vaccine *Vaccine*, **21**, 1776–1779

Hilleman M. R. (2000) Vaccines in historic evolution and perspective: a narrative of vaccine discoveries *Vaccine*, **18**, 1436–1447

Minor P. D. (2004) Polio eradication, cessation of vaccination and re-emergence of disease *Nature Reviews Microbiology*, **2**, 473–482

Noad R. and Roy P. (2003) Virus-like particles as immunogens *Trends in Microbiology*, **11**, 438–444

25

Anti-viral drugs

At a glance

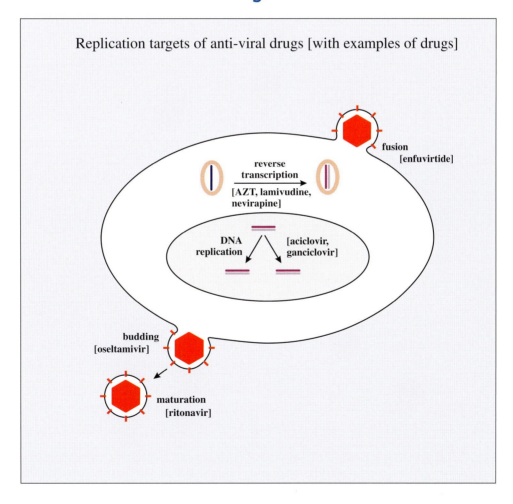

Replication targets of anti-viral drugs [with examples of drugs]

reverse transcription
[AZT, lamivudine, nevirapine]

fusion
[enfuvirtide]

DNA replication   [aciclovir, ganciclovir]

budding
[oseltamivir]

maturation
[ritonavir]

Virology: Principles and Applications   John B. Carter and Venetia A. Saunders
© 2007 John Wiley & Sons, Ltd   ISBNs: 978-0-470-02386-0 (HB); 978-0-470-02387-7 (PB)

25.1 Introduction to anti-viral drugs

For a long time there were very few anti-viral drugs available for clinical use compared with the number of anti-bacterial and anti-fungal drugs. This was because it was difficult to find compounds that interfere specifically with viral activities without causing significant harm to host cell activities. As we saw in Chapters 5–8, however, there are many virus-specific activities that are potential drug targets, some of which are summarized in Figure 25.1.

Most of the early anti-viral drugs were found by screening large numbers of compounds, but this approach has now been largely superseded by designing drugs to inhibit specific virus proteins. Both approaches will be outlined in this chapter.

Drugs are now available for the treatment of diseases caused by a variety of viruses, and in some circumstances drugs are also used for the prevention of virus infections. We shall discuss the clinical uses of several anti-viral drugs and their modes of action. We shall also consider some of the problems associated with anti-viral drug use, especially the emergence of drug-resistant virus strains.

25.2 Development of anti-viral drugs

25.2.1 Screening compounds for anti-viral activity

In this approach to the search for clinically useful anti-viral drugs, large numbers of compounds are tested for

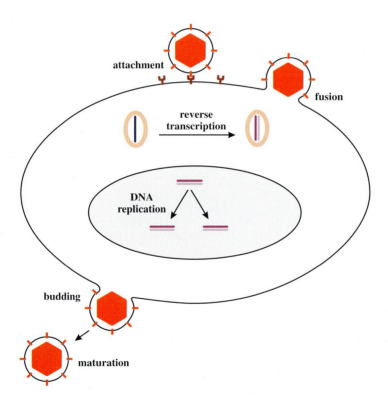

Figure 25.1 *Virus activities that are potential drug targets.* A generalized view of virus replication is depicted; no single virus performs all of these activities! DNA replication provides a target if the virus produces enzymes (e.g. DNA polymerase, thymidine kinase) distinct from those of the host.

possible anti-viral activity. Some of the compounds chosen in the past had already been shown to have anti-tumour-cell activity.

The screening procedure involves testing dilutions of the compounds against a range of viruses growing in cell cultures. Evidence that a compound interferes with the replication of a virus might include inhibition of cytopathic effect (CPE) or of plaque formation. For each potentially useful compound the concentration that inhibits CPE or plaque formation by 50 per cent is determined. This concentration of the drug is known as the 50 per cent inhibitory concentration (IC_{50}).

Screening compounds for anti-viral activity has been largely replaced by the rational design of drugs.

25.2.2 Rational design of anti-viral drugs

The process of designing an anti-viral drug starts with deciding upon a target activity for the drug (Figure 25.1). Once this has been selected, it is necessary to choose a target protein, such as a viral enzyme, that is involved in that activity. A detailed picture of the three-dimensional structure of the protein is derived using techniques such as X-ray crystallography (Section 2.5.3), and a target site in the protein is selected. Computer programs are then used to design compounds that will bind to the target site with the aim of inhibiting the activity of the virus protein.

25.2.3 Safety of anti-viral drugs

If a compound is to have clinical value it must not only inhibit virus replication, but it must cause little or no harm to the host. The research and development procedures therefore include incubating uninfected cell cultures with the compound and examining them for damage such as death or inhibition of cell division.

The potential value of a compound can be assessed by determining the ratio of its cell toxicity to its anti-viral activity. This ratio is known as the selectivity index (SI) and is expressed by the formula

$$SI = \frac{\text{minimum concentration inhibiting cell proliferation (or DNA synthesis)}}{\text{minimum concentration inhibiting virus replication}}$$

A compound with a low IC_{50} and a high SI is most likely to have value as an anti-viral drug. Some examples of IC_{50} and SI values are given in Table 25.1. The data indicate that aciclovir would be preferred to ganciclovir for treatment of herpes simplex virus infections.

25.3 Examples of anti-viral drugs

25.3.1 Nucleoside analogues

A number of anti-viral drugs are synthetic compounds structurally similar to nucleosides such as guanosine

Table 25.1 Some anti-viral drugs and target viruses. IC_{50}: 50 per cent inhibitory concentration. SI: selectivity index. IC_{50} and SI vary depending on the virus strain and the cell type. These figures are ranges determined using several cell types. Data from Kinchington *et al.* (1995)

Drug	Target virus	IC_{50} (μM)	SI
Aciclovir	Herpes simplex virus	0.1–1	>1000
	Varicella-zoster virus	2–10	>1000
Ganciclovir	Herpes simplex virus	0.2–2	100–1000
	Cytomegalovirus	2–10	100–1000
Azidothymidine	HIV-1	0.03–0.5	>1000
Dideoxycytidine	HIV-1	0.5–1.5	100–1000
Dideoxyinosine	HIV-1	2–10	$\geq$100

and 2′-deoxythymidine (Figure 25.2) and they act by interfering with the synthesis of virus nucleic acids. After being taken into a cell a nucleoside analogue, like a nucleoside, is phosphorylated at the 5′ carbon (or its equivalent) to become a nucleotide analogue (Figure 25.3). The 5′ triphosphate derivative of the nucleoside analogue is the active form of the drug and acts as a competitive inhibitor of a viral polymerase such as a reverse transcriptase.

During nucleic acid synthesis, if one of these nucleotide analogues is incorporated into a growing strand, then nucleic acid synthesis is terminated, as in

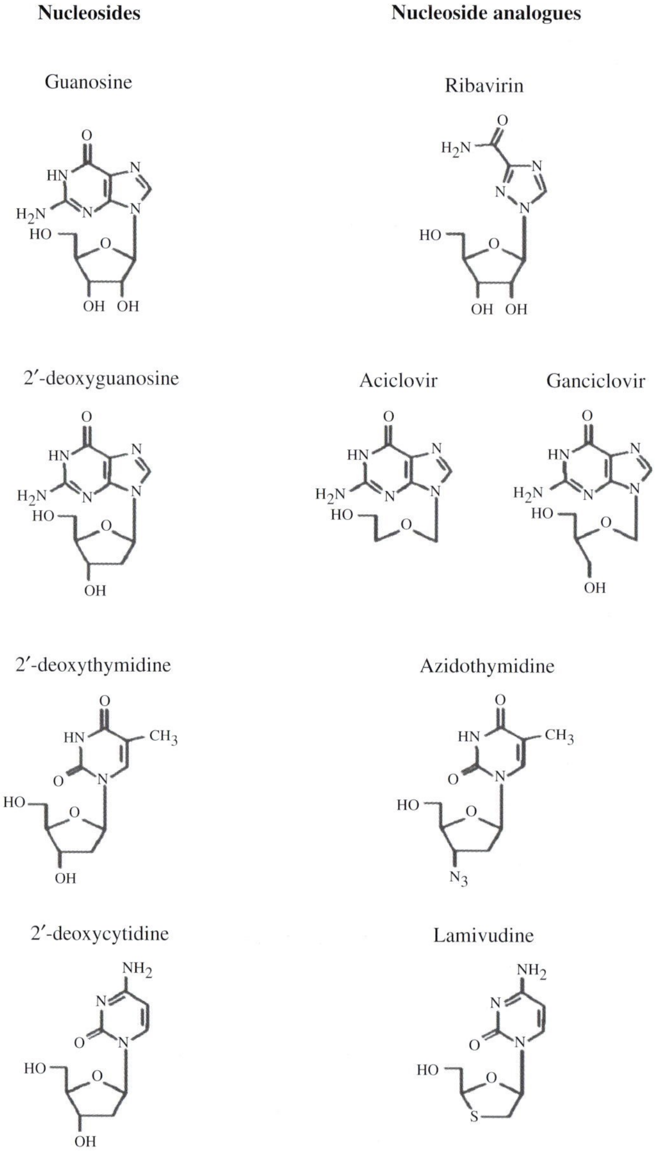

Nucleosides **Nucleoside analogues**

Guanosine

Ribavirin

2′-deoxyguanosine

Aciclovir   Ganciclovir

2′-deoxythymidine

Azidothymidine

2′-deoxycytidine

Lamivudine

Figure 25.2 *Nucleoside analogues used as anti-viral drugs.* The analogues are derived from the nucleosides shown on the left.

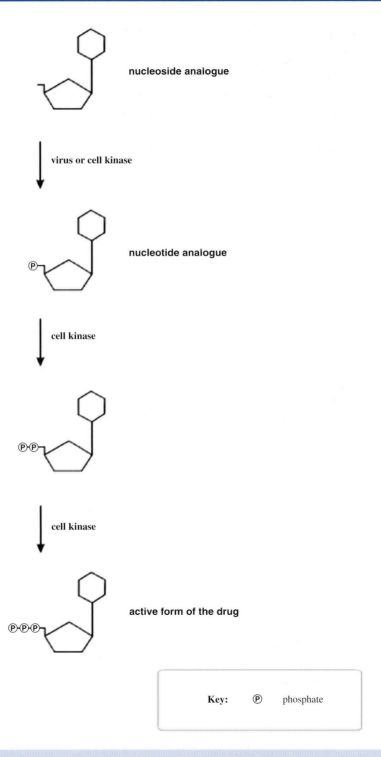

Figure 25.3 *Phosphorylation of a nucleoside analogue.* The active form of the drug is the 5′ triphosphate derivative of the nucleoside analogue.

the dideoxy method of DNA sequencing. The structure of the nucleotide analogue prevents it from accepting the next nucleotide. In some cases the analogue lacks the hydroxyl group on the 3' carbon (or lacks the 3' carbon) necessary for the addition of the next nucleotide.

25.3.1.a Ribavirin

Ribavirin is an analogue of guanosine (Figure 25.2). It is used for the treatment of infection with several RNA viruses, especially persistent infections with hepatitis C virus (HCV; Section 22.11). Combined treatment with ribavirin and α-interferon has eradicated HCV infection in many patients.

Ribavirin has also been recommended for treatment of young children infected with respiratory syncytial virus (Section 15.6), one of the most important causes of severe respiratory tract disease in children. The drug is delivered to the patient in the form of an aerosol for 12–18 hours daily. This treatment reduces the severity of the disease, but the mode of delivery is expensive and inconvenient, so it is not often used.

The mode of action of ribavirin is not completely understood and at least five hypotheses have been proposed to explain its anti-viral activity. The drug has been shown to inhibit several activities, including viral RNA synthesis and mRNA capping, and it is likely that it has more than one mode of action.

25.3.1.b Nucleoside analogues that inhibit herpesvirus DNA synthesis

Aciclovir (also spelt 'acyclovir') is an analogue of guanosine (Figure 25.2). It is a very safe drug, with almost no side-effects, widely used for the treatment and prevention of diseases caused by three of the herpesviruses: herpes simplex viruses 1 and 2 (HSV-1 and HSV-2) and varicella-zoster virus (VZV) (Chapter 11). Uses of aciclovir include

- prevention of cold sores;

- prevention and treatment of genital herpes;

- treatment of HSV encephalitis (early treatment reduces the mortality rate);

- protection of immunocompromised patients after exposure to VZV;

- treatment of shingles (pain is reduced significantly).

Aciclovir strongly inhibits virus DNA synthesis but has very little effect on cell DNA synthesis. In a herpesvirus-infected cell the first phosphorylation of aciclovir (Figure 25.3) is carried out by the virus thymidine kinase (TK) (Section 11.5.3). The cell TK is produced only at certain stages of the cell cycle and it has a much lower affinity for aciclovir than the viral enzyme, so there is very little phosphorylation of aciclovir in uninfected cells.

During DNA synthesis, aciclovir triphosphate competes with deoxyguanosine triphosphate (dGTP) for incorporation into the new strand. When aciclovir becomes incorporated, DNA synthesis is terminated due to the lack of a 3'–OH group.

Ganciclovir, like aciclovir, is an analogue of guanosine, and the first phosphorylation in a cell is carried out by a virus enzyme, in this case a protein kinase. Ganciclovir is more likely than aciclovir to become incorporated into cell DNA. As a result ganciclovir has a lower SI than aciclovir (Table 25.1) and can cause significant side-effects, especially reduced numbers of blood cells because of damage to bone marrow cells.

Ganciclovir is used to treat infections with cytomegalovirus, as aciclovir is less effective against this virus than it is against the other herpesviruses. Examples of cytomegalovirus diseases that are treated with ganciclovir include pneumonia and retinitis in immunocompromised patients, such as those who have received organ transplants and those with AIDS.

25.3.1.c Nucleoside analogues that inhibit reverse transcription

Azidothymidine (AZT), also known as zidovudine, is an analogue of thymidine (Figure 25.2). It had been investigated as a possible anti-cancer drug and it was found to inhibit the reverse transcriptase of HIV. Like other nucleoside analogues, it is phosphorylated to the 5' triphosphate after uptake into a cell (Figure 25.3).

AZT triphosphate binds more strongly to the viral reverse transcriptase than to the cell DNA polymerase, and the reverse transcriptase binds AZT triphosphate in preference to deoxythymidine triphosphate. AZT thus inhibits HIV reverse transcription, but it can also interfere with cell DNA synthesis and, like ganciclovir, it can damage the bone marrow. In the early days of AZT use, when large doses were administered, many patients suffered severe side-effects.

Lamivudine, also known as 3TC, is an analogue of cytidine (Figure 25.2). It inhibits reverse transcriptases, including those of HIV and hepatitis B virus (HBV), and is used, usually in combination with other therapies, to treat infections with these viruses. The drug does not eliminate the infections, but it can control them. For treatment of HBV infection lamivudine is often given in combination with α-interferon. After treatment for a year or more inflammation of the liver is reduced in a proportion of patients.

Several other nucleoside analogues that inhibit reverse transcription are used for treatment of HIV infection; examples are dideoxycytidine and dideoxyinosine.

25.3.2 Non-nucleoside inhibitors of reverse transcription

The approach of rational design was used to develop several HIV reverse transcriptase inhibitors that are not based on nucleosides. These non-nucleoside inhibitors target different sites in the reverse transcriptase to those targeted by the nucleoside analogues. An example of a non-nucleoside inhibitor that is used against HIV is nevirapine (Figure 25.4).

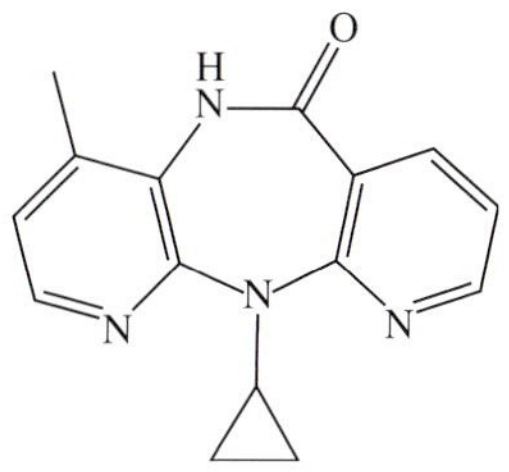

Figure 25.4 Nevirapine (a non-nucleoside inhibitor of reverse transcription).

25.3.3 HIV protease inhibitors

The maturation of a retrovirus virion involves the cleavage by a virus protease of the Gag and Gag–Pol proteins to form the virion proteins. If this processing does not take place the virion does not acquire infectivity (Section 16.3.6). Peptide mimics of the cleavage site in the protein have been developed and these compounds can fit into the active site of the HIV protease. The result is that fewer virions bud from HIV-infected cells and those virions that do bud are non-infectious. An example of an HIV protease inhibitor is ritonavir (Figure 25.5).

25.3.4 HIV fusion inhibitors

After an HIV-1 virion has bound to a cell the transmembrane glycoprotein gp41 must fuse the envelope

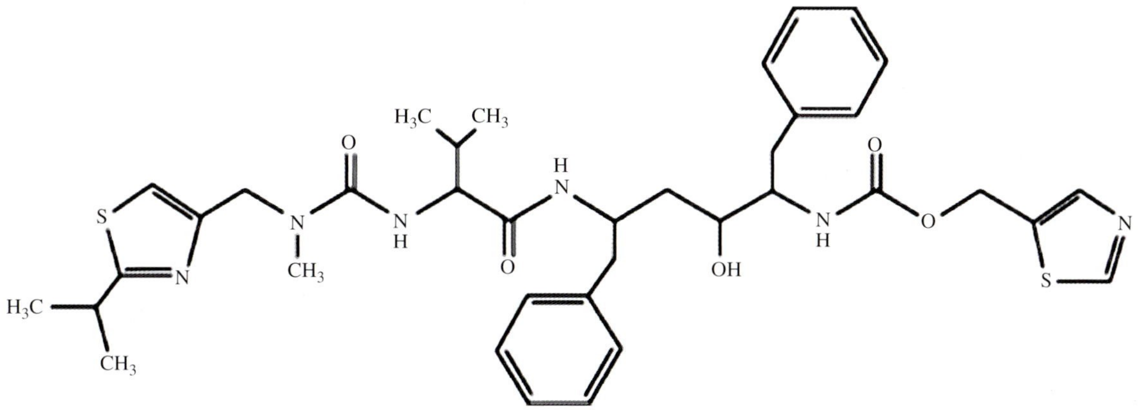

Figure 25.5 Ritonavir (an HIV protease inhibitor).

with the membrane of the cell if the virion contents are to be delivered into the cytoplasm (Section 17.4.1). In order for this to take place gp41 must undergo a conformational change initiated by interaction between different regions of the molecule.

A number of drugs have been developed that can inhibit fusion between the membranes of an HIV-1 virion and a potential host cell. They do this by binding to gp41 and inhibiting the conformational change. An example is enfuvirtide, a 36-amino-acid peptide with the sequence of a gp41 region that interacts with another gp41 sequence to cause the conformational change. Therefore, unlike the other anti-HIV drugs, enfuvirtide prevents infection of a cell with HIV-1. It has much lower activity against HIV-2, which has a trans-membrane glycoprotein only distantly related to that of HIV-1. A disadvantage of enfuvirtide is that it has to be given by subcutaneous injection, whereas the other anti-HIV drugs can be given orally.

25.3.5 Influenza virus neuraminidase inhibitors

One of the surface proteins of influenza A and B viruses is a neuraminidase. This enzyme plays a vital role during the final stages of virion budding from infected cells; if the enzyme is inhibited virions are not released. Each neuraminidase spike on the virion surface is made up of four monomers, each consisting of a 'balloon' on a 'stick'.

In 1983 an influenza virus neuraminidase was crystallized and its structure was determined. It was found that the monomer has a deep cleft and that this forms the active site of the enzyme. Compounds were designed to bind in the cleft and several were found to inhibit neuraminidase activity. One such compound is oseltamivir (Figure 25.6), which is approved for use as an anti-influenza virus drug.

25.4　Drug resistance

Soon after the introduction of anti-viral drugs into clinical practice drug-resistant strains of the target viruses began to emerge. This should not have been surprising, as antibiotic-resistant bacteria, insecticide-resistant insects and rodenticide-resistant rats had

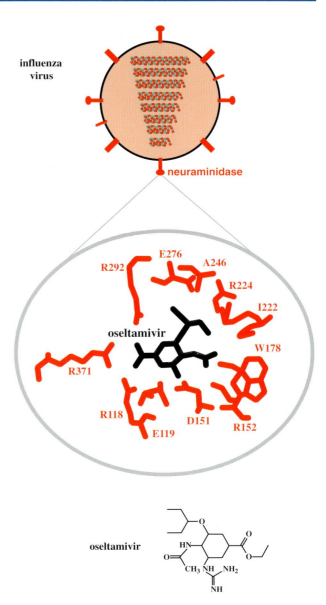

Figure 25.6 *Binding of oseltamivir to influenza virus neuraminidase.* The inset shows the amino acid residues (one-letter abbreviations together with their codon numbers) that bind oseltamivir.

emerged as a result of natural selection. The fact that viruses, especially RNA viruses, can mutate at high frequencies and evolve rapidly (Section 20.3.3.a) means that genotypes encoding drug resistance can arise rapidly. Drug-resistant genotypes may be at an advantage in hosts where the drug is present and may become the dominant genotypes in those hosts.

Drug resistance of viruses is relative, rather than absolute. A measure of the degree of resistance can be obtained by determining the IC_{50} (Figure 25.7). A virus strain is considered to be 'resistant' to a drug if it is able to replicate in the body in the presence of a concentration of the drug that inhibits replication of 'sensitive' strains.

Drug-resistant virus isolates are found to have one or more mutations in the genes encoding the proteins that are the drug targets. For example, the vast majority of aciclovir-resistant HSV isolates have a mutated *tk* gene, while a small number of isolates have mutations in the *DNA pol* gene. Further examples are given in Table 25.2.

Mutations in the HIV-1 *RT* gene that confer resistance to a nucleoside analogue (e.g. AZT) are in different codons to those that confer resistance to a non-nucleoside inhibitor (e.g. nevirapine). This reflects the different target sites in the reverse transcriptase for the two classes of drug. Some mutations confer cross-resistance to other drugs of the same type, as in the examples in Table 25.3.

Table 25.2 Examples of genes mutated in drug-resistant virus strains. *tk*: thymidine kinase. *RT*: reverse transcriptase. *PR*: protease.

Drug	Virus	Gene mutated
Aciclovir	HSV-1	*tk*
AZT	HIV-1	*RT*
Nevirapine	HIV-1	*RT*
Ritonavir	HIV-1	*PR*
Enfuvirtide	HIV-1	*env (gp41)*

Most mutations to drug resistance of HIV-1 are amino acid substitutions, like the examples in Table 25.3, but some mutations are deletions or insertions.

Clinical problems arise when drug-resistant virus strains emerge in patients undergoing treatment and when resistant strains are transmitted to other individuals. When such problems arise patients may be treated with alternative drugs. Infections with aciclovir-resistant HSV strains can be treated with foscarnet or cidofovir, neither of which depends on the HSV TK for their action.

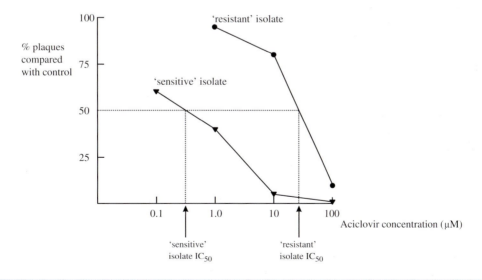

Figure 25.7 *Susceptibility of 'sensitive' and 'resistant' isolates of HSV-1 to aciclovir.* The 'resistant' isolate has an aciclovir IC_{50} approximately 100 times greater than the 'sensitive' isolate.

Table 25.3 Examples of mutations to drug resistance in the *RT* gene of HIV-1.

Mutations conferring resistance to nucleoside analogues	Mutations conferring resistance to non-nucleoside analogues
K65R[1]	K103N
M184V	Y188L

[1]Codon 65 mutated from lysine to arginine. Other mutations are indicated similarly.

When AZT was first used for treatment of HIV infection, AZT-resistant strains of HIV rapidly emerged and a similar problem was encountered when other anti-HIV drugs were used alone. Now the standard treatment for HIV infection involves a combination of drugs. The rationale is the same as that used for chemotherapy of tuberculosis and cancer; there is only a very small probability of a mutant arising that is resistant to all drugs in the combination. Various combinations of anti-HIV drugs are used, such as two reverse transcriptase inhibitors and a protease inhibitor. This form of treatment is known as highly active anti-retroviral therapy (HAART).

The effect of drug treatment on an HIV infection can be monitored by measuring the concentration of HIV RNA (and hence of HIV virions) in the patient's blood. HAART usually results in a rapid decline for about the first 10 days, and then a slower decline for a period of weeks (Figure 25.8). Over a longer period the HIV RNA concentration reaches a steady state in the range 5–50 copies/ml in some patients, while in others it falls to less than 5 copies/ml. HAART also reduces the levels of HIV in the seminal fluid of infected men and in the genital secretions of infected women.

HAART does not eliminate HIV infection from the body; the infection persists in latently infected macrophages and memory CD4 T cells. It may also persist in sanctuary sites, such as the brain and the testes, where infected cells may evade the drugs and/or immune clearance. HAART has, however, markedly reduced death rates from AIDS in developed countries, and treatment of HIV-infected women has significantly reduced the risk of mother-to-child transmission (Section 17.7). It can be difficult for patients on HAART to stick rigidly to their daily routine of taking large numbers of pills, especially when the medication causes unpleasant side-effects, but if patients lapse from their routine they increase the risk of drug-resistant strains of HIV emerging.

25.5 Anti-viral drug research

In addition to the emergence of drug-resistant strains of viruses there are other problems related to the use of anti-viral drugs. Most of the drugs in use do not eliminate virus infections from the body, and many can cause significant side-effects. All of these problems can occur, for example, during treatment of HBV infection

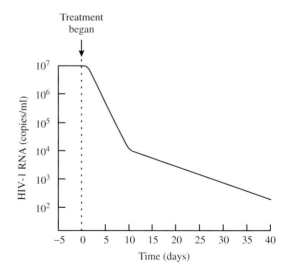

Figure 25.8 *Effect of HAART on the concentration of HIV-1 in the blood.* Redrawn from Di Mascio *et al.* (2004) *Mathematical Biosciences*, **188**, 47. Reproduced by permission of Elsevier Limited and author.

with lamivudine and/or α-interferon. For many virus diseases suitable drugs have not yet been developed.

There is therefore a great need for continued research programmes aimed at extending the range of drugs available. One objective of these programmes is to reduce the severity of the problems discussed above; another objective is to develop drugs for virus diseases that are currently untreatable. In addition to the approaches outlined in this chapter alternative avenues are being explored, including exploitation of some of the body's anti-viral defences, such as RNA silencing (Section 9.2.3).

Learning outcomes

By the end of this chapter you should be able to

- evaluate procedures used to develop new anti-viral drugs;

- describe the modes of action of selected anti-viral drugs;

- evaluate anti-viral drugs in clinical use;

- discuss virus resistance to drugs.

Sources of further information

Books

Cohen J. and Powderly W. G., editors (2004) *Infectious Diseases*, 2nd edn, Chapters 204–207, Mosby

Driscoll J. S. (2002) *Antiviral Drugs*, Ashgate

Flint S. J. *et al.* (2004) Chapter 19 in *Principles of Virology: Molecular Biology, Pathogenesis and Control of Animal Viruses*, 2nd edn, ASM Press

Haaheim L. R. *et al.*, editors (2002) Antiviral drugs. Chapter 4 in *A Practical Guide to Clinical Virology*, 2nd edn, Wiley

Kinchington D. *et al.* (1995) Design and testing of antiviral compounds. Chapter 6 in Desselberger U., editor *Medical Virology: a Practical Approach*, IRL Press

Murray P. R. *et al.* (2003) Chapters 107–109 in *Manual of Clinical Microbiology*, 8th edn, ASM Press

Richman D. D., editor (2003) *Human Immunodeficiency Virus*, Chapters 10–12, International Medical Press

Torrence P. F., editor (2005) *Antiviral Drug Discovery for Emerging Diseases and Bioterrorism Threats*, Wiley Interscience

Journals

De Clercq E. (2004) Antivirals and antiviral strategies *Nature Reviews Microbiology*, **2**, 704–720

De Clercq E. (2005) Antiviral drug discovery and development: where chemistry meets with biomedicine *Antiviral Research*, **67**, 56–75

Feld J. J. and Hoofnagle J. H. (2005) Mechanism of action of interferon and ribavirin in treatment of hepatitis C *Nature*, **436**, 967–972

Jerome K. R. (2005) The road to new antiviral therapies *Clinical and Applied Immunology Reviews*, **5**, 65–76

Shaw T., Bartholomeusz A. and Locarnini S. (2006) HBV drug resistance: mechanisms, detection and interpretation *Journal of Hepatology*, **44**, 593–606

26

Prions

At a glance

PRION (INfectious PROtein)

Normal protein

More α helix than β sheet

Misfolded protein

Mainly β sheet

Prion diseases:
- scrapie (sheep)
- bovine spongiform encephalopathy
- sporadic Creutzfeldt-Jakob disease (humans)
- variant Creutzfeldt-Jakob disease (humans)

Virology: Principles and Applications   John B. Carter and Venetia A. Saunders
© 2007 John Wiley & Sons, Ltd   ISBNs: 978-0-470-02386-0 (HB); 978-0-470-02387-7 (PB)

26.1 Introduction to prions

This chapter is not about viruses, but about some infectious diseases of man and animals that are known as prion diseases. Much of the research on prions is published in virology journals, hence their appearance in this book. The evidence suggests that the causative agents are protein molecules from within the cells of the host; no nucleic acid has been found associated with them.

The prion diseases are characterized by very long incubation periods, measured in years. (The incubation period of an infectious disease is the time that elapses between the infectious agent entering the host and the first appearance of signs and symptoms of the disease.) Signs of prion disease include dementia and loss of co-ordination; the patient gradually deteriorates, and death is inevitable.

This subject assumed particular importance towards the end of the 20th century with the onset in the UK of bovine spongiform encephalopathy (BSE), or mad cow disease as it is popularly known, and its apparent transmission to humans as variant Creutzfeldt-Jakob disease (vCJD).

26.2 Transmissible spongiform encephalopathies

The prion diseases are known as transmissible spongiform encephalopathies (TSEs). Taking these words in reverse order,

- **encephalopathy** means disease of the brain;

- **spongiform** refers to the development of holes in the brain, making it appear like a sponge (Figure 26.1);

- **transmissible** refers to the fact that the causative agent is infectious. It can be transmitted to members of the same species, and sometimes to other species.

26.3 The nature of prions

There is no evidence that the infectious agents that cause TSEs contain any nucleic acid; the agents appear to be misfolded forms of normal cell proteins. This 'protein-only' hypothesis was proposed by Stanley

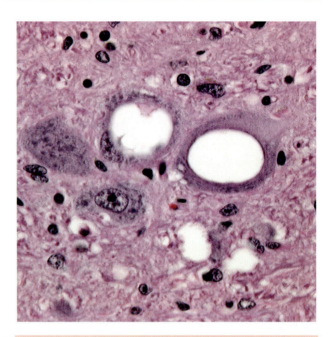

Figure 26.1 *Brain section from a sheep with scrapie.* The spongiform appearance (holes in the tissue) is evident. Magnification × 500. Courtesy of Dr. R. Higgins, University of California, Davis.

Prusiner, who also suggested the term prion (pronounced *pree-on*), derived from *in*fectious *pro*tein. Versions of the normal protein have been found in mammals, birds and reptiles; in humans it is encoded by the *Prnp* gene. The role of the protein is not yet clear. It cycles between endosomes and the cell surface, where it is held in the plasma membrane by a glycosyl-phosphatidyl-inositol anchor at its C terminus. It is found on many cell types, but especially on cells of the central nervous system.

The protein molecule has a loop as a result of a disulphide bond. The loop contains two asparagine residues, either or both of which can be N-glycosylated, so the protein exists as three glycoforms: unglycosylated, monoglycosylated and diglycoslyated; the sugars attached to the protein may be of various types.

The conformation of much of the normal protein is α helix. When the protein misfolds there is a decrease in α helix structure and an increase in β sheet (Figure 26.2). This change in conformation is accompanied by changes in properties of the protein. While the normal protein is completely digested by

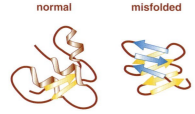

normal   misfolded

Figure 26.2 *Normal and misfolded forms of prion.* The misfolded form has more β sheet (shown in yellow and blue) than the normal form. Redrawn, with permission from Annual Reviews and the author, from Collinge J. (2001) *Annual Review of Neuroscience*, **24**, 519.

proteinase K, the misfolded protein is largely resistant. The enzyme is able to digest only 90–100 amino acid residues from the N terminus; the remaining 27–30 kD portion of the protein remains intact. Misfolding also renders the protein insoluble in nonionic detergents. In prion-diseased tissues molecules of the misfolded protein aggregate as fibrils, rods or other forms, depending on the host and the prion strain.

Various terminologies are used for the normal and the misfolded forms of the protein. The normal protein is commonly designated as PrP^c (PrP = prion protein; c = cell), while the misfolded form is designated as PrP^{Sc} (Sc = scrapie) or PrP^{res} (res = resistant to proteinase).

The mechanism by which prion 'replication' takes place is not understood. It has been suggested that a form of 'evangelism' may be involved, whereby molecules of the misfolded protein 'convert' normal protein molecules and cause them to misfold. In other words, the misfolded protein somehow causes a change in the conformation of the normal protein.

A further suggestion is that initiation of the replication process requires a 'seed' consisting of an aggregate of a number of misfolded protein molecules (Figure 26.3). This 'seed' could be formed after prion inoculation, or after a rare conformational change in the normal molecule, or as the inevitable outcome of the expression of a mutant prion gene.

If the misfolded protein molecules result from normal molecules that have undergone a conformational change, then this is a very different process to virus

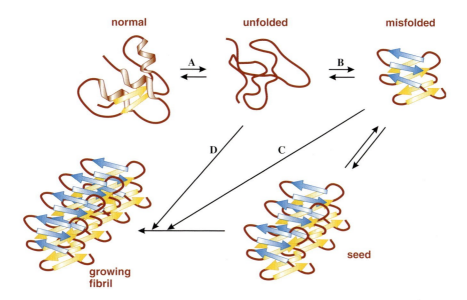

normal   unfolded   misfolded

growing fibril   seed

Figure 26.3 *Model for prion replication.* Copies of the normal protein unfold (A) and refold (B) into a form comprised mainly of β sheet (shown in yellow and blue). Replication may require a critical 'seed' size. Further recruitment of misfolded molecules (C) or unfolded molecules (D) then occurs as an irreversible process. Redrawn, with permission from Annual Reviews and the author, from Collinge J. (2001) *Annual Review of Neuroscience*, **24**, 519.

replication, where new virions are constructed from amino acids and nucleotides.

The misfolded protein accumulates in endosomes and lysosomes, and quantities build up especially in neurones. In some prion diseases the protein can also be found in a number of other organs and tissues, including the spleen and lymph nodes.

Prion infectivity is remarkably heat resistant and some infectivity can survive autoclaving for prolonged periods (Figure 26.4). It is interesting to note that a small subpopulation of the infective material has a higher level of resistance to inactivation. This subpopulation is inactivated at a slower rate, mirroring the situation with many viruses (Section 23.4). Prion infectivity is also very resistant to inactivation by irradiation and by some chemicals that inactivate virus infectivity. Treatments that are used to inactivate prion infectivity include exposure to 2.5–5 per cent sodium hypochlorite solution or 1 M NaOH for 1–2 hours.

It has been difficult for many to accept the idea that an infectious agent that is able to replicate is devoid of any nucleic acid, but all attempts to demonstrate the presence of a nucleic acid have failed. Doubts still remain, but the following work, published in 2004, provides evidence that prions are infectious proteins. Mouse prion protein was produced in recombinant bacterial cells and polymerized into fibrils, which were

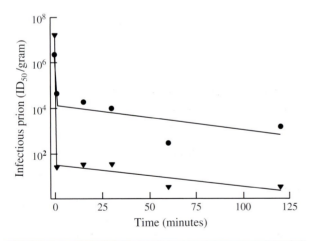

Figure 26.4 *Inactivation of two strains of scrapie prion in an autoclave at 126 °C.* Data from Somerville (2002) *Trends In Biochemical Sciences*, **27**, 606; Reprinted by permission of Elsevier and the author.

injected into the brains of mice. The mice subsequently developed signs of a TSE and extracts of their brains transmitted the disease to other mice.

26.4 Prion diseases

Some cases of prion disease arise spontaneously, some are inherited and some are acquired as a result of a TSE agent entering the body. In the latter case entry may be through the digestive tract, followed by transport to the lymphoreticular system (e.g. spleen, lymph nodes, Peyer's patches), where the molecules are amplified. From these sites they can be transferred into the central nervous system.

Replication takes place slowly but, because the misfolded protein is not degraded, concentrations gradually build up. The molecules form insoluble aggregates, often visible as 'plaques' in sections of central nervous tissue. The accumulation of these aggregates is thought to lead to dysfunction and death of neurones, leading to the development of the holes in the brain (Figure 26.1). The inevitable outcome is the death of the host.

26.4.1 Prion diseases in animals

Scrapie is a disease of sheep and goats that has been known in Britain and other parts of Europe for hundreds of years. Many affected animals scrape against hard objects such as fence posts, hence the name of the disease, and many grind their teeth, stumble and fall; all eventually die. In the 1930s it was demonstrated that scrapie can be transmitted from sheep to sheep by injection of brain tissue.

In the US a TSE in farmed mink was first recognized in 1947, then in the early 1980s chronic wasting disease was described in mule deer and elk in captivity. The latter disease has since been found in wild mule deer, white-tailed deer and elk and is the only TSE known to occur in free-ranging animals.

BSE was first reported in the UK in 1986 and a massive outbreak rapidly developed (Figure 26.5). The disease was spread by feeding meat and bone meal to cattle as a protein supplement. Body parts from animals incubating the disease were incorporated into meat and bone meal and this served to spread the disease on a

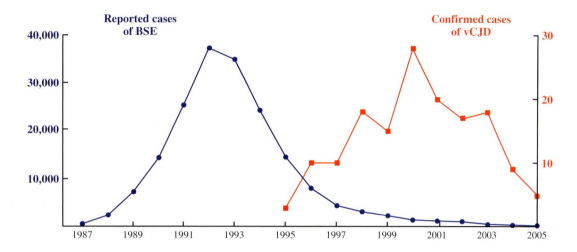

Figure 26.5 *Numbers of cases of BSE and vCJD in the UK.* BSE data from World Organisation for Animal Health. vCJD data from UK National CJD Surveillance Unit.

large scale. Cattle that appeared to be healthy could in fact be incubating BSE and there could be large quantities of misfolded protein in their brains and spinal cords. It is uncertain whether the original infectious material was from a sheep with scrapie or from a cow in which BSE had arisen spontaneously.

BSE was exported from the UK in live cattle, and probably also in the large quantities of meat and bone meal that were exported. BSE is no longer restricted to the UK; it has been reported in many countries around the world.

In the early years of the BSE outbreak many other species of domestic and captive mammal developed spongiform encephalopathies. These animals included domestic cats, big cats (such as puma and tiger) and herbivores (such as bison and eland). It is likely that these cases resulted from feeding animals meat from BSE-infected cattle.

26.4.2 Prion diseases in humans

There are a number of human prion diseases in the three categories; some examples are given in Table 26.1. Sporadic CJD is the most common, occurring through-out the world at an incidence of about 1.7 cases per million people per year. Inherited prion diseases have high

Table 26.1 Examples of human prion diseases

Category of disease	Examples
Spontaneous	Sporadic CJD
Inherited	Familial CJD
	Fatal familial insomnia
Acquired	Kuru
	vCJD

incidences in families whose genomes encode certain amino acids at particular codons in the *Prnp* gene.

Kuru and vCJD are acquired by ingesting prions. In the case of kuru the source of the prions was human brain tissue. This disease used to occur in areas of New Guinea where it was the custom to eat tissue from relatives who had died.

vCJD is a relatively new disease. The first case was reported in the UK in 1995 with the death of a young man called Stephen Churchill, and since then there have been a number of cases each year (Figure 26.5). There is no evidence that these cases represent an inherited form of prion disease, and they have characteristics that distinguish them from sporadic CJD. Amongst these characteristics are

- the relatively young age of the victims;

- the relatively short duration of the disease;

- distinct pathological changes in the brain;

- the presence of infective prion in the spleen, lymph nodes and tonsils.

These features indicated that vCJD was a new disease, and because its appearance coincided with the BSE outbreak in the UK it was suggested that it had resulted from humans ingesting the BSE prion. Further evidence that the BSE prion is the causative agent of vCJD came from laboratory studies, which demonstrated that the BSE prion and the vCJD prion have identical molecular characteristics, and that they induce identical responses in inoculated mice. A few cases of vCJD appear to have resulted from transmission of the agent via blood transfusion, rather than through ingestion.

There have been cases of vCJD in countries other than the UK. Some of these victims had lived in the UK, so may have acquired the disease before moving abroad. Others had never visited the UK, but they could have become infected from meat exported from the UK or from BSE in local cattle.

26.5   Prion strains

For some TSE agents a number of strains can be differentiated, each having phenotypic characters that are consistently transmitted from generation to generation. For example, different strains of the scrapie agent can induce different clinical syndromes in goats. Differences can also be observed in laboratory studies when prion strains may be distinguished by properties such as incubation period in mice after injection into the brain and degree of heat resistance. Figure 26.4 illustrates two strains of scrapie agent that differ in heat resistance.

Virus strains are defined by differences in the sequences of their nucleic acids, but if TSE agents do not contain nucleic acids how can strains with different phenotypic characters be explained? One proposal is that the prion protein can misfold into multiple conformations, one for each strain. Alternatively,

prion strain differences may result from the extent to which the molecules are glycosylated at the two glycosylation sites; in other words, a particular strain may have a particular ratio of the three glycoforms (Section 26.3).

26.6   Prion transmission

As we have already discussed, prions can be transmitted to other members of the same species and often to other species. The efficiency of transmission depends on the route by which the prion molecules enter the body; injection into the brain is much more efficient than ingestion. Dose is also important; the larger the dose, the more likely it is that transmission will occur.

Another factor that affects the efficiency of prion transmission is the existence of small variations in the prion gene sequence, which result in variation in susceptibility between genotypes. The presence of certain amino acids at particular codons can render individuals more or less susceptible to infection. For example, sheep are more susceptible to scrapie if the prion gene has valine rather than alanine at codon 136. Small variations in the prion gene sequence can also have a strong influence on the incubation period and on the clinical course of the disease.

Many attempts to transmit TSE agents to distantly related species have been reported as difficult. When transmission does occur the minimum infective dose is commonly larger than for the same or closely-related species, and incubation periods are commonly longer. This effect has been called the 'species barrier'. Most tests for cross-species transmission have used the development of prion disease in recipient animals as the criterion for successful transmission. There is, however, evidence for subclinical infections, where prions replicate without development of disease during the lifespan of the infected animal. It is therefore possible that the strength of the 'species barrier' has been over-estimated.

If a TSE agent is transmitted to a different species, and then passaged in that species, progressive changes take place; the incubation period becomes shorter over a number of passages and then stabilizes.

The transmission of BSE into humans as vCJD involved a 'species barrier' being crossed. As the BSE outbreak progressed, the UK government introduced increasingly strict measures, to reduce the risk of the BSE prion being present in food. Animal materials likely to constitute a risk were banned; these include brain, spinal cord and mechanically recovered meat from the spine. Governments introduced additional measures, including the testing of cattle and meat for the BSE prion using serological methods.

The possibility of human-to-human transmission is also a matter for concern. Measures are taken to minimize the risk of prion transmission via donated blood and blood products, and via surgical instruments.

The cases of TSE in cats and other species, coincident with the BSE outbreak, are thought to have been transmitted in meat containing the BSE agent (Section 26.4.1). A number of additional species have been infected with the BSE agent under experimental conditions, thus the BSE agent appears to have a wide 'host range'.

26.7 The protein-only hypothesis

If the protein-only hypothesis is correct, and the infectious agent is misfolded prion protein, then transmission of a prion disease implies that the introduction of the misfolded protein into the body of a new host initiates the misfolding of protein molecules in that host. If the agent is 'transmitted' to other species then, because each species makes a specific prion protein, the molecules that become misfolded have the amino acid sequence of the recipient prion, not that of the donor.

The protein-only hypothesis suggests that the agent is derived from one of the body's own proteins, and this can explain the inherited forms of TSEs (Table 26.1). According to the hypothesis, these diseases develop as a consequence of the inheritance of a gene encoding a prion protein with an amino acid sequence that has a high probability of misfolding.

There remain many controversies and there is still much to be learnt about the TSEs and their causative agents. Progress in research is slow as a result of the long incubation periods of the diseases.

Learning outcomes

By the end of this chapter you should be able to

- define the terms 'prion' and 'transmissible spongiform encephalopathy';

- evaluate the 'protein-only' hypothesis;

- describe the characteristics of prions;

- discuss a theory of prion replication;

- describe prion diseases in animals and man;

- discuss the transmission of prions.

Sources of further information

Books

Bishop M. and Ironside J. W. (2004) Genetic susceptibility to prion diseases. Chapter 13 in *Susceptibility to Infectious Diseases*, editor Bellamy R., Cambridge University Press

Dormont D. (2004) Prions. Chapter 223 in *Infectious Diseases*, 2nd edn, editors Cohen J. and Powderly W. G., Mosby

Harris D., editor (2004) *Mad Cow Disease and Related Encephalopathies*, Springer

Manson J. C. and Barron R. M. (2005) The transmissible spongiform encephalopathies. pp. 137–158 in *Molecular Pathogenesis of Virus Infections*, editors Digard P., Nash A. A. and Randall R. E., 64th Symposium of the Society for General Microbiology, Cambridge University Press

Pennington H. (2003) *When Food Kills: BSE,* E. coli *and Disaster Science*, Oxford University Press

Prusiner S. B., editor (2004) *Prion Biology and Diseases*, 2nd edn, Cold Spring Harbor Laboratory Press

Journals

Baylis M. *et al.* (2004) Risk of scrapie in British sheep of different prion protein genotype *Journal of General Virology*, **85**, 2735–2740

Collinge J. *et al.* (2006) Kuru in the 21st century – an acquired human prion disease with very long incubation periods *The Lancet*, **367**, 2068–2074

Henry C. and Knight R. (2002) Clinical features of variant Creutzfeldt-Jakob disease *Reviews in Medical Virology*, **12**, 143–150

Hill A. F. and Collinge J. (2003) Subclinical prion infection *Trends in Microbiology*, **11**, 578–584

Hilton D. A. (2006) Pathogenesis and prevalence of variant Creutzfeldt-Jakob disease *Journal of Pathology*, **208**, 134–141

Legname G. *et al.* (2004) Synthetic mammalian prions *Science*, **305**, 673–676

Mabbott N. A. and MacPherson G. G. (2006) Prions and their lethal journey to the brain *Nature Reviews Microbiology*, **4**, 201–211

Silveira J. R. *et al.* (2005) The most infectious prion protein particles *Nature*, **437**, 257–261

Watts J. C., Balachandran A. and Westaway D. (2006) The expanding universe of prion diseases *PLoS Pathogens*, **2**, e26

Virologists' vocabulary

Brief definitions of terms used in virology, plus a selection of relevant terms from cell biology, molecular biology, immunology and medicine.

Abortive infection	Infection of a cell where the virus replication cycle is not completed and no progeny virions are formed.
Ambisense genome	A virus genome composed of ssRNA or ssDNA that is partly plus sense and partly minus sense.
Antibody	A glycoprotein synthesized in a plasma cell, which is derived from a B cell that has interacted with a specific antigen. The antibody molecule can bind specifically to this antigen.
Antigen	A molecule that (a) triggers synthesis of antibody and/or a T cell response; (b) binds specifically to an antibody or a lymphocyte receptor.
Antiserum (plural **antisera**)	Blood serum from an animal that has been injected with an antigen. The serum contains antibodies specific to that antigen.
Anti-termination	A mechanism involved in the control of transcription in which termination is overcome at specific terminator sites. Anti-terminator proteins allow RNA polymerase to read through these sites into genes downstream.
Apoptosis	Programmed cell death. Cell suicide. A process controlled by a cell that results in the death of the cell.
Archaea	One of two major groups of prokaryotes, the other being the bacteria.
Assembly	The stage in the virus replication cycle when components come together to form virions.
Asymptomatic infection	An infection without symptoms or signs of disease.
Attachment	Binding of a virion to specific receptors on the host cell.
Attenuated strain of a virus	A strain of a pathogenic virus that has much reduced virulence, but can still replicate in its host. The virulence is said to have been attenuated.
Avirulent strain of a virus	A strain of a virus that lacks virulence, but can still replicate in its host.

Virology: Principles and Applications   John B. Carter and Venetia A. Saunders
© 2007 John Wiley & Sons, Ltd   ISBNs: 978-0-470-02386-0 (HB); 978-0-470-02387-7 (PB)

B lymphocyte (B cell)	A cell with surface receptors that can recognize a specific antigen. Antigen binding can trigger a B cell to develop into an antibody-secreting plasma cell.
Bacteriophage (phage)	A type of virus that infects bacteria.
Baltimore classification	A scheme that classifies viruses into seven groups on the basis of the nature of the genome and the way in which it is transcribed.
Bicistronic mRNA	An mRNA with two open reading frames.
Bronchiolitis	Inflammation of the bronchioles.
Burst size	The average number of new virions released from a single infected cell.
Cap	A methylated guanosine triphosphate joined to the 5′ end of an mRNA by a 5′–5′ bond.
Capsid	The protein coat that encloses the nucleic acid of a virus.
Capsomere	A discrete component of a capsid, constructed from several identical protein molecules.
CD (cluster of differentiation)	CD antigens are cell surface molecules recognized by monoclonal antibodies. CD4 is the receptor for HIV.
Chemokine	A cytokine that stimulates the migration and activation of cells in the animal body, especially cells involved in inflammation.
Cirrhosis	Liver disease characterized by scarring and loss of function.
Cohesive end (*cos*)	A single-stranded end of a dsDNA molecule with a base sequence complementary to another single-stranded end of the same or a different DNA molecule, allowing hydrogen bonding between the two ends.
Coliphage	A bacteriophage that infects *E. coli*.
Complement	A series of proteins that is activated when the body is infected. Activation has a number of anti-viral effects, including lysis of infected cells and enhancement of phagocytosis.
Concatemer	A very long DNA molecule composed of multiple repeats of a nucleotide sequence.
Conditional lethal mutant	A mutant virus that expresses wild-type characteristics under certain (permissive) conditions, but mutant (lethal) characteristics under other (non-permissive/restrictive) conditions, e.g. a temperature-sensitive mutant.
Confocal microscopy	The use of a microscope that excludes light from out-of-focus regions of the specimen, producing clear images of thick and fluorescently labelled specimens.
Conservative replication	The replication mode of some dsRNAs. The parental double strand is conserved and both strands of the progeny molecule are newly synthesized.
Continuous cell line	A clone of cells that is immortal and can be subcultured indefinitely.

Co-receptor	A second cell receptor to which some viruses must bind in order to infect a cell.
Cryo-electron microscopy	Examination of frozen specimens in an electron microscope with a cold stage.
Cytokine	A protein that is secreted from a cell and has a specific effect on other cells, including cells of the immune system.
Cytopathic effect (CPE)	Change in the appearance of a cell culture induced by virus infection.
Defective interfering particle	A virus that lacks part of its genome and interferes with the replication of standard virus.
Defective virus	A virus that lacks part of its genome and is unable to complete the replication cycle without the aid of a helper virus.
Diploid	Having two copies of the genome.
DNA-dependent RNA polymerase	An enzyme that synthesizes RNA from a DNA template.
Domain	A discrete portion of a protein or a nucleic acid with its own structure and function.
Elongation factor (EF)	A specific protein required for polypeptide elongation during translation.
Encephalitis	Inflammation of the brain.
Endemic	Describes a disease that is constantly present or commonly present in a geographical area.
Endocytosis	A process whereby eukaryotic cells take in extracellular materials by engulfing them in endosomes.
Endogenous retroviruses	Retrovirus sequences present in the genomes of vertebrate animals.
Endosome	A vesicle formed during endocytosis by membrane pinching off from the plasma membrane.
Enhancer	A short DNA sequence that can increase the frequency of transcription initiation.
Envelope	A lipid bilayer and associated protein forming the outer component of an enveloped virion.
Enzyme-linked immunosorbent assay (ELISA)	A serological method used to assay antigens and antibodies. Positive results detect an enzyme label.
Epidemic	A rapid increase in the number of cases of a disease that spreads over a larger geographical area.
Epitope	The part of an antigen that binds to an antibody or a lymphocyte receptor.
***Escherichia coli* (*E. coli*)**	Gram-negative bacterium that inhabits the human colon.
Exogenous retrovirus	A normal infectious retrovirus, cf. endogenous retrovirus.

Exon	Part of a eukaryotic gene that is transcribed and usually translated into protein.
Exonuclease	An enzyme that digests a DNA or RNA strand one nucleotide at a time from either the 5′ end or the 3′ end.
F pilus	A specialized pilus, encoded by the F plasmid, required for conjugation in *E. coli*.
Fluorescence microscopy	The use of a microscope fitted with an appropriate source of light, e.g. an arc lamp or a laser, to examine fluorescent specimens.
Fusion protein	(a) A virus protein that fuses the membrane of an enveloped virus with a cell membrane. (b) A protein generated by joining two genes together e.g. a foreign protein fused to a coat protein in phage display.
Genome	The DNA or RNA that encodes the genes of an organism or a virus.
Genotype	The complete set of genes of an organism or a virus.
Glycoprotein	A protein with one or more oligosaccharide groups covalently attached.
Glycosylation	The process of adding oligosaccharide groups to proteins.
Golgi complex	A membranous organelle found in most eukaryotic cells. Its primary function is to process and sort proteins.
Green fluorescent protein (GFP)	A jellyfish protein that fluoresces green. The GFP gene can be fused to a gene of interest, producing a 'tagged' fusion protein, and the localization or movement of the protein in living cells can be visualized. The GFP gene can also be used to monitor the expression of a gene of interest in living cells; the control sequences for the gene of interest are linked to the GFP gene.
Guanylyl transferase	An enzyme that adds guanosine 5′-monophosphate to the 5′ end of mRNA when it is capped.
Hairpin	A structure formed by hydrogen bonding within a single-stranded nucleic acid molecule, producing a double-stranded stem and a loop of unpaired nucleotides.
Helical symmetry	A type of capsid symmetry present in many ssRNA viruses where the RNA forms a helix that is coated with protein.
Helicase	An enzyme that unwinds a DNA duplex at replication forks.
Helper virus	A virus that can provide function(s) missing from a defective virus, thereby enabling the latter to complete its replication cycle.
Hepatocyte	The main cell type in the liver.
Heteroploid cell	A cell with an abnormal number of chromosomes.
Hexon	A type of capsomere surrounded by six other capsomeres.
Histones	Proteins comprised mainly of basic amino acids, associated with DNA in the cells of eukaryotes and some archaea. Histones play roles in gene regulation.

Horizontal gene transfer	Transfer of genetic material laterally from one organism to another, mediated in bacteria by conjugation, transduction and transformation.
Host	A cell or an organism in which a virus or a plasmid can replicate.
Host-controlled restriction	A mechanism by which bacteria can cleave unmodified foreign DNA by a restriction endonuclease that recognizes a specific nucleotide sequence.
Icosahedral symmetry	A type of symmetry present in viruses where the capsid is constructed from protein molecules arranged to form 20 triangular faces.
Immunofluorescence	The detection of an antigen using an antibody labelled with a fluorescent dye.
Immunogenic	The ability of a substance to induce an immune response.
Immunoglobulin	A glycoprotein that functions as an antibody.
Incubation period	The time between infection of a host and the appearance of the first signs and/or symptoms of disease.
Induction	Activation of a latent infection or an inactive gene.
Infectivity	The ability of virions to initiate an infection.
Integral membrane protein	A protein that is closely associated with a membrane. Most integral membrane proteins have one or more sequences that span the lipid bilayer.
Integrase	An enzyme that integrates virus DNA into host DNA.
Interferon	Protein produced by animal cells in response to virus infection. There are several types of interferon and they interfere with virus replication in a variety of ways.
Intergenic (IG) region	A sequence of nucleotides between two genes.
Internal ribosome entry site (IRES)	A binding site for eukaryotic ribosomes, present in the RNAs of some plus-strand viruses. The site is internal (near the 5' end of the RNA), in contrast to most eukaryotic mRNAs, which bind ribosomes at the 5' end.
Intron	A non-coding sequence interrupting coding sequences (exons) in a gene. The intron is transcribed into RNA, but is subsequently excised by a splicing reaction during processing of the primary transcript into mRNA. Introns are common in eukaryotes and their viruses, but rare in prokaryotes and their viruses.
Isometric virion	A virion that has the same size from all perspectives. Virions that have icosahedral symmetry are isometric or almost isometric.
Kilobase (kb)	A measure of length for DNA and RNA equal to 1000 bases.
Kinase	An enzyme that catalyses phosphorylation.
Laboratory strain	A virus strain derived from a wild strain after propagation in the laboratory.

Latent infection	Infection of a cell where the replication cycle is not completed, but the virus genome is maintained in the cell.
Ligase	An enzyme that can catalyse bond formation between two similar types of molecule. A DNA ligase can join two DNA molecules by catalysing the formation of a phosphodiester bond between the 5' end of a polynucleotide chain and the 3' end of another (or the same) polynucleotide chain.
Ligation	The joining of two molecules, e.g. DNA.
Lipopolysaccharide (LPS)	A compound comprising lipid and polysaccharide that is a key component of the outer membrane of Gram-negative bacteria.
Lysis	Destruction of a cell caused by the rupture of its membrane and release of contents, e.g. the bursting of a cell and release of progeny virions at the end of the replication cycle.
Lysogen	A bacterium containing the genome of a phage that is in the prophage state and thus repressed for lytic functions. Such a bacterium is said to be lysogenic for (name of the phage).
Lysogenic conversion (phage conversion)	When a cell becomes lysogenized, occasionally extra genes (unrelated to the lytic cycle or lysogeny) carried by the prophage are expressed in the host cell. These genes can change the phenotype of the host.
Lysogeny	Latent infection of a bacterial cell with a phage.
Lysosome	Membrane-bounded organelle in the cytoplasm of eukaryotic cells. Lysosomes contain enzymes, e.g. proteases and nucleases, that can digest virions.
Major histocompatibility complex (MHC)	A region of the vertebrate genome that encodes major histocompatibility proteins. MHC class I and class II molecules are cell surface proteins that play important roles in immune responses.
Meningitis	Inflammation of the meninges (membranes covering the brain and the spinal cord).
Meningoencephalitis	Inflammation of the meninges and the brain.
Methyl transferase	An enzyme that catalyses the transfer of methyl groups from one molecule to another. Methyl transferases play a role in capping eukaryotic mRNA.
Microarray	A substrate with an orderly array of small spots of material (e.g. DNA or protein) attached. Each spot contains a specific probe that can detect specific target molecules in a sample.
Microtubule	Protein structure in eukaryotic cells, forming a component of the cytoskeleton, with roles in intracellular transport and mitosis.
Minus strand (negative strand, (−) strand)	A nucleic acid strand that has the nucleotide sequence complementary to that of the mRNA.
Monocistronic mRNA	An mRNA with one open reading frame.
Monoclonal antibody	A single type of antibody produced by a clone of identical cells.
Monolayer	A layer of cells growing on the surface of a plastic or glass vessel.
Multiplicity of infection (m.o.i.)	The ratio of virions to cells in a virion–cell mixture.

Mutation	An alteration in the sequence of DNA or RNA. Mutations may occur spontaneously or they may be induced.
Myristylation (myristoylation)	The addition of a myristyl (myristoyl) group to a molecule.
Naked virus	A virus that does not have an envelope on the virion.
Negative staining	The use of a compound containing a heavy metal to reveal the structure of specimens in a transmission electron microscope.
Neurone	The main type of cell in the nervous system.
Neurovirulence	A measure of the severity of nervous system disease that a virus (or any micro-organism) is capable of causing.
Neutralization	The inactivation of virus infectivity by reaction with specific antibody.
Non-structural protein	A virus protein that is not a component of the virion, but has one or more roles in the replication cycle.
Northern blotting	The transfer of RNA molecules to a membrane after gel electrophoresis. Specific RNAs can be detected on the membrane using probes.
Nuclear envelope	The structure, composed of two membranes, that separates the nucleus from the cytoplasm in a eukaryotic cell.
Nuclear localization signal	A positively charged sequence in a protein that directs the protein through nuclear pores into the nucleus of a cell.
Nuclear pore	A protein complex in the nuclear envelope of a eukaryotic cell through which materials are transported in and out of the nucleus.
Nucleic acid hybridization	The formation of double-stranded nucleic acids by base-pairing between complementary single strands.
Nucleocapsid	The virus genome enclosed in a protein capsid.
Nucleoside	A nitrogen base (purine or pyrimidine) joined to a sugar (ribose or deoxyribose).
Nucleotide	A nitrogen base (purine or pyrimidine) joined to a sugar, which is joined to one or more phosphate groups.
Okazaki fragment	A short fragment of DNA with the RNA primer attached, produced during discontinuous DNA synthesis. Okazaki fragments are joined by a DNA ligase to form the lagging strand.
Oncogene	A cell or virus gene, the expression of which can cause a cell to become transformed. This may lead to the development of a cancer.
Open reading frame (ORF)	A sequence of nucleotides, starting with an initiation codon and finishing with a termination codon, that encodes the amino acids of a protein.
Origin (*ori*)	A specific sequence of DNA at which replication is initiated.
Packaging signal (PS)	A nucleotide sequence in a virus genome that is recognized by a virus protein during virion assembly.
Pandemic	A disease outbreak throughout the world.

Pararetrovirus	A virus with a DNA genome that is replicated via RNA.
Passage	'Subculture' of a virus in cell culture or in an organism. 'Subculture' of a prion in an organism.
Penton	A type of capsomere surrounded by five other capsomeres.
Perforin	Protein present in CD8 T cells and NK cells. These cells can kill a virus-infected cell by releasing perforins, which form pores in the plasma membrane of the target cell.
Persistent infection	A long-term, possibly life-long, infection of a host. A persistent virus infection may be productive or latent.
Phage display	Display of recombinant proteins on the surface of a phage e.g. M13.
Phagemid	A cloning vector that combines features of a phage and plasmid replicon and can replicate in either mode. The vector carries a plasmid *ori* for replication in plasmid mode, and a filamentous phage *ori*, leading to production of single-stranded copies of the phagemid when the cell is infected with the relevant helper phage.
Phenotype	The observable characteristics of an organism or virus, determined by its genotype and environmental factors.
Phosphoprotein	A protein with one or more phosphate groups covalently attached.
Phosphorylation	The addition of one or more phosphate groups to a molecule.
Phylogenetic tree	A diagram showing evolutionary relationships between viruses or other organisms.
Pilin	The protein subunit that is polymerized into the pilus in bacteria.
Plaque	(a) An area of lysis in a layer of cells, usually initiated by a single virion infecting a cell, followed by the spread of infection to surrounding cells.
	(b) An aggregate of prion fibrils in the central nervous system of a human or animal with a transmissible spongiform encephalopathy.
Plaque-forming unit	The quantity of virus (often a single virion) that can initiate formation of one plaque.
Plaque-purified virus	A clone of virus derived from an individual plaque.
Plasma membrane	The membrane at the surface of a eukaryotic cell.
Plasmid	A self-replicating, extrachromosomal dsDNA molecule that is generally circular, though can be linear. Plasmids are common in prokaryotes and rare in eukaryotes. Artificial plasmids can be created to clone DNA sequences.
Plus strand (positive strand, (+) strand)	A nucleic acid strand that has the same sequence as the mRNA.
Poly(A) polymerase	An enzyme that adds adenylate residues to the 3′ end of a eukaryotic mRNA.
Polyadenylation	The process of adding adenylate residues to the 3′ end of a eukaryotic mRNA.
Polycistronic mRNA	An mRNA with more than one open reading frame.

Polymerase chain reaction (PCR)	An *in vitro* technique for amplifying specific DNA sequences.
Polyprotein	A large protein that is cleaved to form smaller functional proteins.
Post-translational modification	A modification to a protein after it has been translated. Examples: glycosylation, cleavage.
Primary transcript	An RNA molecule synthesized during transcription. Introns may be removed from the primary transcript to form functional RNAs.
Primase	An enzyme that synthesizes RNA primers for DNA synthesis.
Primer	A molecule (often RNA, sometimes DNA or protein) that provides a free –OH group at which a polymerase can initiate synthesis of DNA or RNA.
Primosome	A complex of proteins involved in the priming action that initiates DNA replication on φX-type origins.
Probe	A specific sequence of DNA or RNA used to detect nucleic acids by hybridization. A probe may be labelled, e.g. with a fluorescent molecule, to enable detection and quantification of the hybrid.
Procapsid	A precursor of a capsid.
Processivity factor	A protein that increases the efficiency of DNA synthesis by enhancing the ability of a DNA polymerase to remain associated with, and to process along, its template.
Productive infection	Infection of a cell that results in the production of progeny virions.
Promoter	A specific DNA sequence, usually upstream of the transcription start point of a gene, to which DNA-dependent RNA polymerase binds to initiate transcription.
Proofreading	A mechanism for correcting errors in nucleic acid or protein synthesis.
Prophage	A phage genome that resides in a bacterial host in a latent state. The prophages of most phages are integrated into the bacterial genome, but some are not.
Protease	An enzyme that cleaves protein molecules by breaking peptide bonds.
Proto-oncogene	A cell gene that can become an oncogene (tumour gene) if activated or over-expressed.
Provirus	The dsDNA copy of a retrovirus genome RNA.
Pseudoknot	An ssRNA secondary structure with two loops, formed when a sequence in a loop base-pairs with a complementary sequence outside the loop.
Reading frame	The phase in which the nucleotides of a nucleic acid are read in triplets.
Reassortment	A category of recombination that may occur with those segmented genome viruses that have all the segments packaged in one virion. Reassortment occurs in a cell co-infected with two virus strains, and is the formation of progeny virions containing mixtures of genome segments from the two parental strains.
Receptor	A molecule on a cell surface to which a virus specifically attaches.

Recombinant	An organism or a virus with a new genome, produced as a result of recombination. May also be used to describe the genome of the organism or virus.
Recombination	A process of combining genetic sequences that results in the production of a new genome, derived from two parental genomes. For a cellular organism, recombination can take place between DNA molecules within a cell. For viruses, recombination may occur when a cell is co-infected with two related viruses.
Replicase	An RNA polymerase that replicates the genome of an RNA virus by synthesizing both (+) RNA and (−) RNA.
Replication cycle	The process whereby a virus is replicated. For most viruses the process starts with attachment to a host cell and this is followed by entry into the cell. The process ends with exit of progeny virions from the cell. Some authors use the terms 'infection cycle' and 'life cycle'.
Replicative intermediate	A structure formed during ssRNA replication, consisting of a template RNA associated with nascent RNAs of varying length. The nascent RNAs have the opposite polarity to the template RNA.
Restriction endonuclease	An enzyme, usually from a bacterium, that cuts dsDNA at a specific site.
Retrovirus	A member of the family *Retroviridae*, so named because these viruses carry out reverse transcription.
Reverse genetics	The generation of an RNA virus genome from a cloned copy DNA.
Reverse transcriptase	An enzyme that can synthesize DNA using (a) an RNA template and (b) a DNA template.
Reverse transcription	Synthesis of DNA from an RNA template.
Reverse transcription-polymerase chain reaction (RT-PCR)	An *in vitro* technique for amplifying the data in RNA sequences by first copying the RNA to DNA using a reverse transcriptase. The DNA is then amplified by a PCR.
Ribonuclease (RNase)	An enzyme that hydrolyses RNA.
Ribonuclease H (RNase H)	A ribonuclease that specifically digests the RNA in an RNA–DNA duplex.
Ribosomal frameshifting	A mechanism that allows a ribosome to read two overlapping open reading frames (ORF1 and ORF2) in an mRNA. A ribosome reading ORF 1 shifts into a different reading frame towards the end of the ORF. The ORF 1 stop codon is therefore not recognized and the ribosome now reads ORF 2 to produce an elongated version of the ORF 1 protein.
RNA-dependent RNA polymerase	An enzyme that synthesizes RNA from an RNA template.
RNA interference (RNAi, RNA silencing)	A process that interferes with the expression of a specific gene. The process is induced by a dsRNA and results in the destruction of mRNA that has the same sequence as the dsRNA.
RNA polymerase	An enzyme that synthesizes RNA.

RNA polymerase II	The eukaryotic cell enzyme that synthesizes mRNA.
Rough endoplasmic reticulum	A system of membranes and ribosomes in the cytoplasm of eukaryotic cells.
Satellite virus	A defective virus that depends on a helper virus to provide one or more functions.
Scaffolding protein	A protein that facilitates the assembly of a procapsid. The scaffolding protein is removed in the final stages of assembly and is therefore not present in the mature virion.
Segmented genome	A virus genome that is composed of two or more nucleic acid molecules.
Self-assembly	The ability of a biological structure, e.g. a virus particle, to form spontaneously from its component parts.
Semi-conservative replication	The replication mode of dsDNA and some dsRNAs in which the parental double strand is not conserved. Each progeny molecule consists of one parental strand and one newly synthesized strand.
Serology	The study of antigen–antibody reactions and their use in tests to detect specific antigens and antibodies.
Seropositive	A person or an animal is said to be seropositive for an antigen, e.g. a virus, if their blood contains antibodies specific for that antigen.
Serotype	A strain of a virus or micro-organism distinguished by its antigens.
Sex pilus	A thin protein appendage for bacterial mating.
Shine-Dalgarno (S-D) sequence	A purine-rich sequence in prokaryotic mRNA just upstream of the translation start. The sequence can base-pair with a sequence near the $3'$-end of 16S ribosomal RNA and facilitates the initiation of protein synthesis.
Sigma (σ) factor	A subunit of bacterial RNA polymerase, responsible for the recognition of a specific promoter. Different sigma factors allow recognition of different promoter sequences.
Single-stranded binding (ssb) protein	A basic protein with a high affinity for ssDNA. Ssb protein protects ssDNA from nuclease attack and prevents it from re-annealing into dsDNA.
Southern blotting	The transfer of DNA molecules to a membrane after gel electrophoresis. Specific DNA can be detected on the membrane using probes.
Splicing	The process of removing introns from primary transcripts and joining the exons to form mRNA.
Start (initiation) codon	A codon on mRNA for initiation of protein synthesis: commonly AUG, less commonly GUG and rarely UUG.
Stop (termination) codon	A codon on mRNA for termination of protein synthesis: UAG, UAA or UGA. When a ribosome encounters a stop codon a termination factor interacts with the ribosome, causing polypeptide synthesis to stop and the ribosome to dissociate from the mRNA.

Structural protein	A protein that is a virion component.
Superinfection	Infection by a virus of a cell that is already infected (often as a latent infection) with that virus or a related virus.
Superinfection immunity (homoimmunity)	The immunity of a lysogen to superinfection by a phage with a similar regulatory mechanism.
Synchronous infection	Near simultaneous infection of all cells in a culture with a virus. This can be achieved by e.g. using a high multiplicity of infection, or by limiting the time of attachment then diluting the culture so that virions and cells are unlikely to make contact.
Syncytium (plural **syncytia**)	A multinucleated giant cell formed by the fusion of membranes of a number of individual cells. A number of viruses can cause the formation of syncytia when they infect cells.
T lymphocyte (T cell)	A cell with surface receptors that can recognize a specific antigen. Antigen binding can trigger a T cell to perform one of several roles, including helper T cell or cytotoxic T cell.
Tailed phage	A phage that has a tail attached to the head, which contains the genome.
TATA box	A DNA sequence in a eukaryotic promoter, typically 15–25 nucleotides upstream from the transcription start.
TCID$_{50}$	The dose of virus that infects 50 per cent of tissue cultures inoculated with aliquots of a virus preparation.
Tegument	A layer of protein and RNA between the capsid and the envelope of a herpesvirus particle.
Temperate phage	A phage capable of either establishing a lysogenic state in a susceptible bacterial host or of entering the lytic cycle.
Temperature-sensitive mutant	A virus mutant that is unable to replicate at some temperatures (non-permissive temperatures) at which the wild-type virus can replicate, but is able to replicate at other temperatures (permissive temperatures).
Titre	The concentration of virus, antibody or other material in a preparation, determined by titration.
Transcriptase	An enzyme that carries out transcription.
Transcription	Synthesis of RNA from a DNA or an RNA template.
Transcription factor	A protein that binds specifically to a promoter or enhancer to control gene expression.
Transfection	A process for introducing nucleic acids (e.g. a virus genome) into cells.
Transformation	(a) Changes in an animal cell that result in it developing into a tumour cell. (b) Bacterial gene transfer process involving uptake of naked DNA and acquisition of an altered genotype.
Translation	Synthesis of protein from the genetic information in mRNA.
Transmission electron microscope	A microscope in which the image is formed by electrons transmitted through the specimen.

Transovarial transmission	Transmission of a virus from a female, e.g. insect, to the next generation within eggs.
Uncoating	The release of the virus genome (and associated protein in some cases) from a virion when a cell is infected.
Upstream	The region of a nucleic acid extending in the 5′ direction from a gene; in the opposite direction to that of transcription.
Vector	(a) An organism that transmits a virus from an infected host to an uninfected host. (b) A virus or plasmid DNA, into which foreign DNA can be inserted for the purpose of transferring it to a host cell for amplification/cloning or gene therapy.
Vertical transmission	Transmission of a genetic element, virus or micro-organism to the next generation of the host.
Viraemia	The presence of infectious virus in the bloodstream.
Virion	Virus particle.
Viroplasm	A morphologically distinct region in which virus replication occurs within an infected cell.
Virulence	A measure of the severity of disease that a virus (or micro-organism) is capable of causing.
Virulent strain of a virus	A virus strain that can cause disease when it infects a host.
Virulent phage	A phage that can produce only a lytic infection; it is unable to induce lysogeny.
Virus attachment site	The surface regions of one or more virion proteins, which form a site that attaches to the receptor on a host cell.
Western blotting	The transfer of protein molecules to a membrane after gel electrophoresis. Specific proteins can be detected on the membrane.
Wild strain	A virus strain isolated from a naturally infected host, cf. laboratory strain.
Wild-type virus	The parental strain of a mutant or laboratory-adapted virus.
Zinc finger	A region of a nucleic acid-binding protein that resembles a finger and binds a zinc ion.

INDEX